Smart Nanomaterials Technology

Series Editors

Azamal Husen, Wolaita Sodo University, Wolaita, Ethiopia

Mohammad Jawaid, Department of Chemical and Petroleum Engineering, United Arab Emirates University, Al Ain, United Arab Emirates

Nanotechnology is a rapidly growing scientific field and has attracted a great interest over the last few years because of its abundant applications in different fields like biology, physics and chemistry. This science deals with the production of minute particles called nanomaterials having dimensions between 1 and 100 nm which may serve as building blocks for various physical and biological systems. On the other hand, there is the class of smart materials where the material that can stimuli by external factors and results a new kind of functional properties. The combination of these two classes forms a new class of smart nanomaterials, which produces unique functional material properties and a great opportunity to larger span of application. Smart nanomaterials have been employed by researchers to use it effectively in agricultural production, soil improvement, disease management, energy and environment, medical science, pharmaceuticals, engineering, food, animal husbandry and forestry sectors.

This book series in Smart Nanomaterials Technology aims to comprehensively cover topics in the fabrication, synthesis and application of these materials for applications in the following fields:

- Energy Systems—Renewable energy, energy storage (supercapacitors and electrochemical cells), hydrogen storage, photocatalytic water splitting for hydrogen production
- Biomedical—controlled release of drugs, treatment of various diseases, biosensors,
- Agricultural—agricultural production, soil improvement, disease management, animal feed, egg, milk and meat production/processing,
- Forestry—wood preservation, protection, disease management
- Environment—wastewater treatment, separation of hazardous contaminants from wastewater, indoor air filters.

Rakesh Kumar Bachheti · Archana Bachheti ·
Azamal Husen
Editors

Chitin-Based Nanoparticles for the Agriculture Sectors

Springer

Editors
Rakesh Kumar Bachheti
Department of Industrial Chemistry
Addis Ababa Science and Technology University
Addis Ababa, Ethiopia

Archana Bachheti
Department of Environmental Science
Graphic Era University
Dehradun, Uttarakhand, India

Azamal Husen
Wolaita Sodo University
Wolaita, Ethiopia

ISSN 3004-8273 ISSN 3004-8281 (electronic)
Smart Nanomaterials Technology
ISBN 978-981-96-0919-2 ISBN 978-981-96-0920-8 (eBook)
https://doi.org/10.1007/978-981-96-0920-8

This Springer imprint is published by the registered company Springer Nature Singapore Pte Ltd.
The registered company address is: 152 Beach Road, #21-01/04 Gateway East, Singapore 189721, Singapore

Preface

Agriculture is the backbone of human civilization, and its continuous advancement is essential for addressing the challenges of food security, environmental sustainability, and economic development. In recent years, the integration of nanotechnology into agriculture has opened new avenues for enhancing crop productivity, disease management, soil health, and environmental sustainability. Among the various nanomaterials, chitin-based nanoparticles have shown significant promise due to their unique properties, biocompatibility, and eco-friendliness.

This book, *Chitin-Based Nanoparticles for the Agriculture Sectors*, aims to provide a comprehensive overview of the current state of research, development, and application of chitin-based nanoparticles in agriculture. It brings together contributions from leading experts in the field, presenting the latest advancements, challenges, and future perspectives of this exciting and rapidly evolving area.

The book is organized into twelve chapters, each focusing on a specific aspect of chitin-based nanoparticles and their applications in agriculture: Chapter "Introduction to the Isolation of Chitin and Chitosan and Their Properties" introduces the isolation of chitin and chitosan and their properties, laying the foundation for understanding the subsequent chapters. Chapter "Properties and Synthesis of Chitin-Based Nanoparticles" delves into the properties and synthesis of chitin-based nanoparticles, highlighting the methods and techniques used to produce these nanomaterials. Chapter "Characterization Techniques for Chitin-Based Nanoparticles" discusses various characterization techniques for chitin-based nanoparticles, providing insights into their structural, chemical, and physical properties. Chapter "Role of Agrochemical-Loaded Chitin Nanoparticles for Insect Pest Management" explores the role of agrochemical-loaded chitin nanoparticles for insect pest management, showcasing their potential to enhance the efficacy and reduce the environmental impact of traditional pesticides.

Chapter "Chitin-Based Nanoparticles for Plant Disease Management" focuses on chitin-based nanoparticles for plant disease management, presenting innovative solutions for protecting crops from pathogens. Chapter "Chitin-Based Nanoparticles for Soil Remediation" examines the use of chitin-based nanoparticles for soil remediation, addressing issues of soil contamination and health. Chapter "Role of Chitin-Based Nanoparticles for Crop Nutrition and Fertilization" highlights the role of chitin-based nanoparticles in crop nutrition and fertilization, emphasizing their potential to improve nutrient uptake and utilization. Chapter "Chitin-Based Nanoparticles for Crop Plant Growth Promotion" discusses the application of chitin-based nanoparticles for crop plant growth promotion, revealing their impact on plant development and productivity. Chapter "Application of Chitin/Chitosan-Based Agro Nano Chemicals in Crop Protection" reviews the use of chitin/chitosan-based agro nano chemicals in crop protection, providing a broader perspective on their applications and benefits. Chapter "Challenges and Limitations of Chitin-Based Nanoparticles in Agriculture" addresses the challenges and limitations of chitin-based nanoparticles in agriculture, offering a critical analysis of the current barriers and potential solutions. Chapter "Future Perspectives and Opportunities for Chitin-Based Nanoparticles in Agriculture" explores future perspectives and opportunities for chitin-based nanoparticles in agriculture, envisioning the next steps in research and application. Further, in the last, Chapter "Chitin and Chitosan-Based Hydrogel and Their Use in the Agriculture" investigates chitin and chitosan-based hydrogels and their use in agriculture, presenting novel approaches for improving water retention and soil structure.

We hope this book serves as a valuable resource for researchers, practitioners, and students in the fields of agriculture, nanotechnology, and environmental science. By presenting a holistic view of chitin-based nanoparticles and their agricultural applications, we aim to inspire further research and innovation in this promising area. We extend our gratitude to all the contributors for their expertise and dedication, and to the readers for their interest in advancing the frontiers of agricultural science.

Addis Ababa, Ethiopia — Rakesh Kumar Bachheti
Dehradun, India — Archana Bachheti
Wolaita, Ethiopia — Azamal Husen

Contents

About the Editors

Dr. Rakesh Kumar Bachheti graduated from Hemwati Nandan Bahuguna Garhwal University, India, in 1996. He completed his M.Sc. in Organic Chemistry from the same university in 1998. Later, in 2001, he pursued a one-year Post Graduate Diploma in Pulp and Paper Technology from the Forest Research Institute, Dehradun, India. Subsequently, he obtained his Ph.D. in Organic Chemistry from Kumaun University, Nainital, India, in 2007. Dr. Bachheti has a total of 20 years of research and teaching experience. He is a Visiting Faculty Member at the Forest Research Institute, Dehradun, India, and a Research Fellow at INTI International University, Persiaran Perdana BBN, Putra Nilai, Nilai, Negeri Sembilan, Malaysia. Dr. Bachheti has published more than 150 research papers and edited over 8 books. He is listed among the Top 2% Scientists for the year 2024 by Stanford University, USA, and Elsevier.

Prof. Archana Bachheti completed her B.Sc. in 1997 and M.Sc. in 1999 from Hemwati Nandan Bahuguna Garhwal University. She earned her Ph.D. from the Forest Research Institute, Dehradun, India, in 2006. Currently, she is a Professor at Graphic Era University in Dehradun, India. With over 20 years of extensive experience, she has served in various academic roles in India and has also provided expertise internationally. Her teaching portfolio includes subjects such as ecology and environment, environmental science, freshwater ecology, disaster management, and bryophytes and pteridophytes. Professor Bachheti's research interests span a broad and interdisciplinary field of plant ecology, with a focus on ecorestoration, green chemistry—particularly the synthesis of nanomaterials—and exploring the medicinal properties of plants. Her research has involved ecological restoration of degraded lands, studying the physical and chemical properties of plant oils, and investigating plant-based nanomaterials. Throughout her career, she has been actively involved in mentoring students and researchers. Dr. Bachheti has published more than 130 research papers and edited over 8 books. She is listed among the Top 2% Scientists for the year 2024 by Stanford University, USA, and Elsevier.

Prof. Azamal Husen (B.Sc. from Shri Murli Manohar Town Post Graduate College, Ballia, UP; M.Sc. from Jamia Hamdard, New Delhi, and Ph.D. from Forest Research Institute, Dehra Dun, India) is Foreign Delegate at Wolaita Sodo University, Wolaita, Ethiopia, and Adjunct Professor at Graphic Era (Deemed to be University), Dehradun, Uttarakhand, India. He has served the University of Gondar, Ethiopia, as Full Professor of Biology, and worked there as Coordinator of the M.Sc. program and as Head of the Department of Biology. He was Visiting Faculty Member of the Forest Research Institute and the Doon College of Agriculture and Forest at Dehra Dun, India. Dr. Husen's research and teaching experience of 25 years include nanobiotechnology, plant physiology, environmental botany, medicinal plants, and plant propagation. He is on the advisory board of Cambridge Scholars Publishing, UK. He Fellow of the Plantae group of the American Society of Plant Biologists.

Introduction to the Isolation of Chitin and Chitosan and Their Properties

Payas Arora, Nikita Bahuguna, Manjusha Tyagi, Nishant Rai, and Rakesh Kumar Bachheti

Abstract Chitin and chitosan, two of the most abundant biopolymers existent in a wide variety of natural sources, present a vast array of biological properties that have found wide applicability in several domains of global importance. The dynamic development of such essential use cases of these biologically derived materials can be attributed to the physical, chemical, and biological properties of these biomaterials. Since these properties present such significant importance in governing the success achieved in all possible domains of their usage, extensive research has been conducted, now for a very long time, in an attempt to understand the diverse inherent properties of these substances, the effects of the traditional and modern extraction procedures on these properties, and whether such deviations in the natural properties induced during extraction or intended modifications prove beneficial or detrimental for use by the global communities. In order to delve deeper into these critically important realms intricately associated with these biological products, dedicated attempts must be made to study in considerable detail the scientific information that has been revealed through years of study by the scientific community. Upon establishing a concrete map of every discovered and proven characteristic of these materials can the scientific community hope of developing room for new ground breaking discoveries yet undiscovered that harbour the potential of transforming the current landscape

P. Arora · N. Rai (✉)
Department of Biotechnology, Graphic Era (Deemed to be University), Dehradun 248002, Uttarakhand, India
e-mail: dr_nishantrai.bt@geu.ac.in

N. Bahuguna
Department of Microbiology, Graphic Era (Deemed to be University), Dehradun 248002, Uttarakhand, India

M. Tyagi
Department of Microbiology, SGRR University, Dehradun 248001, Uttarakhand, India

R. K. Bachheti
Department of Industrial Chemistry, College of Natural and Applied Sciences, Addis Ababa Science and Technology University, P.O. Box: 16417, Addis Ababa, Ethiopia

University Centre for Research and Development, Chandigarh University, Gharuan 140413, Punjab, India

R. K. Bachheti et al. (eds.), *Chitin-Based Nanoparticles for the Agriculture Sectors*, Smart Nanomaterials Technology, https://doi.org/10.1007/978-981-96-0920-8_1

of material deployment and reliance on conventional approaches in all sectors of commercial and industrial importance.

Keywords Chitin · Chitosan · Biopolymers · Properties · Importance

1 Introduction

A French botanist, Henri Braconnot, is credited with the discovery of chitin. In 1811, he published his findings on higher fungi from six different varieties, from which he isolated chitin. Another French scientist after Braconnot, A. Odier, in 1823, obtained chitin by isolating it from the elitrium of the May-bug. Following these discoveries, chitosan was obtained for the first time in 1859 by C. Rouget, yet another French scientist. F. Hoppe Seyler, in 1894, named this substance obtained from chitin, chitosan. K. H. Meyer and H. Mark elaborated the repeating unit characterizing chitin's chains in 1928. A. Neuberger and R. Pitt Rivers completed the chemical structure elucidation in 1939, while the crystallographic structure was determined by K. M. Rudall in 1963 [1]. Chitin, a nitrogenous polysaccharide, is a biopolymer with residues that have 1,000–3,000 repeat units. According to the amount generated annually by biosynthesis, chitin is only second to cellulose in the category of the most abundant organic material globally available. Structurally, chitin demonstrates a great degree of similarity to cellulose, with the only difference being the acetamide group in chitin at the C2 position in place of the hydroxyl group present in the structure of cellulose [2]. Chitobiose, 4-*O*-(2-amino-2-deoxy-β-D-glucopyranosyl)-(1 → 4)-2-amino-2-deoxy-D-glucose, constitutes the structural unit found in native chitin. This linear polysaccharide has a degree of deacetylation less than 10% and a molecular weight of about $1–2.5 \times 10^6$ Da. On the contrary, chitosan is a derivative of chitin, produced from the *N*-deacetylation of chitin in the presence of a hot alkali. Being produced by the process of deacetylation, chitosan can have a degree of deacetylation as high as 40–98%, with the molecular weight ranging from 5×10^4 Da to 2×10^6 Da [3].

Chitin and chitosan can be obtained from a plethora of naturally available sources such as Crustaceans (including shrimp shells, mussel shells, crayfish, clams, prawns, lobsters, crab shells, etc.), fungi (including *Aspergillus niger*, *Penicillium chrysogenum*, *Lactarius vellereus*, yeast), molluscs (including squids, oysters, brown seaweed, green algae), and insects (including silk worm, flour moth, hawk moth, comb-clawed beetle, dung beetle, grasshopper, Colorado potato beetle, house cricket, *Lucanus cervus*, *Argynnis pandora*) [4]. Sources of these materials comprise many other organisms in the categories mentioned above, like Isopoda and barnacles in Crustaceans, honey bees, black soldier fly, cockroach, waxworm, ladybug, tobacco hornworm, housefly, and lesser mealworm in insects, and Bryozoa, black coral, horse-shoe worms, sponges, and spiders among other animals [5]. Chitin and chitosan hold importance in various industries like biotechnology, biomedical, food, agriculture, cosmetics, and water treatment [6]. In addition to the aforementioned industries, pharmaceutical, textile, and paper industries also find immense use of these highly valuable biopolymers [7]. This chapter aims to discuss in detail the structural characteristics, compositional features, different methods of isolation adopted for the extraction of chitin and chitosan from varied sources, and the physical and chemical properties of these biopolymers that contribute to their widespread usage in the industrial and commercial domains. Through this elaborate discussion, we hope to provide a broader view of the scientific information available on this topic, enriching the readers with a thorough knowledge of these natural materials before proceeding to more advanced, technical concepts.

2 Chitin and Chitosan: Structure and Composition

2.1 Molecular Structure

2.1.1 Chitin Structure

Chitin exhibits three forms of existence—α-chitin, β-chitin, and γ-chitin. Of these, α-chitin is the commonly occurring and widely known allomorph, followed by β-chitin. Very little is known of γ-chitin and research is in progress to fully characterize it. In α-chitin, the constituent unit cells are orthorhombic with individual chains showing an anti-parallel arrangement. β-chitin, the lesser-known allomorph, consists of a unit cell which is monoclinic and individual chains running in a parallel fashion. The γ-chitin consists of three chain unit cells, with two "up" chains and one "down" chain [8]. Primarily, the 2-fold helix motif is adopted by the solvated chitin chains, with the equilibrium conformation unaffected by ionic strengths reaching a value of 1.1 M. Further, solvated β-chitin nanoparticles have also been shown to contain intermolecular hydrogen bonds existing between the hydroxymethyl and N-acetyl groups. However, these interactions have been observed in low frequency and possess low strength. Furthermore, intrachain hydrogen bonds occur between the HO3 atom of one monomer and the O5 atom belonging to the next monomer [9]. The unit cells of α-chitin possess a space symmetry group $P2_12_12_1$ [10] while the unit cells of β-chitin have a space group of $P2_1$ with a single twofold screw axis [11, 12]. The axes a, b, and c of an α-chitin unit cell possess values 4.76, 10.28, and 18.85 Å respectively [13], with other literature reporting only minor alterations in these values [14, 15]. A, b, and c values for the β-chitin unit cell correspond to 4.85, 9.26, and 10.38 Å, respectively [11]. However, much like with α-chitin, there are also some small differences in the a, b, and c values of β-chitin [16]. Chitin comprises two cellubioses arranged in an antiparallel orientation within a crystallite unit. Whereas β-chitin solely has intrasheet hydrogen bonds, the structures of α- and γ-chitin exhibit both intrasheet and intersheet hydrogen bonding [17].

Similar observations have been made with respect to α-chitin and β-chitin with the former containing two antiparallel molecules per unit cell and intrasheet as well as intersheet hydrogen bonds, while the latter contains a single unit with no intersheet bonds [18].

2.1.2 Chitosan Structure

Two polymorphs of chitosan, annealed (anhydrous) and tendon (hydrated), exist. Both these types are characterized by an extended two-fold helix with intramolecular O(3)-O(5) bonds stabilizing the structure. This is the Type I form. With salts of acetic acid and some other acids like succinic acid, formic acid, fumaric acid, propionic acid, and butyric acid, chitosan assumes the hydrated Type II form, with a relaxed two-fold helix containing asymmetric units of tetrasaccharides. In its hydrogen iodide salt, chitosan takes up 4/1 helix with asymmetric disaccharide units [19]. Upon crystallization, chitosan assumes an orthorhombic unit cell. The values of a, b, and c for this unit cell are 0.828, 0.862, and 1.043 nm, respectively. A space symmetry group of $P2_12_12_1$ characterizes the unit cell. Four glucosamine residues are present in the unit cell. Chain conformation consists of an extended 2-fold helix with intramolecular O(3)-O(5) hydrogen bonds stabilizing the structure and an O6 atom rotated near gt position. Two chains with antiparallel packing pass through the unit cell [20].

2.2 *Chemical Composition*

2.2.1 Monomers and Polymers

The biopolymer chitin is made up of unbranched chains consisting of monomeric units of β–(1 → 4)-linked 2-acetoamido-2-deoxy-D-glucose (N-acetyl-D-glucosamine; GlcNAc) residues, resulting in the consideration that chitin is a derivative of cellulose, resembling cellulose in multiple properties [21]. It has been demonstrated that hydrolysis of chitin using concentrated acid results in a reaction mixture containing a significantly high percentage of N-acetyl-D-glucosamine, amounting to a molar fraction greater than 90% [22].

Chitosan, a derivative of chitin, is made up of D-glucosamine (GlcN) monomers, as opposed to chitin, which consists of monomeric N-acetyl-D-glucosamine units, thus showing that chitosan is formed from the deacetylation of chitin. Chitosan oligomers have been synthesized as GlcN1-GlcN6 [23]. Chemically, chitosan has been described as a copolymer composed of two repeating units: β-(1,4)-2-acetamido-2-deoxy-D-glucose (D-glucosamine) and β- (1,4)-2-anaino-2-deoxy-D-glucose (N-acetyl glucosamine) [24]. Chitosan preparation by deacetylation of chitin fibers has resulted in the determination of the degree of N-deacetylation (DD) by measuring the infrared absorption at 1655 and 3450 cm^{-1}, the ratio of the two absorption parameters, and the use of standard curve obtained from chitosan samples with a known DD. As has already been observed by other approaches, this approach produces satisfactory results for analyzing chitosan DD [25].

2.2.2 Functional Groups

Using Fourier-transform infrared spectroscopy (FTIR) to analyse the biopolymer chitin, distinct signals ranging from 1700 to 1500 cm^{-1} are obtained for the C=O stretching area associated with the amide moiety in both α- and β-chitin. The amide I band recorded for α-chitin occurs as a split into two components—1660 and 1630 cm^{-1}, while for β-chitin, the amide I band occurs at 1630 cm^{-1}. Further, the amide II band occurs in both chitin allomorphs. For α-chitin, its band occurs at 1558 cm^{-1}, while for β-chitin, it corresponds to 1562 cm^{-1}. IR spectra obtained for β-chitin reveals two additional bands of CH_x deformation occurring at around 1455 and 1374 cm^{-1} [26]. Another study states that in the existence of the two amide groups, half of the total number of carbonyl groups are present in a bonded state with the amino group in the same chain through hydrogen bonds. This results in the spectral band at 1660 cm^{-2}. The remaining carbonyl groups have an additional -CH_2OH group in the side chain causing an overall reduction in the original vibration mode, the band occurring at 1627 cm^{-1}. Bands at 3268 and 3106 cm^{-1}, studied in α-chitin, are attributed to the modes of vibration corresponding to the NH present in the amide group [27]. Meanwhile, the GlcN monomeric unit in chitosan has a C2-positioned amine group, and the -OH groups exist at both the C3 and C6 positions [28]. The chitosan chemical structure that has been determined includes acetamido groups in addition to amine and hydroxyl groups [3]. The structure and composition of chitin and chitosan have been illustrated in Fig. 1.

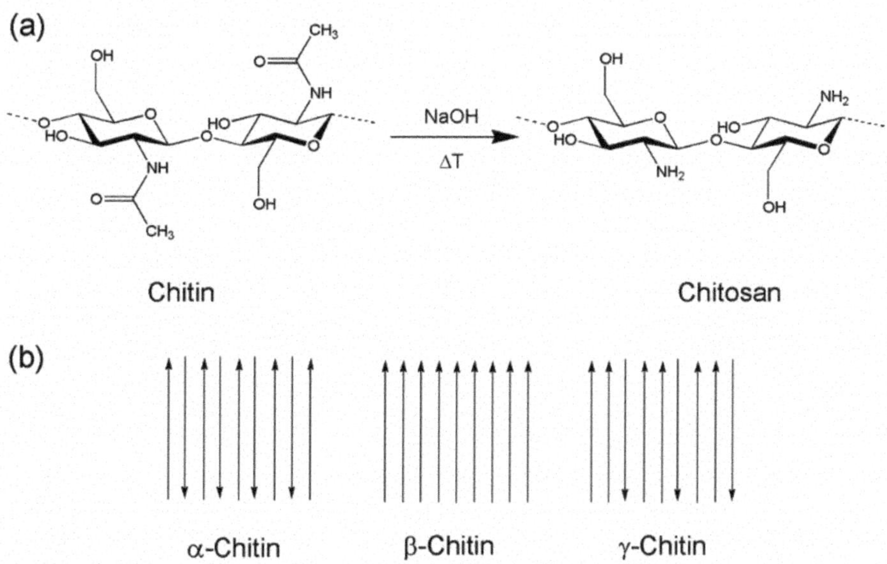

Fig. 1 Representation of the molecular structure and composition of chitin and chitosan [29]

3 Methods of Isolation

3.1 Traditional Extraction Techniques

3.1.1 Acidic Treatment

Demineralization is one of the conventional techniques for extracting chitin. Demineralization is the process of removing inorganic fillers such as calcium carbonate and calcium phosphate from the raw material that is being used to extract chitin. Inorganic acids such as nitric acid, sulfuric acid, and hydrochloric acid are used in this process. Strong organic acids like acetic and formic acid are also utilized for this purpose. The most extensively used acid among these is hydrochloric acid, which is used because it effectively eliminates any existent mineral impurities [30].

In order to carry out demineralization of salted cephalothoraxes obtained from *Parapenaeopsis stylifera* shrimps for amino acid analysis and protein characterization, 1 M HCl was used at ambient temperature. The solution-to-solid ratio used was 10 mL g^{-1} for a duration of 24 h [31]. Based on results from sensitive Inductively Coupled Plasma—Optical Emission Spectrometry (ICP-OES) used in the determination of an optimum process of demineralization for the waste from Omani *Portunidae segnis*, 2 M HCl used for a duration of 1 h with the treatment carried out at room temperature was the best optimized method for chitin extraction from the waste with minimum concentration obtained at the end of the process for calcium and phosphorus [32]. Lactic acid and acetic acid were found to have an effectiveness similar to HCl in demineralizing *Pandalus borealis* shells (Northern Pink Shrimp). A shells to acid ratio of 1:20, 24 °C as the reaction temperature, and a retention time of 2 h were determined as appropriate for the process. The efficiencies of removal of total minerals and calcium were found to be 97.4 and 99.11% for lactic acid, and 86.36 and 85.33% for acetic acid [33]. Acidic treatment has also been employed for the purpose of chitosan extraction from various sources. From the derivatives of cell wall prepared from *Rhizomucor pusillus*, chitosan extraction was carried out using different concentrations of hot sulphuric acid (0.5, 1.0, and 2.0% (v/v)) at 95 °C overnight, followed by treatment carried out at a high temperature of 121 °C with a time duration of 20 min in an autoclave. The acid/chitosan ratio was maintained at 100-, 150-, and 200 mL g^{-1}. Sulphuric acid of concentration 1.0% (v/v) was chosen as the satisfactory candidate for chitosan extraction [34]. The alkali-insoluble material (AIM) obtained from freeze-dried mycelia (FDM) of *Absidia glauca* was autoclaved in 1 M HCl solution with a treatment temperature of 121 °C for 15 min for chitosan extraction. This procedure yielded the highest amount of chitosan lacking N-deacetylation, but the recorded molecular weight was 3–4 times lower than other treatments. Another approach, which involved treatment with 2% acetic acid for 15 min at a temperature of 121 °C, yielded significant amounts of chitosan with a low degree of degradation and low N-deacetylation [35]. Demineralization of the milled adult beetles *Holotrichia parallela* has been carried out by the use of 1 M HCl for chitin extraction [36]. Demineralization of the exoskeleton of mud crabs for both chitin and chitosan extraction has been performed using 300 ml 2 mol dm^{-3} H_2SO_4 in which 10 g of dried crab shell powder was treated for 4 h at room temperature [37].

3.1.2 Alkaline Hydrolysis

Deproteinization is another conventional extraction technique that can be applied either before or after demineralization. It includes using alkalis such as potassium and sodium hydroxide to remove the protein content present in the source material. The most common alkali used in the industrial chitin synthesis process is sodium hydroxide [30]. The ratio of the alkali solution to the source material, processing temperature, and base concentration are some variables that affect how effective deproteinization is. Treatment of crustacean shell wastes is typically carried out using dilute NaOH with concentrations anywhere between 1 and 10% (w/v) with increased processing temperatures (65–100 °C) [38]. Extraction of chitin from *Agaricus bisporus* was carried out by treatment of 4 g each of lyophilized stipes, gills, and pileus with 120 mL of 1 M NaOH solution with the temperature maintained at 80 °C for 2 h in order to carry out the reaction [39]. From the beetle *Holotrichia parallela*, chitin extraction was carried out by deproteinization of milled adult beetles using 1 M NaOH following decolorization [36]. From chitin isolated from brown and pink shrimp shells, crab shells, two different varieties of squid pens, and shells sourced from *Procambarus clarkii* (Crayfish), chitosan was obtained following treatment of the chitin obtained from these sources with strong NaOH solution. The degree of deacetylation increased with an elevation in NaOH concentration [40]. The process of deproteinization of the exoskeleton of mud crabs was carried out using 300 ml 2 mol dm^{-3} NaOH, into which the sample was dissolved at room temperature for 4 h [37].

3.2 *Modern Extraction Methods*

3.2.1 Enzymatic Methods

Enzymatic extraction of chitin has been proven capable of yielding high-grade chitin by using white shrimp (*Litopenaeus vannamei*) waste under relatively mild and more eco-friendly conditions when compared to the production process of commercial chitin. Chitin isolated by trypsin and ficin in the presence/absence of sodium metabisulfite under mild alkaline hydrolysis yielded a smooth microfibrillar crystalline structure, a higher crystalline index (CI), and a higher degree of acetylation (DA) percent amounting to 81–83% as compared to commercial chitin [41]. Using a Box-Behnken design employing three variables and three levels, it was determined that *Bacillus mojavensis* A21 crude proteases can effectively perform deproteinization, removing 88 $\pm$ 5% of the shrimp (*Metapenaeus monoceros*) shell proteins with the enzyme/substrate ratio being 7.75 U mg^{-1}, incubation carried out for a period of 6 h, and a temperature of 60 °C [42]. Nine different crude enzymes were tested for their degrees of deproteinization (DDP) achieved with shrimp shell wastes as a measure to devise a process for the extraction of chitin plagued by less problems than the chemical treatment. Among the microorganism-derived enzymes, *B. mojavensis* A21 crude protease showed the highest DDP (77 ± 3%). Although the fish crude proteases also showed considerable efficiency in deproteinization, the protease from *Balistes capriscus* (Grey Triggerfish) provided a higher DDP (78 ± 2%) as compared to other enzymes derived from fishes [43]. Chitin recovery from deproteinized shells of *Litopenaeus vannamei* was performed by hydrolysis with chitinase rChit46. With a substrate (shrimp shell powder) concentration of 100 g L^{-1}, the highest chitin recovery rate (CRR) was obtained, with the CRR reaching 88.9% based on calculations from weight reductions, thus reporting a higher recovery as compared to other methods deduced for this purpose [44]. Comparison of various commercially available enzymes like Econase CEPi®, Maxazme™ NNP®, Cytolase PCL5®, Econase MP 1000®, Ccllupulin MG®, and Delvolase® revealed that Delvolase® possesses the highest deproteinization activity among these enzymes with 1% Delvolase resulting in an 85% deproteinization of crab shell waste within a single day. This deproteinization escalated to 90% upon further treatment [45]. An alkaline protease, the viscera of *Portunus segnis* (blue crab) as its source, was used to carry out the deproteinization of *P. segnis* as well as *Penaeus kerathurus* (shrimp) shells. With an enzyme/substrate ratio maintained at 5 U mg^{-1} of proteins at a pH 8.0 with temperature 50 °C for 3 h, the degrees of deproteinization recorded with *P. segnis* and *P. kerathurus* were 85 and 91%, respectively. The chitin extracted was identified as α-chitin [46]. Further chitin extraction has also been performed with shrimp head waste using the enzyme alcalase, with 0.5% of the enzyme giving a high protein recovery of 64% when used with 200 mmol L^{-1} sodium sulfite [47]. Like chitin, extraction of chitosan can also be carried out by enzymatic treatment. Chitosan extraction was carried out with the mycelia of *Gongronella butleri* acting as the source and the enzyme being utilized was α-amylase. This enzyme-mediated chitosan extraction resulted in a 2–5 times better yield as compared to the extraction carried out by demineralization and deproteinization [48]. Papain hydrolysis has also been performed to extract chitosan from *Penaeus vannamei* shells. Although the chitosan obtained upon treatment was partially purified, this method enabled amino acid isolation. Also, the easy availability of papain, as well as its low cost, make this method suitable for chitosan extraction [49].

3.2.2 Biological Methods

Chitin and chitosan have been biologically extracted from *Penaeus semisulcatus* waste by the usage of three different probiotic lactic acid bacteria, namely *Lactobacillus plantarum* PTTC 1058, *L. acidophilus* PTTC 1643, and *L. rhamnosus* PTTC 1637. Among these bacteria, *L. plantarum* is an efficient candidate for the microbe-mediated extraction process devised for chitin and chitosan from shell wastes [50].

Since demineralization and deproteinization constitute essential steps in the extraction procedure of chitin, these processes may also be carried out by biological treatment apart from the conventional chemical treatments. *Lactobacillus plantarum, L. salivarius, L. paracasei*, and *Serratia marcescens*, which are organic acid producing bacteria, and *Pseudomonas aeruginosa, P. acidilactici, Bacillus subtilis*, and *B. firmus*, which represent protease producing bacteria, can be employed for demineralization and deproteinization, respectively [51]. From the cephalothoraxes of *Litopenaeus vannamei*, biological extraction of chitosan has been carried out using *L. plantarum*, where the lactic acid fermentation (LAF) was performed by mixing the shrimp waste with 10% (w/w) sucrose along with 5% (v/w) bacterial inoculum and placing the mixture in a 5 kg column reactor for incubation at a temperature of 35 °C for 144 h [52]. A combinational treatment of lactic acid fermentation by *Lactobacillus* and proteolytic bacteria performing demineralization and deproteinization can be investigated as a possible alternative for higher yield of chitosan during the extraction procedure [53]. The different methods of isolation of chitin and chitosan have been shown conclusively in Fig. 2.

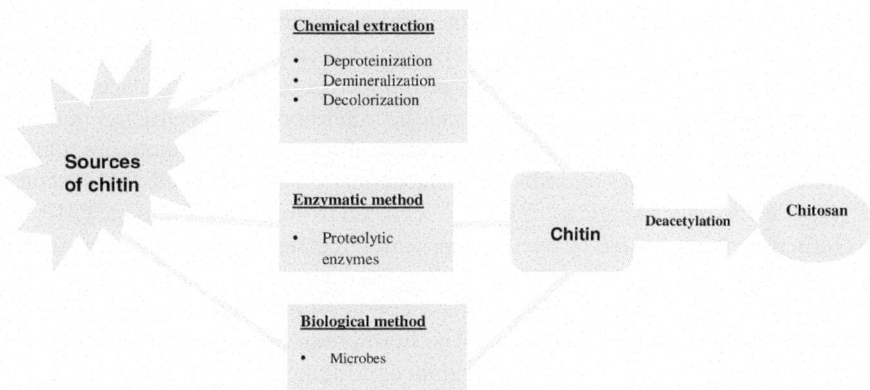

Fig. 2 Conventional techniques for isolating chitin and chitosan, as covered in this book chapter, include demineralization, deproteinization, enzymatic, and biological approaches [4]

4 Properties of Chitin and Chitosan

4.1 Physical Properties

4.1.1 Morphology and Surface Characteristics

According to scanning electron microscopy (SEM) studies, the surface morphological features of γ-chitin are composed of microfibers, while the surface structure of both α- and β-chitin is composed of nanofibers. Further investigations using atomic force microscopy (AFM) reveal that the rod-shaped microfibers of γ-chitin are made up of densely connected nanofibers [54]. However, chitin extracted from cockroach wings (*Periplaneta americana*) revealed a new morphology of chitin different from previous morphological observations. Upon observation with environmental scanning electron microscopy (ESEM), the absence of nanofibers and the presence of only oval nanopores ranging from 230 to 510 nm was revealed on the surface. The pores show a certain degree of ordered arrangement with a central pore surrounded by six or seven pores. Between the pores, structures bearing similarity to cell walls exist [55]. SEM analysis of chitin obtained from silkworm chrysalides shows the surface consisting of various fine, loosely united leaves. This observation corresponds to previous surface observations of chitin recorded with crustaceans [56]. Further, chitin obtained from the wings and the other body parts (OBP) of butterfly (*Argynnis pandora*) revealed a source-dependent alteration in morphologies. Out of five different morphologies recorded from the butterfly, chitin from OBP contributed four types, while only a single morphology occurred in the wings. From the OBP, the first chitin was characterized by overlapping scales, rectangular in shape with pointed tips and ridges. These scales contained embedded pores about 20 μm in size. Some smooth porous areas were observed in the second chitin with small pores scattered all over the surface of the porous area. The third chitin possessed some embedded tubular structures that provided the appearance of big pores, while the fourth chitin had a surface morphology defined by a plane area that was exposed and did not contain pores. Upon evaluation, the wing chitin showed a tough and rough surface, with fibers clearly visible under increased magnification [57]. Furthermore, the α-chitin obtained from six different species, *Asellus aquaticus* (belonging to Crustacea) and *Agabus bipustulatus*, *Anax imperator*, *Hydrophilus piceus*, *Notonecta glauca*, and *Ranatra linearis* (belonging to Insecta) had a similar morphology—presence of microfibrillar structure and an absence of porous structure [58]. Also, chitin obtained from two sources, one being *Geolycosa vultuosa* and the other *Hogna radiata*, also revealed characteristically different morphologies between the two species as studied by ESEM. Chitin from both species was identified as α-chitin. The chitin from *G. vultuosa* shows two contrasting pore structures with the type one pore being rarely sequenced, with size in the range of 190–240 nm, and the type two pore being tightly sequenced with a size of 11–32 nm. The chitin from *G. vultuosa* thus reveals a morphology not previously documented for chitin. The chitin studied from *H.*

radiata being the source material has typical morphology consisting of nanofibrillar structures measuring 11–17 nm with pores of size 195–260 nm [59]. Chitosan derived from *Thenus unimaculatus* after initial chitin extraction and analysed using SEM revealed a rough surface with pores and voids, block irregularities, and crystalline structures [60]. Transmission electron microscope (TEM) and low-voltage high-resolution scanning electron microscope (LVHRSEM) studies conducted on chitosan films demonstrated a surface characterized by smoothness and uniformity. The structure appeared dense with a few defects. AFM surface images also support a homogenous, smooth film structure with low micro-roughness [61]. Chitosan acetate containing 0.40 g of ethylene carbonate (a plasticizer) presents a smooth surface morphology compared to the softer and spongy surface characteristics present in fatty acid plasticized films when observed using SEM [62].

4.1.2 Solubility and Swelling Behaviour

While dilute acidic mediums at a pH less than the pKa of chitosan (pH = 6.5) can readily dissolve chitosan, chitin does not dissolve in organic and regular solvents. The solubility of chitin depends on various factors like fraction of the number of N-acetyl-D-glucosamine present, solution pH, molecular weight of the chitin being studied, ionic strength, and temperature [63]. NaOH/urea aqueous solution is capable of dissolving chitin. 8 wt% NaOH/4 wt% urea was deemed an adequate medium of solvation, with -20 °C as the optimal treatment temperature. The chitin aqueous solution was a pseudoplastic fluid at high concentration and a Newtonian fluid at low concentration [64]. Although its insolubility nullifies the usage of many ordinary and commonly used solvents, chitin is soluble in highly concentrated inorganic acids including sulphuric acid, hydrochloric acid, and phosphoric acid. Concentrated formic acid dissolves β-chitin. Other solvents such as lithium thiocyanate, hexafluoroacetone, $CaBr_2 \cdot H_2O$ saturated methanol, *N*-methyl-2-pyrrolidone, hexafluoroisopropyl alcohol, and methanol saturated with calcium chloride dihydrate are also capable of dissolving chitin [65]. Formic acid, although capable of dissolving chitin, becomes effective in its functionality only at concentrations as high as 98% or greater. In order to bring the solubility parameters (δ) of standard formic acid (95–97%, $\delta = 13.0$) within the range of that of chitin (δ in range of 12.1–12.5), a cosolvent with low δ, low viscosity, and poor (P) to moderate (M) hydrogen bonding potential can be used along with the acid in the mixture. Such cosolvents include methylene chloride and 2-chloroethanol. Further, as solvents, LiCl, trichloroacetic acid (TCA) and dichloroacetic acid (DCA) can also be used. Cosolvents like dimethylformamide, dimethyl sulfoxide, low molecular weight chlorohydrocarbons, acetonitrile, and nitro-paraffins can be used along with LiCl [65, 66]. Swelling of β-chitin from *Tevnia jerichonana* presents three levels of swelling: (a) inter-sheet swelling induced by water and methanol forming crystalline β-chitin cystallosolvates without any effect on the constitutive sheets, (b) combinational inter- and intra-sheet swelling caused by 6 N HCl or DCl leading to the formation of a paracrystalline gel-like product from β-chitin crystals, and (c) a severe inter- and intra-sheet swelling by

8 N HCl or DCl causing irreversible obliteration of the sheet organization and the parallel nature of the sheets [67]. A previous study similarly based on β-chitin from *T. jerichonana* also demonstrated identical results. HCl, with a normality of 6–7, could bring about swelling of the β-chitin crystal lattice, forming a decrystallized HCl-chitin association with the decrystallization being reversible. Swelling of the β-chitin in HCl having a normality 8 or above caused β-chitin to completely convert into α-chitin with irreversible damage to the microfibrillar structure [68].

Chitosan has been dissolved in carbonic acid using CO_2 gas to dissolve chitosan in water [69]. Chitosan can also be dissolved in 1-ethyl-3-methylimidazolium acetate (EMIM Ac), which is an ionic liquid. Freezing carried out at −20 °C follows this step of dissolution, with solvent exchange utilizing water as the subsequent step [70]. Chitosan dissolves well in diluted inorganic acids, such as perchloric, phosphoric, nitric, hydrochloric, and carbonic acids. Further, organic acids like acetic, formic, and lactic acids are also appropriate solvents. Different chitosan salts also suitably dissolve in dimethyl sulfoxide (DMSO) [71]. The swelling behavior of chitosan hydrogels shows both pH- and temperature dependence. Uncrosslinked chitosan (NCLChs) has the highest swelling ratio, increasing with temperature. However, the swelling behaviour of crosslinked chitosan hydrogels (CLChs1 and CLChs2) remained relatively constant over the studied temperature range. Also, the highest amount of swelling was recorded in the hydrogels at acidic pH compared to treatment at basic pH [72]. Chitin, chitosan, and microcrystalline chitin have a water-binding capacity ranging from 230 to 440% (w/w), with chitosan displaying the highest water-binding capacity [73].

4.2 Chemical Properties

4.2.1 Reactivity and Chemical Modifications

It has been shown that the rate of the deacetylation process of chitin is dependent on the crystallinity index of the chitin being used, with a lower crystallinity index being associated with a faster deacetylation process. Also, hydrolysis of α-chitin occurs with a rate constant different from that of β-chitin during the process of N-deacetylation of the sample chitin being used [74]. The reactivity of β-chitin isolated from squid pens has been found to be considerably higher than α-chitin, with degrees of occurrence of reactions such as N-acetylation, full acetylation, tosylation, and tritylation holding greater values as compared to those recorded for α-chitin [75]. Moreover, chitosan can participate in electrophilic and nucleophilic substitution processes [76]. The chemical alterations that chitin nanofibers can undergo include acetylation, surface maleylation, *N*-phthaloylation, naphthaloylation, and partial deacetylation, which results in chitin nanowhiskers. Additional modifications that have been accomplished include surface *N*-halamine chitin nanofibers, chitin nanocrystals made from 2,2,6,6-tetramethylpiperidine-1-oxyl radical (TEMPO)-mediated oxidation of α-chitin, and graft polymerization of acrylic acid on the surface chitin nanofibers [77]. Apart from acetylation of chitosan and deacetylation of chitin, other possible chemically-induced modifications exist in literature. Carbohydrate branched chitin and chitosan derivatives with α-mannoside at the C6 position and another with β-galactoside have been documented. Further, chitosan derivatives with lactose, glucose, galactose, and N-acetylglucosamine branches have also been documented. Graft polymers of chitin and chitosan with polyethylene glycol (PEG) have also been synthesized [78]. In addition, surface modification of chitin nanowhiskers by alkenyl succinic anhydride (ASA) and by a chemical reaction between hydroxyl and isocyanate groups belonging to phenyl isocyanate (PI) and isopropenyl-α,α'-dimethylbenzyl isocyanate (TMI) have also been carried out in different experimental studies. Furthermore, β-(1 → 4)-linked polyuronic acids like chitouronic acid can be synthesized from TEMPO oxidation. The sodium salt of *N*-acetylglucosaminuronic acid constitutes the repeating unit present in these acids [79]. Also, Schiff reaction occurring between chitosan and aldehydes or ketones forming aldimines and ketimines, respectively, *O*-carboxymethylation, *N*-acylation, sulfation, and cyanoethylation of chitin and chitosan have been recorded as well [80].

4.2.2 pH Sensitivity

The adsorption capacity of chitin demineralized with HCl recorded at pH 3.0 was 1.9- to 2.3-fold higher as compared to the values recorded without pH adjustment. Chitin demineralized with HCl and deproteinized with KOH also exhibits increased adsorption after pH adjustment to 3.0. This adsorption capacity was measured for helactine, polactine, and remazol, all of which are reactive dyes [81]. Chitosan films demonstrated changes occurring with alterations in pH. As the pH increased, water vapour permeability (WVP) and the total soluble matter (TSM) also increased. A considerable decrease was observed in tensile strength (TS). Also, the chitosan film spectra showed an absence of the NH_3^+ band when the chitosan was maintained at pH 5.0 [82]. Between pH 2.0–5.0, chitosan-organic acid solutions presented an increase in apparent viscosities when pH was decreased, while an opposite behaviour was observed with chitosan-HCl solutions. Between pH 5.0–6.0, apparent viscosities of chitosan-NaOH solutions demonstrated a decreasing trend with increasing pH. However, these apparent viscosities increased when the pH was further raised to the range of 6.0–6.4 [83]. Chitosan derivatives with an acetyl content of more than 50% are stable even in alkaline conditions without the occurrence of gelation or precipitation of the solutions [84].

5 Applications of Chitin and Chitosan

Chitin and chitosan have been shown to accelerate the wound-healing process. Shrimp chitin and chitosan show activities of up to +30%, while chitin derived from lobster and king-crab shells had activities reaching +75%. Different forms of chitin, like regenerated chitin fibers, sponges, films, and non-woven mats had wound-healing activities of +30% or above [85]. While both chitin and chitosan induced an increase in prolyl hydroxylase (PHL) activity which is concerned with collagen synthesis required for efficient wound healing, higher concentrations of chitin were found to have a profound impact on stable collagen synthesis during the initial stages of wound recovery [86]. Chitin and chitosan find diverse applications in several prevalent drug delivery systems. Oral, gastric-specific, colon-specific, buccal, periodontal, topical and transdermal, ocular, nasal, pulmonary, vaginal, rectal, and parenteral drug delivery can be carried out using chitin and chitosan in combination with different polymers loaded with a variety of drugs in different dosage forms [87]. Gastrointestinal, ophthalmic, and sublingual drug delivery systems have also been developed using chitosan. Chitosan-alginate polyelectrolyte complex (PEC), chitosan aspartate, chitosan lactate, and chitosan glutamate have been employed as drug carriers for targeted delivery of the drugs at the desired locations [88]. Chitosan coatings have been shown to enhance the shelf life of carrots, apples, mangoes, and strawberries. Also, chitosan films, in conjunction with chemical compounds like glycerol or Tween 80, are used for wetting surfaces of products like apple skins. Anti-microbial properties have also been associated with the beneficial properties linked with these edible films and coatings [89]. Chitin significantly inhibits the occurrence of Fusarium yellows of celery and also increases populations of bacteria and potential chitin-degrading actinomycetes in the soil, which, in turn, can also inhibit the growth of pathogenic *Fusarium oxysporum* f. sp. *apii*. In vitro studies also showed that chitosan inhibited the development of five isolates of *F. oxysporum* [90]. For water treatment, chitosan is generally preferred over chitin owing to its high degree of deacetylation and low crystallinity. Highly deacetylated chitosan shows a higher adsorption efficiency for azo-dyes. Also, polycyclic aromatic hydrocarbons, anionic dyes, metal ions, and polychlorinated biphenyls (PCBs) are more efficiently removed by chitosan [91]. These materials are also employed in the paper industry. Chitosan is used in the manufacturing of paper due of its resemblance to cellulose. Also, addition of chemical in the paper manufacturing process is reduced along with an increase in output. Further, the paper produced by the use of chitosan is characterized by a smoother surface and the paper is also moisture resistant. Using these biopolymers, packaging materials become environmentally friendly, as also seen among other products derived from paper [92].

6 Conclusion

The use and importance of chitin and chitosan has rapidly escalated in the recent industrial and commercial exploitation of these bioresources. Available from various sources discussed previously, a number of methods of extraction and isolation of chitin and chitosan have been formulated and rigorously tested in a number of experimental studies. However, chemical methods of performing demineralization and deproteinization for the extraction of chitin and finally chitosan demand considerable supply of these chemicals which can prove to be expensive as well as dangerous in the handling procedures. More recently, enzymatic and biological methods have been devised which enable chitin and chitosan extraction from shell wastes, exoskeletons, and other sources in an environmental-friendly approach, being cost-effective and less harmful at the same time. Properties of chitin and chitosan, like molecular structure, monomeric units, polymers, morphological characteristics, solubility, and pH sensitivity, have proven highly beneficial for applying these materials in various sectors of public importance. With abundant scientific knowledge of these properties, uses can be tailored according to the needs of particular industries. These uses can belong to various prevalent domains in the current global scenario. These highly valuable materials have found significant applications in many aspects of the pharmaceutical industry, water treatment protocols, food preservation, biological control for the protection of plant populations, and the production of biodegradable materials that provide durability, flexibility, and non-toxicity. However, such elaborate employment of these biopolymers has been accompanied by ongoing scientific challenges. These include determining a highly appropriate isolation method for these materials, eliminating all possible issues existing in current isolation methods, and ensuring maximized efficacy. Further, more concrete breakthrough solutions need to be created in order to utilize chitin and chitosan as an irreplaceable component of major industries. Upon solving these questions of attention, chitin and chitosan present a stupendous horizon illuminated with revolutions in the clinical practice of drug administration enabling increased life expectancy and effective treatment of diseases, reduction in suffering induced by trauma and accidents in humans, prevention of damage to food grains thus solving global hunger issues, and maintaining and enhancing global sustainability for future generations.

References

1. Yeul VS, Rayalu SS (2013) Unprecedented chitin and chitosan: a chemical overview. J Polym Environ 21:606–614
2. Kardas I, Struszczyk MH, Kucharska M, van den Broek LA, van Dam JE, Ciechańska D (2012) Chitin and chitosan as functional biopolymers for industrial applications. In: The European polysaccharide network of excellence (EPNOE) research initiatives and results, pp 329–373
3. Mourya VK, Inamdar NN (2008) Chitosan-modifications and applications: opportunities galore. React Funct Polym 68(6):1013–1051
4. Iber BT, Kasan NA, Torsabo D, Omuwa JW (2022) A review of various sources of chitin and chitosan in nature. J Renew Mater 10(4):1097
5. Bastiaens L, Soetemans L, D'Hondt E, Elst K (2019) Sources of chitin and chitosan and their isolation. In: Chitin and chitosan: properties and applications, pp 1–34
6. Philibert T, Lee BH, Fabien N (2017) Current status and new perspectives on chitin and chitosan as functional biopolymers. Appl Biochem Biotechnol 181:1314–1337
7. Hamed I, Özogul F, Regenstein JM (2016) Industrial applications of crustacean by-products (chitin, chitosan, and chitooligosaccharides): a review. Trends Food Sci Technol 48:40–50
8. Tharanathan RN, Kittur FS (2003) Chitin—the undisputed biomolecule of great potential

9. Franca EF, Lins RD, Freitas LC, Straatsma TP (2008) Characterization of chitin and chitosan molecular structure in aqueous solution. J Chem Theory Comput 4(12):2141–2149
10. Mogilevskaya EL, Akopova TA, Zelenetskii AN, Ozerin AN (2006) The crystal structure of chitin and chitosan. Polym Sci, Ser A 48:116–123
11. Gardner KH, Blackwell J (1975) Refinement of the structure of β-chitin. Biopolym: Orig Res Biomol 14(8):1581–1595
12. Salmon S, Hudson SM (1997) Crystal morphology, biosynthesis, and physical assembly of cellulose, chitin, and chitosan. J Macromol Sci Part C: Polym Rev 37(2):199–276
13. Carlstrom D (1957) The crystal structure of α-chitin (poly-N-acetyl-D-glucosamine). J Cell Biol 3(5):669–683
14. Minke RAM, Blackwell J (1978) The structure of α-chitin. J Mol Biol 120(2):167–181
15. Sikorski P, Hori R, Wada M (2009) Revisit of α-chitin crystal structure using high resolution X-ray diffraction data. Biomacromol 10(5):1100–1105
16. Dweltz NE (1961) The structure of β-chitin. Biochem Biophys Acta 51(2):283–294
17. Moussian B (2019) Chitin: structure, chemistry and biology. In: Targeting chitin-containing organisms, pp 5–18
18. Younes I, Rinaudo M (2015) Chitin and chitosan preparation from marine sources. Structure, properties and applications. Mar Drugs 13(3):1133–1174
19. Ogawa K, Yui T, Okuyama K (2004) Three D structures of chitosan. Int J Biol Macromol 34(1–2):1–8
20. Yui T, Imada K, Okuyama K, Obata Y, Suzuki K, Ogawa K (1994) Molecular and crystal structure of the anhydrous form of chitosan. Macromolecules 27(26):7601–7605
21. Foster AB, Webber JM (1961) Chitin. In Advances in carbohydrate chemistry, vol 15. Academic, pp 371–393
22. Einbu A, Vårum KM (2008) Characterization of chitin and its hydrolysis to GlcNAc and GlcN. Biomacromol 9(7):1870–1875
23. Okamoto Y, Watanabe M, Miyatake K, Morimoto M, Shigemasa Y, Minami S (2002) Effects of chitin/chitosan and their oligomers/monomers on migrations of fibroblasts and vascular endothelium. Biomaterials 23(9):1975–1979
24. Tyliszczak B, Drabczyk A, Kudłacik-Kramarczyk S, Sobczak-Kupiec A (2020) Sustainable production of chitosan. In: Sustainable production: novel trends in energy, environment and material systems 45–60
25. Sabnis S, Block LH (1997) Improved infrared spectroscopic method for the analysis of degree of N-deacetylation of chitosan. Polym Bull 39:67–71
26. Kumirska J, Czerwicka M, Kaczyński Z, Bychowska A, Brzozowski K, Thöming J, Stepnowski P (2010) Application of spectroscopic methods for structural analysis of chitin and chitosan. Mar Drugs 8(5):1567–1636
27. Cárdenas G, Cabrera G, Taboada E, Miranda SP (2004) Chitin characterization by SEM, FTIR, XRD, and 13C cross polarization/mass angle spinning NMR. J Appl Polym Sci 93(4):1876–1885
28. Jaworska MM, Antos D, Górak A (2020) Review on the application of chitin and chitosan in chromatography. React Funct Polym 152:104606
29. Hahn T, Tafi E, Paul A, Salvia R, Falabella P, Zibek S (2020) Current state of chitin purification and chitosan production from insects. J Chem Technol Biotechnol 95(11):2775–2795
30. Broquá J, Zanin BG, Flach AM, Mallmann C, Taborda FGD, Machado LEL, Alves SML, Silva MM, Dias RJSP (2019) Methods of chitin production a short review. Am J Biomed Sci Res 3:307–314
31. Percot A, Viton C, Domard A (2003) Characterization of shrimp shell deproteinization. Biomacromol 4(5):1380–1385
32. Al Shaqsi NHK, Al Hoqani HAS, Hossain MA, Al Sibani MA (2020) Optimization of the demineralization process for the extraction of chitin from Omani Portunidae segnis. Biochem Biophys Rep 23:100779
33. Mahmoud NS, Ghaly AE, Arab F (2007) Unconventional approach for demineralization of deproteinized crustacean shells for chitin production. Am J Biochem Biotechnol 3(1):1–9

34. Zamani A, Edebo L, Sjöström B, Taherzadeh MJ (2007) Extraction and precipitation of chitosan from cell wall of zygomycetes fungi by dilute sulfuric acid. Biomacromol 8(12):3786–3790
35. Hu KJ, Yeung KW, Ho KP, Hu JL (1999) Rapid extraction of high-quality chitosan from mycelia of absidia glauca. J Food Biochem 23(2):187–196
36. Liu S, Sun J, Yu L, Zhang C, Bi J, Zhu F, Qu M, Jiang C, Yang Q (2012) Extraction and characterization of chitin from the beetle Holotrichia parallela Motschulsky. Molecules 17(4):4604–4611
37. Ali M, Shakeel M, Mehmood K (2019) Extraction and characterization of high purity chitosan by rapid and simple techniques from mud crabs taken from Abbottabad. Pak J Pharm Sci 32(1):171–175
38. Synowiecki J, Al-Khateeb NA (2003). Production, properties, and some new applications of chitin and its derivatives
39. Hassainia A, Satha H, Boufi S (2018) Chitin from Agaricus bisporus: extraction and characterization. Int J Biol Macromol 117:1334–1342
40. Abdou ES, Nagy KS, Elsabee MZ (2008) Extraction and characterization of chitin and chitosan from local sources. Bioresour Technol 99(5):1359–1367
41. Marzieh MN, Zahra F, Tahereh E, Sara KN (2019) Comparison of the physicochemical and structural characteristics of enzymatic produced chitin and commercial chitin. Int J Biol Macromol 139:270–276
42. Younes I, Ghorbel-Bellaaj O, Nasri R, Chaabouni M, Rinaudo M, Nasri M (2012) Chitin and chitosan preparation from shrimp shells using optimized enzymatic deproteinization. Process Biochem 47(12):2032–2039
43. Younes I, Hajji S, Frachet V, Rinaudo M, Jellouli K, Nasri M (2014) Chitin extraction from shrimp shell using enzymatic treatment. Antitumor, antioxidant and antimicrobial activities of chitosan. Int J Biol Macromol 69:489–498
44. Deng JJ, Mao HH, Fang W, Li ZQ, Shi D, Li ZW, Zhou T, Luo XC (2020) Enzymatic conversion and recovery of protein, chitin, and astaxanthin from shrimp shell waste. J Clean Prod 271:122655
45. Jo GH, Park RD, Jung WJ (2011) Enzymatic production of chitin from crustacean shell waste. In: Chitin, chitosan, oligosaccharides and their derivatives, pp 37–45
46. Hamdi M, Hammami A, Hajji S, Jridi M, Nasri M, Nasri R (2017) Chitin extraction from blue crab (Portunus segnis) and shrimp (Penaeus kerathurus) shells using digestive alkaline proteases from P. segnis viscera. Int J Biol Macromol 101:455–463
47. Mizani AM, Aminlari BM (2007) A new process for deproteinization of chitin from shrimp head waste. In:Proceedings of European congress of chemical engineering (ECCE-6), September 2007. Copenhagen, pp 16–20
48. Nwe N, Stevens WF (2002) Production of fungal chitosan by solid substrate fermentation followed by enzymatic extraction. Biotechnol Lett 24:131–134
49. Gartner C, Peláez CA, López BL (2010) Characterization of chitin and chitosan extracted from shrimp shells by two methods. E-Polymers 10(1):069
50. Khanafari A, Marandi R, Sanati SH (2008) Recovery of chitin and chitosan from shrimp waste by chemical and microbial methods
51. Pal J, Verma HO, Munka VK, Maurya SK, Roy D, Kumar J (2014) Biological method of chitin extraction from shrimp waste an eco-friendly low cost technology and its advanced application. Int J Fish Aquat Stud 1(6):104–107
52. Pacheco N, Garnica-Gonzalez M, Gimeno M, Bárzana E, Trombotto S, David L, Shirai K (2011) Structural characterization of chitin and chitosan obtained by biological and chemical methods. Biomacromol 12(9):3285–3290
53. Hisham F, Akmal MM, Ahmad F, Ahmad K, Samat N (2023) Biopolymer chitosan: potential sources, extraction methods, and emerging applications. Ain Shams Eng J 102424
54. Kaya M, Mujtaba M, Ehrlich H, Salaberria AM, Baran T, Amemiya CT, Galli R, Akyuz L, Sargin I, Labidi J (2017) On chemistry of γ-chitin. Carbohydr Polym 176:177–186
55. Kaya M, Baran T (2015) Description of a new surface morphology for chitin extracted from wings of cockroach (Periplaneta americana). Int J Biol Macromol 75:7–12

56. Paulino AT, Simionato JI, Garcia JC, Nozaki J (2006) Characterization of chitosan and chitin produced from silkworm crysalides. Carbohydr Polym 64(1):98–103
57. Kaya M, Bitim B, Mujtaba M, Koyuncu T (2015) Surface morphology of chitin highly related with the isolated body part of butterfly (Argynnis pandora). Int J Biol Macromol 81:443–449
58. Kaya M, Baran T, Mentes A, Asaroglu M, Sezen G, Tozak KO (2014) Extraction and characterization of α-chitin and chitosan from six different aquatic invertebrates. Food Biophys 9:145–157
59. Kaya M, Seyyar O, Baran T, Erdoğan S, Kar M (2014) A physicochemical characterization of fully acetylated chitin structure isolated from two spider species: with new surface morphology. Int J Biol Macromol 65:553–558
60. Arasukumar B, Prabakaran G, Gunalan B, Moovendhan M (2019) Chemical composition, structural features, surface morphology and bioactivities of chitosan derivatives from lobster (Thenus unimaculatus) shells. Int J Biol Macromol 135:1237–1245
61. Jiang H, Su W, Caracci S, Bunning TJ, Cooper T, Adams WW (1996) Optical waveguiding and morphology of chitosan thin films. J Appl Polym Sci 61(7):1163–1171
62. Yahya MZA, Harun MK, Ali AMM, Mohammat MF, Hanafiah MAKM, Ibrahim SC, Mustaffa M, Darus ZM, Latif F (2006) XRD and surface morphology studies on chitosan-based film electrolytes. J Appl Sci 6(15):3150–3154
63. Roy JC, Salaün F, Giraud S, Ferri A, Chen G, Guan J (2017) Solubility of chitin: solvents, solution behaviors and their related mechanisms. Solub Polysacch 3:20–60
64. Hu X, Du Y, Tang Y, Wang Q, Feng T, Yang J, Kennedy JF (2007) Solubility and property of chitin in NaOH/urea aqueous solution. Carbohydr Polym 70(4):451–458
65. El Knidri H, Laajeb A, Lahsini A (2020) Chitin and chitosan: chemistry, solubility, fiber formation, and their potential applications. In: Handbook of chitin and chitosan. Elsevier, pp 35–57
66. Austin PR (1984) Chitin solvents and solubility parameters. In: Chitin, chitosan, and related enzymes. Academic, pp 227–237
67. Saito Y, Okano T, Gaill F, Chanzy H, Putaux JL (2000) Structural data on the intra-crystalline swelling of β-chitin. Int J Biol Macromol 28(1):81–88
68. Saito Y, Putaux JL, Okano T, Gaill F, Chanzy H (1997) Structural aspects of the swelling of β chitin in HCl and its conversion into α chitin. Macromolecules 30(13):3867–3873
69. Sakai Y, Hayano K, Yoshioka H, Yoshioka H (2001) A novel method of dissolving chitosan in water for industrial application. Polym J 33(8):640–642
70. Li B, Wang J, Moustafa ME, Yang H (2019) Ecofriendly method to dissolve chitosan in plain water. ACS Biomater Sci Eng 5(12):6355–6360
71. Lizardi-Mendoza J, Monal WMA, Valencia FMG (2016) Chemical characteristics and functional properties of chitosan. In: Chitosan in the preservation of agricultural commodities. Academic, pp 3–31
72. Rohindra DR, Nand AV, Khurma JR (2004) Swelling properties of chitosan hydrogels. South Pac J Nat Appl Sci 22(1):32–35
73. Knorr D (1982) Functional properties of chitin and chitosan. J Food Sci 47(2):593–595
74. Arrouze F, Desbrieres J, Rhazi M, Essahli M, Tolaimate A (2019) Valorization of chitins extracted from North Morocco shrimps: comparison of chitin reactivity and characteristics. J Appl Polym Sci 136(30):47804
75. Kurita K, Ishii S, Tomita K, Nishimura SI, Shimoda K (1994) Reactivity characteristics of squid β-chitin as compared with those of shrimp chitin: high potentials of squid chitin as a starting material for facile chemical modifications. J Polym Sci Part A: Polym Chem 32(6):1027–1032
76. Akman F (2017) Prediction of chemical reactivity of cellulose and chitosan based on density functional theory. Cellul Chem Technol 51(3–4):253–262
77. Ifuku S (2014) Chitin and chitosan nanofibers: preparation and chemical modifications. Molecules 19(11):18367–18380
78. Morimoto M, Saimoto H, Shigemasa Y (2002) Control of functions of chitin and chitosan by chemical modification. Trends Glycosci Glycotechnol 14(78):205–222

79. Mincea M, Negrulescu A, Ostafe V (2012) Preparation, modification, and applications of chitin nanowhiskers: a review. Rev Adv Mater Sci 30(3):225–242
80. Kurita K (1986) Chemical modifications of chitin and chitosan. Chitin in nature and technology. Springer US, Boston, pp 287–293
81. Klimiuk E, Filipkowska U, Wojtasz-Pajak A (2003) The effect of pH and chitin preparation on adsorption of reactive dyes. Pol J Environ Stud 12(5):575–588
82. Kim KM, Son JH, Kim SK, Weller CL, Hanna MA (2006) Properties of chitosan films as a function of pH and solvent type. J Food Sci 71(3):E119–E124
83. Chen RH, Lin WC, Lin JH (1994) Effects of pH, ionic strength, and type of anion on the rheological properties of chitosan solutions. Acta Polym 45(1):41–46
84. Aiba SI (1989) Studies on chitosan: 2 Solution stability and reactivity of partially N-acetylated chitosan derivatives in aqueous media. Int J Biol Macromol 11(4):249–252
85. Balassa LL, Prudden JF (1978) Applications of chitin and chitosan in wound-healing acceleration. In: Proceedings of the first international conference on chitin/chitosan, May 1978. Massachusetts Institute of Technology, Cambridge, MA, pp 296–305
86. Kojima K, Okamoto Y, Kojima K, Miyatake K, Fujise H, Shigemasa Y, Minami S (2004) Effects of chitin and chitosan on collagen synthesis in wound healing. J Vet Med Sci 66(12):1595–1598
87. Parhi R (2020) Drug delivery applications of chitin and chitosan: a review. Environ Chem Lett 18(3):577–594
88. Mitra A, Dey B (2011) Chitosan microspheres in novel drug delivery systems. Indian J Pharm Sci 73(4):355
89. De Azeredo HMC, Britto DD, Assis OB (2011) Chitosan edible films and coatings-a review 179–194
90. Bell AA, Hubbard JC, Liu L, Davis RM, Subbarao KV (1998) Effects of chitin and chitosan on the incidence and severity of Fusarium yellows of celery. Plant Dis 82(3):322–328
91. Kaur S, Dhillon GS (2014) The versatile biopolymer chitosan: potential sources, evaluation of extraction methods and applications. Crit Rev Microbiol 40(2):155–175
92. Dutta PK, Dutta J, Tripathi VS (2004) Chitin and chitosan: chemistry, properties and applications. J Sci Ind Res 63:20–31

Properties and Synthesis of Chitin-Based Nanoparticles

Nisha Sharma, Kalpana Kushwaha, Swarnakshi Upadhyay, Priyanka Mishra, Sweta Rai, Jay Kumar, and Prakash Chandra Gupta

Abstract Chitin, a biopolymer, has drawn a lot of interest as a flexible precursor for creating nanoparticles with various uses. Chitin is unique in that it is biocompatible, biodegradable, and non-toxic. These qualities have sparked research into chitin-based nanoparticles as potential solutions in diverse industries, including agriculture, medicine, and environmental remediation. The innate characteristics of chitin play a vital role in the development of its nanoparticles. The distinctive structure of chitin, which consists of N-acetylglucosamine units that are -(14)-linked, makes it a perfect scaffold for nanoparticle formation. Chitin-based nanoparticles have been created using a variety of synthesis techniques, including chemical, physical, and biological ones. These methods make it possible to precisely control the size, shape, and surface properties of nanoparticles, making it easier to design them for particular uses. Through cross-linking or grafting processes, chitin is frequently modified chemically to produce nanoparticles with improved functionality and durability. Biotechnological methods use chitin's enzymatic breakdown to create nanoparticles, matching with sustainable and environmentally beneficial practices. Chitin and supporting components interact dynamically in the complex processes driving chitin-based nanoparticle formation. Understanding synthesis pathways is crucial for accurate nanoparticle engineering since these interactions affect the form and characteristics of nanoparticles. Additionally, the extraordinary physicochemical properties of chitin-based nanoparticles, such as their large surface area, porosity, and potential for functionalization, expand the range of applications for these materials, including drug delivery, wound healing, agricultural formulations, and pollutant sequestration. This chapter provides a thorough insight into chitin-based nanoparticles' properties, synthesis, characterization, and applications.

Keywords Chitin · Nanoparticles · Biopolymer · Biocompatibility · Biodegradability · Drug delivery · Pollutant sequestration · Nanotechnology

N. Sharma · K. Kushwaha · S. Upadhyay · P. Mishra · S. Rai · J. Kumar · P. C. Gupta (✉)
School of Pharmaceutical Sciences, Chhatrapati Shahu Ji Maharaj University, Kanpur 208024, India
e-mail: prakashchandragupta@csjmu.ac.in

R. K. Bachheti et al. (eds.), *Chitin-Based Nanoparticles for the Agriculture Sectors*, Smart Nanomaterials Technology, https://doi.org/10.1007/978-981-96-0920-8_2

1 Introduction

Chitin is a naturally occurring organic compound that is the second most abundant polysaccharide found in the planet's biosphere after cellulose. The word Chitin is derived from the word '*Chiton*', which in Latin language means Mollusc, which can be so interpreted as an exoskeleton of a mollusc. In the Greek language, it is connected to the word '*Khiton*', which means shirt [12]. The discovery of chitin is commonly attributed to Henri Braconnot, a French naturalist from Nancy, who identified this biopolymer in 1811 in extracts of mushrooms. However, it is worth noting that an English scientist named A. Hachett reported in 1799 the existence of a material that exhibited exceptional resistance to typical chemicals, which may have been chitin.

Nonetheless, Hachett did not pursue further investigations into this material. Consequently, Braconnot's identification of chitin in mushroom extracts and subsequent naming it as "fungine" is widely recognized as the pivotal discovery in the field. Several years later, namely in 1823, Auguste Odier made the discovery of the identical biopolymer within the exoskeleton of the insect. Consequently, Odier named "chitin" to this substance, drawing inspiration from the Greek term "chiton," which signifies a "tunic." Chitin is a prevalent component observed in the exoskeletons of many insects, the shells of crustaceans, the cell walls of coralline algae, and numerous marine species, fungi, plants, and bacteria. The presence of iridophores reflecting material has been seen in the epidermis and eyes of specific arthropods and cephalopods. Chitin derived from marine sources offers several advantages compared to terrestrial animal sources. Chitin forms an important part of cell walls in fungi, whereas it is absent in plants and vertebrates. Chitin in its crystallite, i.e., fibrillar form, provides protective properties to the organism's cell walls in the form of mechanical support and barrier, which is protective in nature. These advantages include minimal regulatory compliance and quality control issues, a much-reduced presence of biological pollutants and toxins, a low inflammatory reaction, and enhanced metabolic compatibility.

Chemically, chitin is a β (1,4) linked homopolymer of N acetyl-glucosamine, a simple polysaccharide type. Chitin is a heavy molecule with a molecular weight of 1000 KDa. Chitin shares similar chemical structure with cellulose as it has hydroxyl group at the C-2 position replaced by the acetamido group [21]. Chitin is usually insoluble in water. Chitin is generally extracted from the exoskeleton of crustaceans via the process of demineralization and deproteination.

Chitin and its derivatives are one of the most prevalent compounds due to various types of advantages like non-toxicity, bio-renewability, bio-degradability and bio-compatibility, due to which it finds numerous applications in a wide array of industries like agriculture, pharmacy, cosmetics and food etc. Among the numerous chitin derivatives, chitosan is one of the most valuable as it is used in many industries like agriculture and pharmacy due to its properties like less toxicity, bio-renewability, and biodegradability.

2 Properties of Chitin Nanoparticles (CNP)

2.1 *Structural Properties*

Chitin is a ubiquitous biopolymer that forms the backbone of crustacean (crab, shrimp, and lobster) shells, insect and molluscs exoskeletons, and the cell walls of certain fungi [3]. In addition to other macromolecules, chitin is often found in many different organisms. However, neither chitin nor its derivative, chitosan, is present in the composition of higher animals and plants [44]. Researchers have shown that marine creatures have 20–30% chitin, 30–40% proteins, 0–14% lipids, and 30–50% minerals [13]. But ascomycetes, zygomycetes, basidiomycetes, and deuteromycetes all have chitin in their cell walls and septa. The chitosan and chitin content of the cell walls of zygomycetes is high. 10–20% chitin, 50–60% glucans, 20–30% glycoproteins, and traces of lipids, pigments, and inorganic salts comprise the fungal cell wall [16]. There are three crystalline forms of chitin found in nature: α, β, γ chitin, with distinct physicochemical characteristics dependent on hydration level, unit cell size, and chitin chain density, respectively [64].

Chitin can be classified based on the arrangement of the polymer chains into three different polymer forms, i.e., α-chitin, β-chitin, and γ-chitin [54]. In α-chitin polymer chains are antiparallel to each other, whereas in the β-chitin polymer chains are parallel to each other, and in γ-chitin there is a combination of α + β-chitin arrangement, i.e., some polymer chains are parallel while some are antiparallel [60] (Fig. 1).

2.2 *Physicochemical Properties of Chitin*

The -(1–4) linkage holds hundreds of D-glucosamine residues that make up chitin. The number of N-acetyl-D-glucosamine units in chitosan ranges from 15 to 50%, depending on the degree of deacetylation of chitin. Since chitin and chitosan differ in nitrogen concentration (5–8%, respectively), they serve different biological functions. Chitosan is made by converting acetamide groups in chitin into primary amino groups, making it an N-deacetylated derivative of chitin. Because of the presence of primary and secondary hydroxyl groups on each repeat unit and the amine group on each deacetylated unit, chitosan is more water-soluble and chemically reactive than chitin [1].

2.2.1 Solubility

Chitin can be dissolved in an acidic solvent but not in a basic. Deacetylating the chitin produces soluble chitosan with main amino groups with a PKa of 6.5, despite chitin's

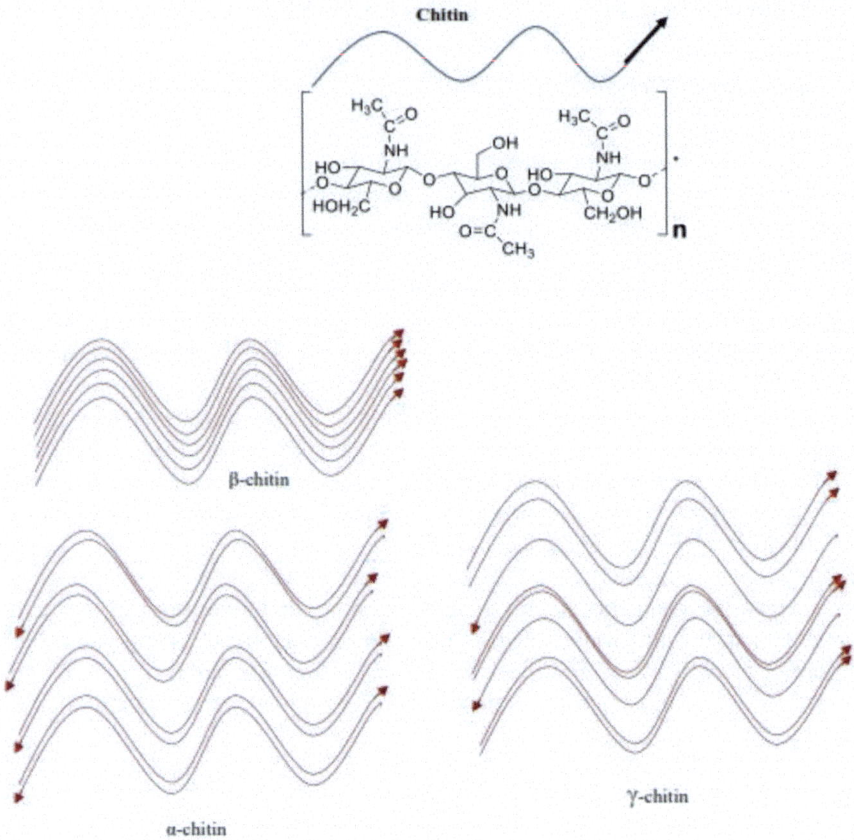

Fig. 1 Structure of α-chitin, β-chitin and γ-chitin

inherent insolubility in solvents. Solubility in acidic solutions is due to the protonation and positive charge of the amine in chitin. On the other hand, they become uncharged and insoluble at pH levels of 6 or higher [75]. Chitin solubility is influenced by a number of factors besides pH, including molecular weight, degree of deacetylation, temperature, and polymer crystallinity. α-Chitin has moderate solubility in many solvents, including aqueous thiourea solution, aqueous alkaline urea solution, 5% LiCl/DMAC, some ionic liquids, hexafluoroacetone, hexafluoro-2-propanol, and methane-sulfonic acid. In contrast, the β-chitin structure exhibits the ability to undergo swelling in aqueous environments, forming a suspension. Additionally, it demonstrates solubility in formic acid [53].

2.2.2 Molecular Weight

The molecular weight of chitin has a significant impact on its physicochemical and biological qualities. The molecular weight of chitin varies based on the starting material and processing methods. Based on its molecular weight distribution, chitin may be categorized as high, medium, or low [41]. Increasing the molecular weight of chitosan makes it less soluble and more viscous, which is undesirable in many industrial settings. Low molecular weight chitin is favoured because of its superior solubility and durability in biological and commercial settings [77].

2.2.3 Degree of Deacetylation

The degree of deacetylation of chitin is an additional component that impacts its physicochemical qualities, activity, and potential uses. The degree of deacetylation is measured in terms of amino group distribution throughout the polymer chain. Since the amino group in the polymer chain is cationic, chitin behaves as a cation in basic environments. Consequently, the degree of deacetylation significantly impacts both solubility and viscosity [5].

2.2.4 Viscosity

The degree of deacetylation and the molecular weight of the chitin are two of the most critical criteria in determining the economic viability of chitin. As molecular weight drops and deacetylation increases, viscosity rises. Chitin's particle size and storage period may play a role in this [7]. When compared to a solution of regular chitosan, nano chitin has a viscosity that is around 30% lower. When stored for 24 h, conventional chitin has a loss in viscosity of around 10%, whereas nano colloids experience a drop of about 17%.

2.3 *Thermal Properties*

The primary focus of existing literature revolves around composite materials. However, Tran et al. conducted a study that specifically assessed the thermal degradation properties of different kinds of nano chitin [65]. The chemicals that were investigated in their research included ChNFs (nanofibers), ChNCs (nanocrystals), and carboxylated ChNCs. The respective melting points (Td5) of these substances were found to be 301, 277, and 256 °C. The heat resistance of chitin is observed to decrease with its transformation into ChNCs, whereas a modest increase is observed upon its transformation into ChNFs. These findings align with the thermal decomposition temperature (Td5) of bulk chitin, measured at 285 °C. By comparing the heat stability of nanochitin substances to that of nanocellulose (250 °C), it can be inferred that nanochitin exhibits greater thermal stability. This observation implies that nanochitin can be utilized as high-quality nanofillers in composite materials [39].

The initial experiments in 2001 were carried out by Paillet and Dufresne, who employed composites containing ChNC as thermoplastics [47]. A latex-colloidal nanocomposite film was fabricated using ChNCs at varying weight percentages ranging from 0 to 20%. No significant effect on the shear modulus was seen when the temperatures fell below the glass transition temperature (Tg). Nevertheless, when the temperature exceeded the glass transition temperature (Tg), incorporating 10 weight percent (w/w) ChNC resulted in a substantial increase of 160 times in the relaxed modulus. The authors explained this outcome by attributing it to the significant aspect ratio of the ChNC.

The inverse relationship between Young's modulus of a material and the coefficient of thermal expansion is typically observed. This relationship was also investigated in the context of nanocomposites that incorporate ChNF. In contrast to orderly chitosan films, [29] observed a decrease in the coefficient of thermal expansion in chitosan-based films upon incorporating ChNFs. In a study conducted by [56], a direct comparison was made between the thermal degradation temperatures of Chitosan Nanofibers (ChNF) and Chitosan Nanocrystals (ChNCs). The researchers observed that using ChNFs as nano-fillers in a starch-based nanocomposite led to an increase in temperature, whereas using ChNCs resulted in a drop in temperature.

2.4 *Antibacterial and Antioxidant Properties*

Deacetylation is well known to improve the antibacterial properties of chitin in the bulk phase thus researchers recently looked at how deacetylation at different levels at the nanoscale affected these capabilities [51]. The degree of deacetylation (DDA) values range from 16 to 47%, and the bactericidal capabilities of ChNCs have tested [72]. It was reported that pH and DDA were both significant characteristics of antibacterial activity in their study monitoring the growth of *E. coli* bacteria, with low pH and high DDA producing the best circumstances. Increased cationic amino groups in both factors most likely inhibited the growth of bacteria. To measure the effectiveness of an antibacterial agent against an organism, the minimum inhibitory concentration (MIC) was utilized as a generic criterion [4]. The MIC was seen to decrease with increased DDA. Additionally, the MIC dropped concerning the number of cationic amino groups present on each sample. This was computed using the DDA value and the pH in which the research was carried out. It was known before this experiment that positive amino groups in bulk chitosan might communicate electrostatically with the negatively charged microbial membrane, causing leakage of internal electrolytes like potassium and effectively suppressing cell development [27].

Another study also highlighted the significance of the positively charged protonated amino group, which lends confirmation to these findings [65]. They discovered that partially deacetylated ChNCs exhibit excellent antibacterial behavior based on the presence of positive charges when compared to other forms of nanochitin material, such as ChNF, ChNCs, and carboxylated ChNCs generated through TEMPO-mediated oxidation.

Using these nanomaterials to create polymer nanocomposite materials, most of the literature on the antibacterial characteristics of ChNCs and ChNFs has focused on their combination with other polymers. The addition of ChNCs as nanofillers for carrageenan to make a uniform nanocomposite film in this study enhanced antibacterial action against the microorganisms *L. monocytogenes* and *E. coli*. The scientists proposed a potential pathway in which the ChNFs caused the bacteria to flocculate, inhibiting development due to a lack of nutrients and dioxygen, even though a definite mechanism for antibacterial behaviour has not yet been confirmed [59].

Other uses for ChNCs' antibacterial and antioxidant capabilities include food packaging materials, where microbial decomposition is a significant problem for food preservation and quality [17]. According to a research, Lysozymes immobilized on ChNCs have been shown to have more excellent antibacterial action than free lysozymes [31]. This work encouraged the potential for ChNCs to serve as a crucial nanomaterial for usage as an antibacterial in food preservation.

2.5 *Mechanical Properties*

Chitin, a biopolymer characterized by its remarkable mechanical properties, is able to provide structural support for diverse organisms and has been engineered for many applications. In nature, several notable examples of robust structures employ chitin as a protective armor for defense and safeguarding purposes. These include insect cuticles, crustacean shells, and molluscan nacre [15, 63].

Chen et al. conducted a study whereby they investigated the incorporation of ChNCs into hydrogels for drug delivery systems and cell scaffolds [69]. Injectable hydrogels have garnered significant attention as a viable approach for fabricating tissue scaffolds, primarily owing to their capacity to effectively minimize the need for invasive surgical interventions [6]. This endeavor's prominent objective is pursuing hydrogels possessing elevated elastic modulus and tensile strength. The gelation period of the hydrogel decreased significantly from 6038 to 25 s upon incorporating 5% (w/w) ChNCs into a beta-glycerophosphate-based hydrogel, as reported in this study. Furthermore, in a study, an enhancement in the mechanical durability of the hydrogels was observed [25].

The mechanical toughness of ChNFs was described and experimentally evaluated across many potential applications. A ChNF-(meth) acrylic resin nanocomposite film was fabricated, exhibiting optical clarity (as shown in Fig. 2a) and possessing a satisfactory tensile strength of no less than 30 MPa. As depicted in Fig. 2b, the incorporation of ChNF into the (meth) acrylic resin resulted in a notable increase of up to 3.0 GPa in the Young's modulus of the film [28].

It has been reported that ChNFs have been utilized to enhance the mechanical properties of gelatin-based nanocomposites. In a study conducted by Rolandi et al., it

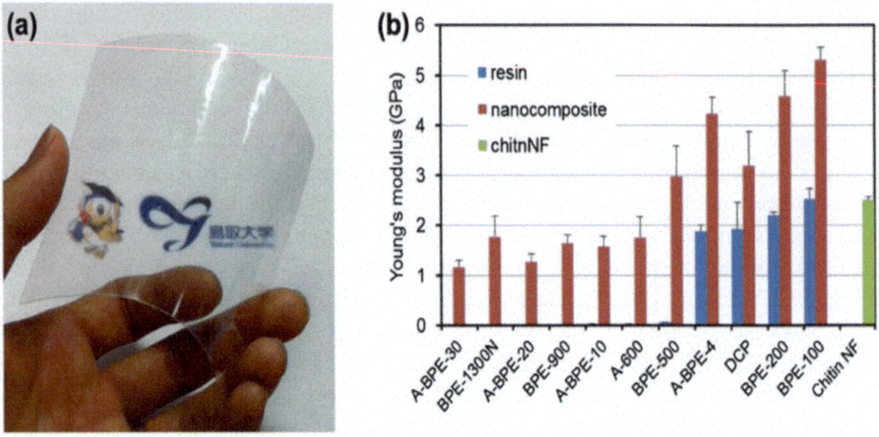

Fig. 2 **a** Picture of the Tricyclodecane-dimethanol dimethacrylate (DCP) film with ChNF and **b** Young's moduli of different meth(acrylic)-ChNF films and related nanocomposites. Standard deviation is shown by the error bars. Based on Ref. [28]

was demonstrated that incorporating ChNFs into gelatin methacryloyl significantly enhanced the hydrogel's elastic modulus by a factor of 1,000. Additionally, this modification led to a substantial increase of 200% in the stress-to-strain failure of the hydrogel [22].

Li and coworkers showed how to create gelatin-based nanocomposites utilizing ChNFs with surface deacetylation, demonstrating this was a crucial characteristic [8]. In particular, the gelatin containing surface-deacetylated ChNF had a 2.5-fold higher Young's modulus than gelatin alone. Neat ChNFs that weren't subjected to a deacetylation process had less noticeable effects than surface-deacetylated ChNFs. According to the authors, the increased hydrogen bonding provided by the amines was crucial in raising the nanocomposite's tensile contraction strength. Additionally, ChNFs have been employed directly to improve CNC-based packaging film. The Meredith group has demonstrated that adding ChNFs to CNC-based films can significantly increase the film's ultimate tensile strength (UTS), with the optimal ratio of CNC to ChNF being 1: 3 [57].

3 Synthesis of Chitin

Due to its abundant availability in biosphere, chitin is being studied to understand its production methods and potential as a biomaterial. The principal raw material for chitin production is the crustaceans' cuticle like crabs and shrimps. In shellfish, chitin is usually found in the shell of a complex network of proteins on which calcium carbonate is deposited. At present, the production of chitin is generally done by the extraction method in which the chitin is extracted from the shell of the crustaceans by the chemical or biological method. In the chemical method, the chitin is extracted from the shells of the crustaceans with the help of the acid/alkali treatment. In contrast, in the case of the biological method, it usually employs bacteria or enzymes for chitin production. The chemical method, compared to the biological method, is more commonly used at the industrial level because it is cost-effective and because chitin is produced quickly in less time. However, there are various disadvantages to the chemical method, as it is not eco-friendly and employs various harsh and toxic chemicals. The primary purpose of the extraction process is to obtain pure chitin free from proteins, lipids, and any other organic compound. Conventionally or traditionally, chitin is synthesized from chemical methods, which are fast in production but also have many disadvantages. For example, it employs toxic chemicals, solvents, and crosslinkers, and they also create environmental problems. To cope and to rise as an alternative to the chemical method, green extraction methods have been developed, which use deep eutectic solvents (DESs) and natural deep eutectic solvents (NADESs) as a solvent for the extraction of chitin. Deep eutectic solvents (DESs) and natural deep eutectic solvents (NADESs) are eco-friendly chemicals as they are biodegradable and can be recycled, making them environment-friendly and economical efficient. Out of the many methods used to extract chitin from crustacean's shells, green extraction method is proving to be best and superior over other methods including chemical because of its environment friendly nature. Also, a high-quality chitin with a very high yield can be obtained with the help of it without causing any harm or pollution to the environment cost-effectively [36].

3.1 Chemical Method

The chemical method is one of the most commonly used methods for chitin extraction commercially. The chemical method for chitin extraction is usually done in three steps, i.e., Demineralization, Deproteination, and decoloration. In the first step, i.e., demineralization, there is an acidic treatment with the HCl of the crustacean's shell to remove the calcium compounds like calcium carbonate and calcium phosphate and increase the surface area for the deproteination. In the second step, i.e., deproteination alkali treatment is done on the demineralized shells, in which proteins are removed.

3.1.1 Advantages of Chemical Method

1. Low cost
2. Easy to apply in case of scale-up process
3. Fast in production
4. Increased demand for chitin can be met by it

3.1.2 Disadvantages of Chemical Method

1. Usage of toxic chemicals.
2. Not eco-friendly
3. Changes in physicochemical properties of chitin
4. Removal of important proteins and minerals

3.1.3 Demineralization

Demineralization involves subjecting the crustacean's shell to acidic conditions to dissolve the crust and remove minerals like calcium phosphate and calcium carbonate. Two primary kinds of acids are employed in demineralization: mineral and organic. While HCl is a popular mineral acid, safer and more environmentally friendly alternatives like citric and acetic acid are increasingly favored. Calcite may be demineralized simply by dissolving it in acid, producing calcium chloride and water, and then releasing carbon dioxide. In addition to the calcium compound, the other minerals present in the shellfish cuticle readily react with HCl to give water-soluble salts; once these salts have been produced, they can be easily separated from the chitin via filtration and subsequent washing with deionized water, with only minor adjustments in concentration and acid selection typically required.

The manufacture of chitin from the shrimp shells required a two-step extraction technique, which Zhao et al. devised utilizing citric acid and deep eutectic solvents. More than 98% of the minerals were removed from the shrimp shells using citric acid in the first phase of the demineralization process. More than 80% deproteinization was achieved in the second stage, including deep eutectic solvents and microwaves [80].

Khanafari et al. recovered chitin from waste *Penaeus semisulcatus* using a chemical and microbiological process. A 510 mg/g yield was discovered when chitin was extracted chemically using NaOH and acetic acid. The shells of shrimp that had been microbially deproteinized were demineralized using an organic acid (lactic acid) generated by bacteria. Microbial production of chitin was more productive than chemical production [35].

3.1.4 Deproteinization

In this process, protein in the shell is isolated from the chitin. In this, strong bases like NaOH/KOH and high temp are used so that proteins become water-soluble and can be easily separated from the chitin. Deproteinization is difficult to carry out as compared to the demineralization step due to the disturbance of chemical bonds between chitin and proteins. The deproteinization step is a very important step as it removes the proteins from the chitin, which is essential for the biomedical view of applications. The chemical approach is usually commonly used for deproteinization, and a variety of bases are used for it, like NaOH, KOH, Na_2S, and $NaHCO_3$, etc. NaOH is the most commonly used base for deproteinization; however, due to its toxic effects and also environmental problems, KOH is being referred to. The reaction temperature involved during deproteinization is usually 65°C to 100°C; however, deproteinization is also carried out at room temperature.

Soon et al. extracted and evaluated chitin from superworm (*Zophobas morio*) larvae using varying concentrations of NaOH in the range of 0.5–2.0 M during deproteinization. It was found that extraction with 0.5 M resulted in the highest yield of chitin at about 5.43% with a high degree of acetylation. The rate of deproteinization increases as there is an increase in the concentration of NaOH, and the yield of chitin decreases as the concentration of NaOH increases [61].

Mizani et al. [43] conducted a study in which they investigated the effect of the application of Sodium sulfite, Ascalase, Triton X-100, and a combination of reagents on the quality of chitin that was produced. They found that the highest protein recovery of protein about 64% was obtained from shrimp head waste when sodium sulphite and alcalase were used in concentrations of 200 mmol/l and 0.5%, respective after this, the extracted chitin was purified with 0.5% NaOH at 60 °C for 0.5 h. The chitin was found to have good quality with little protein [43].

3.2 *Biological Method*

The biological method is an eco-friendly method of extraction of chitin, which is employed as an alternative to the chemical method for the production of chitin as there is no use of toxic chemicals and usage of natural enzymes. In biological methods there are two stages for chitin extraction i.e., demineralization and deproteination [58]. Demineralization is usually carried out with the help of the acid-producing bacteria, which produces acid that precipitates calcium carbonate in the form of calcium lactate, which can be removed. Other than this, acid also creates an acidic environment that activates the protease enzyme and the deproteination stage occurs. The biological method mainly employs bacteria or enzymes for the extraction of chitin. There are two types of biological methods: an enzymatic method and the fermentation method.

3.3 *Enzymatic Method*

In the enzymatic method, proteases enzymes are used to remove proteins present in the crustacean's shell. The primary sources of proteases are bacteria and marine organisms. Proteases, through their high specificity, isolate the chitin with minimal deacetylation and structural changes. Many proteases can be used for eliminating proteins from the crustaceans shells like trypsin, pepsin and alcalase, etc. In the enzymatic extraction, both types of crude and purified proteases are used in the deproteinization step. Crude is mainly used compared to purified proteases because they are cheaper than purified ones and contain other existing proteases. The main sources of crude proteases are bacteria and fish viscera, in which bacteria-produced proteases are commonly used.

Many studies have been done to represent the application of bacterial proteases in enzymatic deproteinization. For instance, Glyose et al., used the enzyme protease produced by the *Erwinia chrysanthemi* to extract the pure chitin from the crustacean's waste by using the standard methods; for this, they used two types of proteases, i.e., crude and pure type. Protein removal by the pure proteases was similar in mineralized and demineralized waste and was found to be about 75 and 80% in 16 h. Protein removal efficiency was very high about 95% in crude protease [76].

Hamdi et al. [20] conducted a study in which they extracted chitin from the blue crab (*Portunus segnis*) and shrimp (*Penaeus kerathurus*) shells with the help of the alkaline proteases derived from the *P. segnis* viscera. Their findings showed that crude protease extract is very effective in the deproteinization of crab and shrimp waste [20].

3.4 Fermentation Method

The fermentation method consists of the use of bacteria instead of enzymes for the extraction of the chitin. Compared to the enzymatic method, it is more ideal for producing chitin in mass quantity due to less cost involved. The downside of the fermentation method is the long duration of time, which elapses in extracting chitin. The Fermentation method incorporates demineralization with the help of endogenous microorganisms like bacteria or by the addition of selected strains of acid-producing bacteria.

Lactic and non-lactic acid fermentation are the two main categories of fermentation processes.

3.4.1 Lactic Acid Fermentation

Fermenting crustacean shells requires an inoculum of a strain of Lactobacillus sp., which produces lactic acid and proteases. Silage has a lower pH than other forages because lactic acid is created during glucose conversion, preventing the growth of rotting bacteria. Calcium lactate, a byproduct of lactic acid's interaction with calcium carbonate, is separated from the lighter shells and then recovered for further processing steps, including water washing and processing. This process may be utilized with cleaned crustacean shells or complete shrimp waste (including heads and viscera). The effectiveness of lactic acid fermentation is usually dependent upon the following:

a. Type of species of bacteria used
b. Quantity of inoculum used

c. Source of carbohydrate
d. Temperature
e. pH
f. Duration of fermentation

In order to preserve shrimp waste to recover chitin, Cira et al. did research on lactic acid fermentation of the waste. Shrimp waste was fermented in 100 g flasks inoculated with varying concentrations of lactobacilli bacteria and fed a diet of sucrose, lactose, and milk whey powder. Column reactors increased the scale, and sucrose was used as the carbohydrate supply. Acid and alkali concentrations of 0.5 and 0.4 M in the chitin fraction, respectively, were shown to reduce mineral and protein content by the most after 6 days of insulation [11].

For several days, [58] incubated waste at 30 °C with 100 rpm and varied the concentrations of glucose (0, 10, 15, 20) and inoculation (10, 15, 20) to extract chitin from chitin from shrimp waste using the biological treatment using *Pseudomonas aeruginosa*. Protease activity was measured at several temperatures and found to be highest at 50 °C. Their findings show a direct correlation between parameter concentration and demineralization and deproteinization. At ideal conditions of 20% glucose, 20% inoculation, and fermentation for around 6 days, the chitin output was about 47%, with 82% demineralization and 92% deproteinization.

3.4.2 Non-lactic Acid Fermentation

Non-lactic acid fermentation usually employs both bacteria and fungi to carry out the fermentation of crustacean shells.

To remove chitin, fermented powdered shrimp shells in two stages using Serratia marcescens B742 and Lactobacillus plantarum ATCC 8014 were studied. Deproteinization and demineralization rates of 95.5 and 93.0% are achieved in a two-step fermentation process that, under ideal circumstances, yields 18.9% chitin. The isolated chitin had the same physiochemical and structural properties as commercial chitin [76].

In contrast to chemical processes, the different biological techniques by microorganisms used to extract chitin are more accessible, more efficient, and ecologically benign. However, there are disadvantages to microbial fermentation, such as a longer processing time than chemical techniques and worse accessibility of proteases (due to the presence of minerals that result in high residual proteins). Nevertheless, the pace of deproteinization may be increased depending on the needs of the final product, particularly for biomedical applications. Using concurrent or sequential procedures, such as the co-fermentation of microorganisms or two-step fermentation, could be used to achieve this. The biotechnology procedure must be finished with a minor chemical treatment to get highly purified chitin by removing the remaining protein and minerals.

3.5 Green Extraction of Chitin

Green extraction or extraction of chitin by green chemistry is nowadays becoming more and more popular as compared to the chemical synthesis methods because in green extraction there is an eco-friendly process in which there is a use of non-toxic chemicals like organic acids, no use of toxic solvents or crosslinkers. There is also recycling and reusing of chemicals, which makes green extraction of chitin an ideal process that is not only eco-friendly but also an economically feasible process for chitin production.

In conventional chemical methods for extraction of chitin, there is a use of non-environment friendly chemicals like strong mineral acids and bases, which not only create environment and ecological problems but also are very toxic in nature hence, to cope with this, a new type of synthesis is widely employed which is known as green extraction in which there is a use of organic chemicals which are non-toxic and doesn't create any environment pollution and there is also fast production of chitin as compared to the conventional chemical method which is very slow and take long duration of time.

Green chitin extraction often uses either a deep eutectic solvent (DES) or a natural deep eutectic solvent (NADES) due to their low environmental impact.

An emerging class of ionic liquids, deep eutectic solvents (DESs) are made up of both hydrogen-bond-donating (HBD) and accepting (HBA) cations. When compared to the melting points of the hydrogen bond donor (HBD) and hydrogen bond acceptor (HBA), deep eutectic fluids have a lower melting point.

Zhang et al. [78, 79] extracted high purity chitin from shrimp shells by using a green, eco-friendly, and well-structured one-pot approach by employing deep eutectic solvent consisting of glycerol and choline chloride combined with a small amount of acetic acid. Extraction of chitin was done under different types of conditions, and the conditions in which there is an acetic acid concentration of 7.5% and the temperature of reaction was 120 °C, the purity of chitin was found to be 96.1% which was way higher than that of conventional chemical method at 87.7% other than this eutectic solvent was also found to be recycled at three times without degrading the quality of obtained chitin [78].

A fast, eco-friendly, and easy method of chitin extraction from lobster shells using deep eutectic solvents consisting of a mixture of choline chloride-urea (CCU), choline chloride-glycerol (CCG), and choline chloride-malonic acid (CCMA) was developed [81]. The various physicochemical properties of isolated chitin compared those prepared with chemical methods with commercial ones from shrimp shells. A yield of isolated chitin using choline chloride-malonic acid (CCMA) DES was obtained at about ($20.63 \pm 3.30\%$) which was high as compared to that one which was prepared with the chemical method ($16.53 \pm 2.35\%$) [67].

The term "natural deep eutectic solvent" (NADES) refers to deep eutectic solvents composed of two or more molecules that are often primary metabolites of plants. Natural deep eutectic solvents are also widely utilized since they are simple to recycle and have a low environmental impact. The chitin in crab shells was extracted using a green method developed by Mao et al., who employed choline chloride and malic acid as a natural deep eutectic solvent (NADES) before evaluating its demineralization and deproteinization efficacy. Natural deep eutectic solvents (NADES) were effective in removing minerals and proteins from shrimp shells, and the grade of the chitin extracted from the shells was likewise high [24].

Wang et al. developed a new green and efficient method for extraction of chitin from the crustaceans' waste with a new type of ternary natural deep eutectic solvent (NADES) in which they attached N-acetyl-D-glucosamine to the choline chloride. In their process, they obtained chitin having a high molecular weight of 3.92×10^5 Da, purity of 90.2%, and yield of 85.6%. As compared with traditional acid/alkali and binary natural deep eutectic solvent (NADES) maximum yield of chitin increased by 1.57 and 1.39 times [70].

4 Characterization of Chitin Based Nanoparticles

Understanding the process of CNP production and their physicochemical and therapeutic characteristics need a thorough characterization. Several research have explored characterizing CNPs to understand their physicochemical features better.

4.1 *Measurement of the Size of Chitin Nanoparticles*

The size of CNPs is crucial in controlling their numerous features, including their ability to bind medicines, their crystallinity, and their total charge. Dynamic light scattering (DLS) or transmission electron microscopy (TEM) are typically used to determine the size of CNPs. The significant polydispersity of CNPs makes using DLS to determine size challenging. The morphological, optical, and structural characteristics of CNPs can be characterized using a variety of complementary approaches. This comprises atomic force microscopy (AFM), X-ray diffraction (XRD), Raman spectroscopy, TEM, and scanning electron microscopy (SEM). CNP solution concentrations are occasionally determined using UV–Visible spectroscopy.

4.2 *X-ray Diffraction*

XRD may be utilized to learn about the crystallinity of CNPs and what kind they are. XRD is the go-to analytical method for identifying the crystalline phase(s) of materials. XRD may be used to assess structural factors and the mean diameter of the nanoparticles [26]. After correcting for the widening of the diffraction peaks due to instrument, strain, and lattice faults, the mean crystallite size of the material may be calculated using the Scherrer equation from powder X-ray diffraction data. Once these adjustments have been made, the Scherrer equation can be used to determine the average size of nanoparticles from the FWHM of a specific XRD peak for a given crystalline phase, assuming that the size of a crystallite and a typical nanoparticle are the same [9].

4.3 Raman Spectroscopy

As determined by the activation of the related Raman-active vibrational modes, Raman spectroscopy is a sensitive method used to detect structural alterations in macromolecular complexes [66]. The Raman effect is based on the scattering of light, which comprises inelastic scattering at a shifted wavelength caused by the stimulation of a particular molecular vibration, as well as elastic scattering at the same wavelength as the input light.

4.4 Transmission Electron Microscopy (TEM)

TEM is useful for studying nanoparticle size, shape, and crystallinity. On the opposite side of the impinging electron beam is an imaging system (i.e., phosphor screen) displaying the picture TEM created. To move the item, TEM directs a beam of high-energy electrons at it. The picture is formed due to the superposition of the electron beam that goes through the sample and the beam that is refracted from the sample. It is common for TEM imaging to have a resolution of roughly 2 nm or better. Extremely thin samples (500 or less) are required for high-resolution TEM evaluation [23]. TEM specimens are made by either applying a negative staining solution (such as uranyl acetate) or putting a diluted sample suspension on a carbon-coated copper grid. Ghadi et al. [18] conducted TEM analysis of magnetic chitin nanoparticles to determine their shape and size.

4.5 Scanning Electron Microscopy (SEM)

SEM uses a concentrated, high-energy electron beam to photograph a specimen's surface. Samples are analyzed after drying, except an ambient SEM that will not be covered in this article. It is necessary to sputter-coat non-conducting materials with a thin layer of metal or carbon. Nanoparticle surface shape and size can be directly seen using SEM. The collected emission X-ray spectra from SEMs with element dispersive spectroscopy (EDS) detection capabilities can be used to analyze the elemental composition of materials. The Saharan et al. [55] Ionic gelation was used to create chitin nanoparticles, chitin-saponin nanoparticles, and Cu-CNPs, which were then tested for each of their specific antifungal properties. The organisation of their chitin nanoparticles on the nanoscale scale was verified in their investigation using SEM. Additionally, SEM imaging revealed that Cu-CNPs possessed the structure of a compact polyhedron, whereas chitin-saponin nanoparticles were spherical in shape [55]. SEM was employed for a morphological examination of TPP-cross-linked CNPs loaded with salicylic acid and gentamicin. The research revealed that these CNPs had an average size of about 200 nm and a almost spherical shape [33].

4.6 Dynamic Light Scattering (DLS)

The DLS method detects the intensity of scattered light to calculate the dynamic fluctuations of particles in a solution due to their Brownian motion. A laser is often used as the monochromatic light source, and the particles in the solution generate dynamic light diffraction, resulting in a speckle pattern. The typical particle size inside a sample may be determined using the Stokes–Einstein equation. DLS may also be used to calculate the Zeta potential of nanoparticles in a sample and the polydispersity index (PDI) or dispersity according to the current IUPAC convention. When utilizing the DLS method to assess the size of nanoparticles, it is typically necessary to monitor intensity fluctuations at various detector angles for numerous polarisations relative to the incoming light [10]. The Zeta potential, a measurement of the electric potential between the attached and non-attached fluid concerning the nanoparticle surface, determines the nanoparticles' stability in the flui [19]. Because of their higher Zeta potential values, CNPs and Cu-CNPs are more stable than saponin-CNPs in water [50].

5 Applications of Chitin Nanoparticles

Since chitin and chitosan, a derivative of chitin, are highly biocompatible, they have found several applications in the pharmaceutical field. This is mostly because enzymes like lysozyme may biodegrade chitosan by splitting the acetylated units apart and breaking the molecule down into simpler sugars [73]. Chitin has mechanical strength and immunogenicity in addition to its degradability. To bind and form complexes with other negatively charged molecules, such as growth factors, nucleic acids, and cytokines, chitosan has the unusual capacity to become protonated. When bound to these negatively charged molecules, they are protected against degradation. Instead, a higher concentration of chemicals beneficial for tissue regeneration is seen [40]. Tissue engineering is a method of replacing diseased or missing tissues and organs and aiding in wound healing and dressing by using modern medical technology and molecular and cellular biology [71]. The numerous beneficial biological qualities of the biopolymer, such as its cheap cost of separation and quick manufacturing, are attracting the attention of scientists interested in tissue regeneration.

5.1 Drug Delivery

Developments in the biomedical sector employing chitin have led to the biopolymer's potential use in medication delivery. Beads, microparticles, nanoparticles, micelles, hydrogels, tablets, and capsules are some of the form's biopolymers that may take advantage of their versatility to provide [52]. Nanogels are hydrogels only present on the nanoscale and may be created from biopolymers like chitin. These nanogels can be tuned in size, have a high surface area for bioconjugation, can be loaded with drugs for controlled release, and are very responsive to external stimuli [48]. These features have led to the creation of innovative chitin-based drug delivery systems, such as polylactic acid (PLA)/chitosan (CS) nanoparticles and water-soluble carboxymethyl chitin (CMC).

5.2 Wound Healing

Researchers are also exploring the possibility of using chitin in place of standard treatments for wound healing. Wound healing is an ongoing and intricate process that involves five distinct phases: homeostasis, inflammation, migration, proliferation, and maturation [2]. An untreated bacterial infection is one of the numerous issues that may arise during wound healing and can have serious consequences [32]. Many considerations go into selecting the best candidate for wound dressing material, such as allowing for gaseous exchange, protecting against bacteria, and ensuring no damage is done to the patient [37]. The healing process may be controlled and initiated risk-free with the help of bioactive agents, which are now widely available on the market.

Interleukin 8, prostaglandin E, and interleukin 1 ß are all regulated by chitin and its derivatives, making them effective wound healers. In addition to chitin's versatility, this quality makes it an attractive material for wound dressing [30]. Scaffolds made of chitin and chitosan have been used with encouraging results in treating severe burns and wounds [74]. In addition to chitin, both metallic silver and silver sulphadiazine ointments are very resistant to germs. Separately, these materials have the potential to improve wound healing significantly, but recently, scientists have been exploring the prospect of combining them. Antibacterial activity and blood clotting capacity against various microorganisms, including Staphylococcus aureus and *Escherichia coli*, have been reported to be significantly enhanced when chitin is combined with nano-silver in scaffold composites. Similarly, a composite scaffolding was created utilizing ß-chitin and nano-silver, demonstrating antimicrobial activity, whole blood clotting, and excellent cell adhesion. Dissimilarities may explain the small difference between the two forms of chitin in their intermolecular interactions. To put it another way, ß-chitin's intermolecular connections are much weaker than those typically seen in chitin, allowing for easier dissolving in many different solvents and making them more reactive and flexible.

5.3 *Tissue Engineering*

Using active bio-materials for tissue regeneration requires a knowledge of how chitin interacts with injured tissues to promote recovery. The concept of a biomaterial scaffolding that provides a cell with the appropriate form and structure is essential to tissue engineering. Chitin, for instance, is being used in new ways as a biomaterial scaffolding material by being coupled with smaller molecules like sugars and peptides. Hybrid studies have shown that human keratinocytes and fibroblasts may link and spread over binders made of chitin fibres, silk-fibroins, and polyglycolic acid. Important cell types involved in the innate inflammatory response during epidermal healing and regeneration include keratinocytes and fibroblasts [62, 68].

Based on the above, studies are investigating whether chitin may aid in cartilage regeneration. This may be explained by the fact that articular cartilage is composed mainly of glycosaminoglycans, with N-acetyl-ß-D-glucosamine serving as the building block for other glycosaminoglycans such as keratin sulphate and hyaluronate. Because chondrocytes can be extracted from cartilage so simply by enzymatic digestion, they are a leading candidate for tissue engineering studies. Cartilage grafts have been shown to thrive on biodegradable scaffolds made from polylactic and polyglycolic acids [46]. Chitin's capacity to connect with the glycosaminoglycans of cartilage gives it an edge over these synthetic polymers. Previous research [38] has also shown that chondrocytes produced on chitosan films have a distinct spherical shape and display type II collagen and aggrecan.

5.4 *Cancer Therapy*

Worldwide, 13% of fatalities are attributed to cancer, and the number of people diagnosed with the disease is on the rise. Cancer is expected to affect 12 million people worldwide by the year 2030. The most prevalent forms are breast cancer, colorectal cancer, lung cancer, and prostate cancer. Carriers for medications used in anti-cancer treatment must be non-toxic and capable of targeting the malignant tissue for effective drug delivery [45]. Polymeric vehicles have attracted a lot of interest among the numerous kinds of drug delivery vehicles because they not only boost drug targeting ability but also improve blood retention duration by preventing pharmaceuticals from being eliminated in the urine [14]. Both chitosan and chitin have found use in chemotherapy as drug delivery vehicles.

Mi et al. [42] used an emulsion cross-linking method to incorporate the anti-cancer medication 5-fluorouracil (5-Fu) into water-soluble carboxymethyl chitin (CMC) nanoparticles, which were generated by cross-linking with $CaCl_2$ and $FeCl_3$. Solubility in water, physiological characteristics, biodegradability, and reactivity with other molecules are all drastically altered by the chemical alteration of chitin and chitosan. As a result, improving the solubility of chitin and chitosan in everyday solvents is crucial for their practical use. Previous research has shown that carboxy methylating chitin increases its water solubility, is biocompatible, and does not induce antibodies [34]. They use a condensation reaction of chitin powder and monochloroacetic acid in isopropyl alcohol as a solvent [49].

6 Conclusion

The diverse characteristics of chitin fulfill the demands of consumers seeking cost-effective and naturally derived products, therefore exerting a favorable influence on human health. There is unanimous consensus regarding the significance of these polysaccharides due to their appealing attributes of biocompatibility, biodegradability, non-toxicity, and solubility, which are essential for the processing of polysaccharides into innovative materials that have the potential for various creative applications. In various fields of application, there has been a recent focus on utilizing functional nanomaterials to enhance and explore polysaccharides. These nanoparticles include desirable characteristics such as biocompatibility, hydrophilicity, and the ability to increase water swelling ratio through electrostatic interactions. This method, known as nanotechnology, shows promise in advancing the potential of polysaccharides.

This chapter summarizes the properties, types of production, and characterization techniques employed in the investigation of chitin nanoparticles. Chitin emerges as a highly favorable choice for the field of nanomedicine due to its exceptional properties and characteristics. Chitin nanoparticles exhibit significant promise as carriers for encapsulating and integrating pharmaceutical compounds. Therefore, with their mucoadhesive characteristics, chitosan nanoparticles demonstrate promising prospects for efficient drug delivery, regulated drug release, and heightened therapeutic effectiveness of pharmaceutical agents.

References

1. Abdel-Gawad KM, Hifney AF, Fawzy MA, Gomaa M (2017) Technology optimization of chitosan production from Aspergillus niger biomass and its functional activities. Food Hydrocoll 63:593–601
2. Ahmed S, Ikram S (2016) Chitosan based scaffolds and their applications in wound healing. Achieve Life Sci 10:27–37
3. Aranaz I, Alcantara AR, Civera MC, Arias C, Elorza B, Heras Caballero A, Acosta N (2021) Chitosan: an overview of its properties and applications. Polymers (Basel). 13(19):3256
4. Benhabiles MS, Salah R, Lounici H, Drouiche N, Goosen MFA, Mameri N (2012) Antibacterial activity of chitin, chitosan and its oligomers prepared from shrimp shell waste. Food Hydrocoll 29:48–56
5. Bhardwaj S, Bhardwaj NK, Negi YS (2020) Effect of degree of deacetylation of chitosan on its performance as surface application chemical for paper-based packaging. Cellulose 27(9):5337–5352

6. Brandl F, Sommer F, Goepferich A (2007) Rational design of hydrogels for tissue engineering: impact of physical factors on cell behaviour. Biomaterials 28:134–146
7. Chattopadhyay D, Inamdar MS (2010) Aqueous behaviour of chitosan. Int J Polym Sci 7:939536
8. Chen C, Wang Y, Yang Y, Pan M, Ye T, Li D (2018) High strength gelatin-based nanocomposites reinforced by surface-deacetylated chitin nanofiber networks. Carbohydr Polym 195:387–392
9. Chicea D, Indrea E, Chicea D, Indrea E, Cretu CM (2012) Assessing Fe_3O_4 nanoparticle size by DLS, XRD and AFM. J Optoelectron Adv Mater 14:460–466
10. Chu B (1985) Light scattering studies of polymer solutions and melts. Polym J 17:225–238
11. Cira LA, Huerta S, Hall GM, Shirai K (2002) Pilot scale lactic acid fermentation of shrimp wastes for chitin recovery. Process Biochem 37(12):1359–1366
12. Cohen E (2010) Chitin biochemistry: synthesis, hydrolysis and inhibition. Adv Insect Physiol 38:5–74
13. Crognale S, Russo C, Petruccioli M, D'Annibale A (2022) Chitosan production by fungi: current state of knowledge, future opportunities and constraints. Fermentation 8(2):76
14. Dev A, Binulal N, Anitha A, Nair S, Furuike T, Tamura H, Jayakumar R (2010) Preparation of poly (lactic acid)/chitosan nanoparticles for anti-HIV drug delivery applications. Carbohydr Polym 80:833–838
15. Fernandez JG, Ingber DE (2013) Bioinspired chitinous material solutions for environmental sustainability and medicine. Adv Func Mater 23:4454–4466
16. Garcia-Rubio R, de Oliveira HC, Rivera J, Trevijano-Contador N (2019) The fungal cell wall: candida, cryptococcus, and aspergillus species. Front Microbiol 10:2993
17. Gayan E, Condon S, Alvarez I (2014) Biological aspects in food preservation by ultraviolet light: a review. Food Bioprocess Technol 7:1–20
18. Ghadi A, Mahjoub S, Tabandeh F, Talebnia F (2014) Synthesis and optimization of chitosan nanoparticles: potential applications in nanomedicine and biomedical engineering. Casp J Intern Med 5:156–161
19. Gugliotta LM, Vega JR, Meira GR (2000) Latex particle size distribution by dynamic light scattering: computer evaluation of two alternative calculation paths. J Colloid Interface Sci 228:14–17
20. Hamdi M, Hammami A, Hajji S, Jridi M, Nasri M, Nasri R (2017) Chitin extraction from blue crab (Portunus segnis) and shrimp (Penaeus kerathurus) shells using digestive alkaline proteases from P. segnis viscera. Int J Biol Macromol 101:455–463
21. Hamed I, Fatih O, Joe MR (2016) Industrial applications of crustacean by-products (chitin, chitosan, and chitooligosaccharides): a review. Trends Food Sci Technol 48:40–50
22. Hassanzadeh P, Kazemzadeh-Narbat M, Rosenzweig R, Zhang X, Khademhosseini A, Annabi N, Rolandi M (2016) Ultrastrong and flexible hybrid hydrogels based on solution self-assembly of chitin nanofibers in gelatin methacryloyl (gelma). J Mater Chem B 4:2539–2543
23. Hembram KC, Prabha S, Chandra R, Ahmed B, Nimesh S (2016) Advances in preparation and characterization of chitosan nanoparticles for therapeutics. Artif Cells Nanomed Biotechnol 44:305–314
24. Huang WC, Zhao D, Guo N, Xue C, Mao X (2018) Green and facile production of chitin from crustacean shells using a natural deep eutectic solvent. J Agric Food Chem 66(45):11897–11901
25. Huey DJ, Hu JC, Athanasiou KA (2012) Unlike bone, cartilage regeneration remains elusive. Science 338:917–921
26. Hull AW (1919) A new method of chemical analysis. J Am Chem Soc 41:1168–1175
27. Hwang KT, Jung ST, Lee GD, Chinnan MS, Park YS, Park HJ (2002) Controlling molecular weight and degree of acetylation of chitosan by response surface methodology. J Agric Food Chem 50:1876–1882
28. Ifuku S, Morooka S, Norio Nakagaito A, Morimoto M, Saimoto H (2011) Preparation and characterization of optically transparent chitin nanofiber/(meth)acrylic resin composites. Green Chem 13:1708–1711
29. Ifuku S, Ikuta A, Egusa M, Kaminaka H, Izawa H, Morimoto M, Saimoto H (2013) Preparation of high-strength transparent chitosan film reinforced with surface-deacetylated chitin nanofibers. Carbohydr Polym 98:1198–1202

30. Jayakumar R, Prabaharan M, Nair S, Tokura S, Tamura H, Selvamurugan N (2010) Novel carboxymethyl derivatives of chitin and chitosan materials and their biomedical applications. Prog Mater Sci 55:675–709
31. Jiang S, Qin Y, Yang J, Li M, Xiong L, Sun Q (2017) Enhanced antibacterial activity of lysozyme immobilized on chitin nanowhiskers. Food Chem 221:1507–1513
32. Jimenez-Gomez CP, Cecilia JA (2020) Chitosan: a natural biopolymer with a wide and varied range of applications. Molecules 25:3981
33. Jingou J, Shilei H, Weiqi L, Danjun W, Tengfei W, Yi X (2011) Preparation, characterization of hydrophilic and hydrophobic drug in combine loaded chitosan/cyclodextrin nanoparticles and in vitro release study. Colloids Surf B Biointerfaces 83:103–107
34. Karagozlu MZ, Kim SK (2014) Anticancer effects of chitin and chitosan derivatives. Adv Food Nutr Res 72:215–225
35. Khanafari A, Marandi R, Sanatei Sh (2008) Recovery of chitin and chitosan from shrimp waste by chemical and microbial methods. Iran J Environ Health Sci Eng 5:19–24
36. Kozma M, Bishnu A, Rabin B (2022) Chitin, chitosan, and nanochitin: extraction, synthesis, and applications. Polymers 14(19):3989
37. Kumar PS, Abhilash S, Manzoor K, Nair S, Tamura H, Jayakumar R (2010) Preparation and characterization of novel-chitin/nanosilver composite scaffolds for wound dressing applications. Carbohydr Polym 80:761–767
38. Kuo YC, Ku IN (2008) Cartilage regeneration by novel polyethylene oxide/chitin/chitosan scaffolds. Biomacromol 9:2662–2669
39. Lam E, Hrapovic S, Majid E, Chong JH, Luong JHT (2012) Catalysis using gold nanoparticles decorated on nanocrystalline cellulose. Nanoscale 4:997–1002
40. Madhumathi K, Kumar PS, Kavya K, Furuike T, Tamura H, Nair S, Jayakumar R (2009) Novel chitin/nanosilica composite scaffolds for bone tissue engineering applications. Int J Biol Macromol 45:289–292
41. Matica T, Aachmann H (2019) Sletta and Ostafe, Chitosan as a wound dressing starting material: antimicrobial properties and mode of action. Int J Mol Sci 20(23):5889
42. Mi FL, Shyu SS, Lin YM, Wu YB, Peng CK, Tsai YH (2003) Chitin/PLGA blend microspheres as a biodegradable drug delivery system: a new delivery system for protein. Biomaterials 24:5023–5036
43. Mizani AM, Aminlari BM (2007) A new process for deproteinization of chitin from shrimp head waste. Proc Eur Cong Chem Eng 6:16–20
44. Mostafa M, Elsoud A, El Kady EM (2019) Current trends in fungal biosynthesis of chitin and chitosan. Bull Natl Res Centre 43:1–12
45. Narayanan D, Jayakumar R, Chennazhi KP (2014) Versatile carboxymethyl chitin and chitosan nanomaterials: a review. Wiley Interdiscip Rev-Nanomed Nanobiotechnol 6:574–598
46. Nettles DL, Elder SH, Gilbert JA (2002) Potential use of chitosan as a cell scaffold material for cartilage tissue engineering. Tissue Eng 8:1009–1016
47. Paillet M, Dufresne A (2001) Chitin whisker reinforced thermoplastic nanocomposites. Macromolecules 34:6527–6530
48. Parhi R (2020) Drug delivery applications of chitin and chitosan: a review. Environ Chem Lett 18:577–594
49. Peter S, Lyczko N, Gopakumar D, Maria HJ, Nzihou A, Thomas S (2021) Chitin and chitosan based composites for energy and environmental applications: a review. Waste Biomass-Valoriz 12:4777–4804
50. Qi L, Xu Z, Jiang X, Hu C, Zou X (2004) Preparation and antibacterial activity of chitosan nanoparticles. Carbohydr Res 339:2693–2700
51. Raafat D, von Bargen K, Haas A, Sahl HG (2008) Insights into the mode of action of chitosan as an antibacterial compound. Appl Environ Microbiol 74:3764
52. Rejinold NS, Chennazhi KP, Tamura H, Nair SV, Rangasamy J (2011) Multifunctional chitin nanogels for simultaneous drug delivery, bioimaging, and biosensing. ACS Appl Mater Interfaces 3:3654–3665

53. Roy JC, Salaun F, Giraud S, Ferri A, Chen G., Guan J (2017) Solubility of chitin: solvents, solution behaviors and their related mechanisms. In: Solubility of polysaccharides. Intech Open, pp 109–127
54. Ruiz-Herrera J, San-Blas G (2003) Chitin synthesis as a target for antifungal drugs. Curr Drug Targets Infect Disord 3(1):77–91
55. Saharan V, Mehrotra A, Khatik R, Rawal P, Sharma SS, Pal A (2013) Synthesis of chitosan based nanoparticles and their in vitro evaluation against phytopathogenic fungi. Int J Biol Macromol 62:677–683
56. Salaberria AM, Labidi J, Fernandes SCM (2014) Chitin nanocrystals and nanofibers as nano-sized fillers into thermoplastic starch-based biocomposites processed by melt-mixing. Chem Eng J 256:356–364
57. Satam CC, Irvin CW, Coffey CJ, Geran RK, Ibarra-Rivera R, Shofner ML, Meredith JC (2020) Controlling barrier and mechanical properties of cellulose nanocrystals by blending with chitin nanofibers. Biomacromol 21:545–555
58. Sedaghat F, Yousefzadi M, Toiserkani H, Najafipour S (2017) Bioconversion of shrimp waste Penaeus merguiensis using lactic acid fermentation: an alternative procedure for chemical extraction of chitin and chitosan. Int J Biol Macromol 104:883–888
59. Shankar S, Reddy JP, Rhim JW, Kim HY (2015) Preparation, characterization, and antimicrobial activity of chitin nanofibrils reinforced carrageenan nanocomposite films. Carbohydr Polym 117:468–475
60. Singh R, Shitiz K, Singh A (2017) Chitin and chitosan: biopolymers for wound management. Int Wound J 14(6):1276–1289
61. Soon CY, Tee YB, Tan CH, Rosnita AT, Khalina A (2018) Extraction and physicochemical characterization of chitin and chitosan from Zophobas morio larvae in varying sodium hydroxide concentration. Int J Biol Macromol 108:135–142
62. Souren JEM, Ponec M, VanWijk R (1989) Contraction of collagen by human fibroblasts and keratinocytes. In Vitro Cell Dev Biol-Anim 25:1039–1045
63. Sun J, Bhushan B (2012) Structure and mechanical properties of beetle wings: a review. RSC Adv 2:12606–12623
64. Tejas Namboodiri MM, Pakshirajan K (2020) Valorization of waste biomass for chitin and chitosan production. In: Waste biorefinery. Elsevier, pp 241–266
65. Tran TH, Nguyen HL, Hwang DS, Lee JY, Cha HG, Koo JM, Hwang SY, Park J, Oh DX (2019) Carbohydr Polym 205:392–400
66. Tuma R (2005) Raman spectroscopy of proteins: from peptides to large assemblies. J Raman Spectrosc 36:307–319
67. Vicente Filipa A, Hus M, Likozar B, Novak U (2021) Chitin deacetylation using deep eutectic solvents: ab initio-supported process optimization. ACS Sustain Chem & Eng 9(10):3874–3886
68. Wan AC, Tai BC (2013) CHITIN-A promising biomaterial for tissue engineering and stem cell technologies. Biotechnol Adv 31:1776–1785
69. Wang Q, Chen S, Chen D (2017) Preparation and characterization of chitosan based injectable hydrogels enhanced by chitin nano-whiskers. J Mech Behav Biomed Mater 65:466–477
70. Wang Y, Yongqing Y, Rui W, Yanlin Z, Pengbo Y, Zhangnan L, Zhihui W, Wei H (2022) Efficient extraction of chitin from crustacean waste via a novel ternary natural deep eutectic solvents. Carbohydr Polym 286:119281
71. Wojtowicz AM, Oliveira S, Carlson MW, Zawadzka A, Rousseau CF, Baksh D (2014) The importance of both fibroblasts and keratinocytes in a bilayered living cellular construct used in wound healing. Wound Repair Regen 22:246–255
72. Xu J, Liu L, Yu J, Zou Y, Wang Z, Fan Y (2019) DDA (degree of deacetylation) and pH-dependent antibacterial properties of chitin nanofibers against Escherichia coli. Cellulose 26:2279–2290
73. Yang TL (2011) Chitin-based materials in tissue engineering: applications in soft tissue and epithelial organ. Int J Mol Sci 12:1936–1963
74. Yusof NLBM, Wee A, Lim LY, Khor E (2003) Flexible chitin films as potential wound-dressing materials: wound model studies. J Biomed Mater Res Part A 66:224–232

75. Zargar V, Asghari M, Dashti A (2015) A review on chitin and chitosan polymers: structure, chemistry, solubility, derivatives, and applications. Chem Bio Eng Rev 2(3):204–226
76. Zhang H, Jin Y, Deng Y, Wang D, Zhao Y (2012) Production of chitin from shrimp shell powders using Serratia marcescens B742 and Lactobacillus plantarum ATCC 8014 successive two-step fermentation. Carbohydr Res 362:13–20
77. Zhang J, Feng M, Lu X, Shi C, Li X, Xin J, Yue G, Zhang S (2018) Base-free preparation of low molecular weight chitin from crab shell. Carbohydr Polym 190:148–155
78. Zhang J, Xu WR, Yu-Cang Z (2022) Facile production of chitin from shrimp shells using a deep eutectic solvent and acetic acid. RSC Adv 12(35):22631–22638
79. Zhang Q, Xiang Q, Li Y (2022) One-step bio-extraction of chitin from shrimp shells by successive co-fermentation using *Bacillus subtilis* and *Lactobacillus plantarum*. Innov Food Sci Emerg Technol 80:103057
80. Zhao D, Huang WC, Guo N, Zhang S, Xue C, Mao X (2019) Two-step separation of chitin from shrimp shells using citric acid and deep eutectic solvents with the assistance of microwave. Polymers 11(3):409
81. Zhu P, Gu Z, Hong S, Lian H (2017) One-pot production of chitin with high purity from lobster shells using choline chloride–malonic acid deep eutectic solvent. Carbohydr Polym 177:217–223

Characterization Techniques for Chitin-Based Nanoparticles

Ritika Puri, Ruchika Kabra, Vimal Arora, Atul Kabra, and Ritika Shyam

Abstract Nanotechnology plays a key role in many different fields, including electronics, energy, and biology. Drug delivery systems rely heavily on nanoparticles, such as nanotubes, dendrimers, and lipid-based carriers, since they enable targeted distribution with the least number of adverse effects. Polymers or polymer science is primarily responsible for the unique physicochemical properties of nanoparticles. The second most common polysaccharide, chitin, and its derivative, chitosan, are highlighted in this chapter because they are becoming more and more viable materials for the formulation of nanoparticles. Because of its many uses and well-known biodegradability, chitosan sticks out in drug delivery systems by shielding bioactive substances. This chapter emphasizes the various preparation methods of chitin and chitin-based nanoparticles, majorly the ionic gelation technique, coacervation method, spray drying, and emulsification solvent evaporation method. Further, various characterization methods for the elucidation of chitin-based nanoparticles have been discussed, such as spectroscopy techniques like UV–visible and FTIR, dynamic light scattering technique, microscopy, and zeta potential. These characterization methods provide important insights into the physicochemical characteristics necessary to understand chitin-based nanoparticles' behaviour and possible uses.

Keywords Characterization methods · Polysaccharide · Preparation methods of chitin and chitin-based nanoparticles

1 Introduction

Nanotechnology, a branch of science and technology, has gained enormous attention in the last two decades and shows its rapid transition from academia to industry in order to fulfill the present demand. According to one report published by [42], it is predicted that the world economy will see a surge of approximately three trillion dollars due to the potential demand for the advancement of nanotechnology

R. Puri · R. Kabra · V. Arora · A. Kabra (✉) · R. Shyam
University Institute of Pharma Sciences, Chandigarh University, Mohali, Punjab, India
e-mail: atul.kbr@gmail.com

R. K. Bachheti et al. (eds.), *Chitin-Based Nanoparticles for the Agriculture Sectors*, Smart Nanomaterials Technology, https://doi.org/10.1007/978-981-96-0920-8_3

techniques, indicating it is a highly lucrative industry. Polymers or polymer science is primarily responsible for the unique physicochemical properties of nanoparticles at the nexus of engineering, physics, medicine, environmental science, agriculture, and chemistry [49]. For instance, to build faster, smaller, and more energy-efficient devices, researchers in electronics communication are investigating the utilization of nanoscale electronic transistors and their related components. Nanotechnology is being used in the energy sector for the development of advanced materials and machinery for the storage of energy, conversion of solar energy, and many other uses [16, 26, 63]. According to [58], nanotechnology is also used in biomedicine to create new tissue engineering methods, therapies, and diagnostic tools. Therefore, nanotechnology's current status is tremendously dynamic, rapidly evolving, and full of fascinating new findings and applications in various disciplines.

Various nanosystems, including different types of nanoparticles, liposomes, ethosomes, nanotubes, and dendrimers, are being used. These systems determine the pharmacokinetics profile and distribution of active drugs by protecting the drug against degradation, allowing more efficient concentrations in the target tissues, and decreasing undesirable toxic side effects [32]. The potential medical applications of several nanosystems, specifically different types of nanoparticles, are being investigated more and more.

Nanoparticles possess nanosized sub-colloidal structures comprised of semisynthetic or synthetic polymers, having a size range of 10–100 nm and available in amorphous or crystalline form [23, 29]. Researchers have prepared various types of nanoparticles, as shown in Fig. 1, to oversee the confinements of traditional therapy, to release the active drug at a specific site to achieve targeted delivery. Therapeutic drugs can be embedded, encapsulated, dissolved, or connected to the nanoparticle matrix. To achieve the targeted delivery of drugs, a drug delivery system is one of the most promising application areas, as medications can be delivered to particular human regions or cells using nanoparticles acting as carriers [18]. Among the many benefits of nanoparticles are their unique surface characteristics, which enable them to target diseased cells only while ignoring healthy ones, increasing efficacy and lowering the side effects of medicament [56].

Among various nanoparticulate delivery systems, nanoparticles formulated using different polymers have gained prominence since it is biodegradable and the availability of a wide range of manufacturing procedures [6, 17, 24]. According to [62], these systems possess the ability to traverse biological barriers and deliver bioactive materials to the intended location without triggering the degradation of proteins, oligonucleotides, or peptides in the biological milieu. Polylactic acid (PLA), polyhydroxyalkanoates (PHAs), starch blends, and cellulose derivatives are a few examples of biodegradable substitutes [9, 32].

Chitin is a hydrophobic, non-toxic, biodegradable polysaccharide that is the second most common polysaccharide available globally [31, 46, 54]. An estimated one billion tonnes of chitin are formulated naturally each year, it is present in the exoskeletons of arthropods, including shrimp shells, and the cell walls of fungi [3, 13, 38, 48, 52]. As seen in Fig. 2, partial deacetylation brought on by acid treatment can result in the formation of chitin nanoparticles by exposing amino moieties.

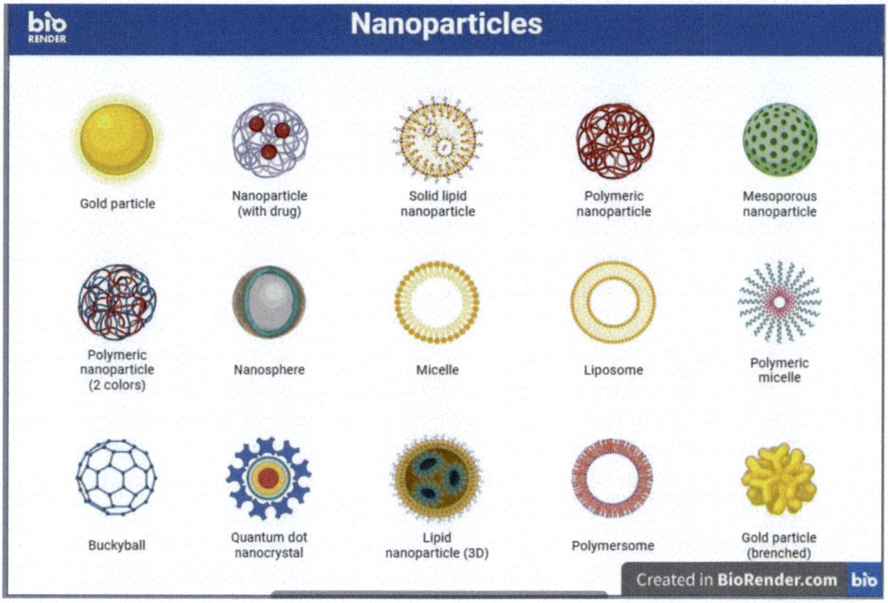

Fig. 1 Classification of nanoparticles

These moieties receive a positive charge from protonation in an acidic environment, stabilizing the colloidal suspension because of repulsive forces [35]. A portion of the chitin produced worldwide is easily accessible, such as in the form of prawn production waste, whose regular disposal is expensive [51].

Research has demonstrated that nanosized chitin can enhance barrier characteristics when enmeshed in a polymer matrix [43]. This is particularly helpful for food packaging applications since barrier qualities are associated with food shelf life, which is associated with a possible reduction in food waste (now ~135 kg per capita annually) [41, 60]. In this regard, the impact of reducing food waste may be far more significant than the impact of reducing plastic waste, which would further support a sustainable society.

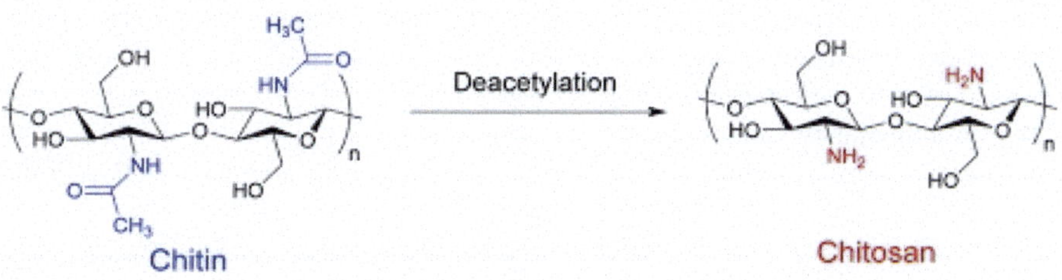

Fig. 2 Deacetylation reaction (chitin to chitosan)

2 Chitosan

Chitosan is the name for the deacylated derivative of chitin. Additionally, chitosan is the fundamental component of the exoskeleton of various arthropods. Chitosan is a linear polymer made up of β (1–4) connected to D-glucosamine, which displays N-acetylglucosamine groups at random positions primarily based on the polymer's degree of deacetylation [21, 39, 59]. In its unaltered state, chitosan is a solid and solubilized in water. Because chitosan dissolves easily in very acidic conditions, it creates cationic polymers with high charge density.

Moreover, they can interact with different anionic polymers to create complexes of polyelectrolytes [57].

The pharmaceutical industry uses several polymeric drug carriers based on chitosan to deliver antibiotics, growth hormones, proteins, peptides, and other drugs [55]. They are also used as a strategy in gene therapy and vaccination delivery.

Enzymes like lysozyme, chitosanase, and papain assist in the in vitro breakdown of chitosan [30, 53]. Lysozyme also causes the acetylated residues to hydrolyze, which initiates the in-vivo degradation of chitosan. There is reportedly an inverse relationship between the rate of chitosan breakdown and the polymer's degree of crystallization and acetylation. The highly deacetylated forms—which can persist in vivo for several months—have lower breakdown rates [5, 28]. Apart from that, any changes in the chemical structure of chitosan can impact its solubility and rate of degradation.

3 Preparation of Chitosan Nanoparticles:

Chitosan can get dissolved, entrapped, and encapsulated and allow bioactive molecules to adhere to the matrix to formulate a stable formulation of nanoparticles. Further, Active pharmaceutical ingredients or bioactive substances can adsorb on the surface of these nanoparticulate systems. Additionally, their nano size facilitates steady circulation and effective passage through cell membranes. These nanoparticles help in the vast delivery of negatively charged, low to high molecular-weight drugs to the target cells, tissues, and specific organ delivery.

Due to their characteristics and functions, chitosan nanoparticles are also appropriate for parenteral and mucosal routes of delivery, such as the mucosa of the mouth, nose, and eyes [37, 47, 64].

By simultaneously ionic crosslinking chitosan and tripolyphosphate, [22] were able to create a hybrid chitosan-tripolyphosphate-alginate nanoparticle that could attach and release insulin in vitro with efficiencies as high as 50.7%.

Wilson et al. [61] used spontaneous emulsification to create chitosan nanoparticles loaded with tacrine. The in vitro release assays revealed that all of the drug-loaded batches produced a continuous and steady release of the drug following the initial

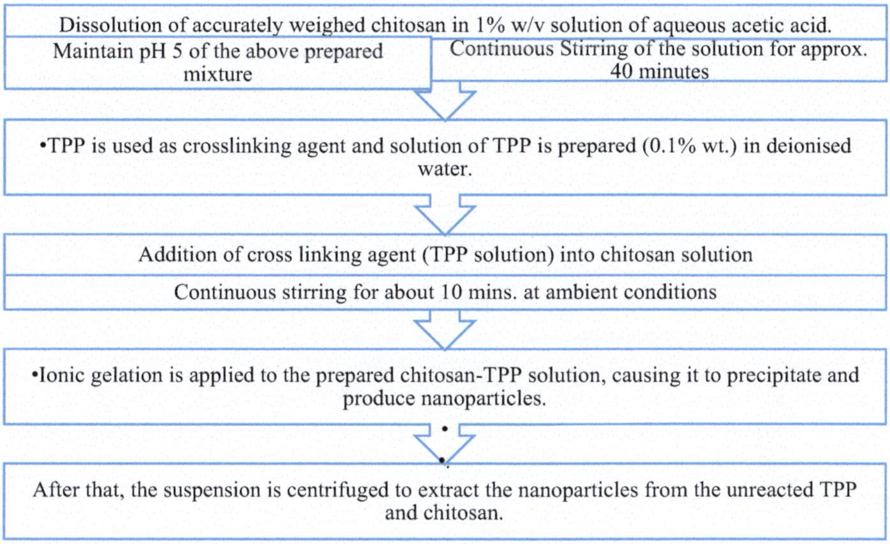

Fig. 3 Ionic gelation method to formulate chitosan nanoparticles

burst, and the nanoparticles made using this process demonstrated good drug-loading capacity.

There are several ways to create chitosan nanoparticles; the method will rely on the intended use and the characteristics of the final product [4]. These are a few typical techniques for the formulation of chitosan nanoparticles, as discussed below:

3.1 Ionic Gelation

This method involves mixing a chitosan solution with an anionic polyanion-like sodium tripolyphosphate (TPP) [25, 44, 59]. The interaction between chitosan and the polyanion leads to the formation of nanoparticles as depicted in Fig. 3:

3.2 Coacervation

This technique is based on the formation of chitosan nanoparticles through the phase separation of a chitosan solution brought about by adding a non-solvent. To make a chitosan solution and then prepare chitosan nanoparticles using the coacervation method, first dissolve chitosan in acetic acid. Depending on the required properties of the nanoparticles, one can adjust the concentration of chitosan solution. Using a pH meter, adjust the chitosan solution's pH. Generally speaking, chitosan dissolves in an acidic solution. Usually, the pH is adjusted to a range of 4–5.

Keep stirring the chitosan solution while gradually adding the non-solvent dropwise resulting in the phase separation brought on by the non-solvent and formation of chitosan nanoparticles. The duration of mixing can impact the nanoparticles' size and characteristics. Thus, if necessary, probe sonication should come after a certain amount of time of continuous mixing. Use a syringe or Millipore filter to filter the nanoparticle suspension in order to control the particle size and remove large aggregates. To separate the nanoparticles from the supernatant, centrifuge the mixture if needed. Drying methods for chitosan nanoparticles include using a desiccator or other suitable drying apparatus.

To ensure the resulting chitosan nanoparticles meet the needs of your application, you can characterize their size, surface charge, and other properties. The characteristics of nanoparticles produced by the coacervation method can be controlled by adjusting the chitosan concentration, selecting different non-solvents, and altering other process variables.

3.3 Emulsification Solvent Evaporation

This method involves emulsifying the polymer solution into an aqueous phase, followed by solvent evaporation, which precipitates the polymer as nanospheres. By adjusting concentrations, surfactants, and other variables, the procedure can be customized to achieve the desired particle size and drug release profiles.

3.4 Reverse Micellar Method

Formulating chitosan nanoparticles via the reverse micellar approach involves creating nanoparticles within the aqueous core of reverse micellar droplets, followed by crosslinking with glutaraldehyde. This technique begins by dissolving a surfactant in an organic solvent and then adding a crosslinking agent, like glutaraldehyde, under vigorous stirring [33]. As noted by [14], this method helps enhance the stability and control the size of the nanoparticles.

3.5 Electrostatic Deposition

To formulate nanoparticles with specific characteristics, chitosan is introduced into an electrolyte solution, and an electric field is applied.

3.6 *Spray Drying*

Spray-drying methods are effective for formulating chitosan nanoparticles, allowing the preparation of drug-loaded free-flowing powders or granules. In this process, a solution of the drug and excipient mixture is atomized into tiny droplets and introduced into a heated, air-filled chamber. The droplets dry, forming drug-laden particles [7, 10, 19].

4 Characterization of Chitin and Chitin Based Nanoparticles

Characterizing the prepared nanoparticles is crucial for comprehensively understanding their physicochemical properties and assessing their therapeutic potential. The following section briefly overviews the various methods used to characterize chitin-based nanoparticles.

(1) **Fourier Transform Infrared Spectroscopy (FTIR)**

Fourier-transform infrared (FTIR) spectroscopy is the most practical method by which chemical identities can be detected based on the existence of functional groups or interactions between these groups. Gaseous forms can be analyzed in unknown mixtures. Fourier Transform Infrared (FTIR) spectroscopy records data as interference patterns, which are then converted into spectra to measure frequencies as wave numbers, typically ranging from 4000 to 400 cm^{-1} [11]. FTIR is a powerful tool for identifying unknown compounds due to its ability to produce spectra with narrow, distinct lines that clearly differentiate the bands corresponding to specific bonds present in the sample. Materials are mixed with KBr for sample preparation and pressed into disks using a hydraulic press. In the case of chitosan, a band at 3420 cm^{-1} is attributed to the combined stretching vibrations of -OH and -NH2 groups. Furthermore, bands near 1649, 1385, and 892 cm^{-1} correspond to the -CONH2 group, C-N stretching of amines, and anhydrous glucosidic bonds, respectively [8].

(2) **X-Ray Diffraction Studies**

XRD can be used to quantify the mean diameter of the nanoparticles and identify their structural features, like the crystalline state of the material [27]. When measured in powder X-ray diffraction, the Scherrer equation can calculate the material's mean crystallite size after corrections for strain, lattice, and instrumental flaws that cause the diffraction peaks to broaden [45].

(3) **UV–Visible Spectroscopy**

In UV–visible spectroscopy, a range of 200–800 nm is selected to determine the absorbance of electromagnetic radiations, and as a result, electrons are excited to higher energy levels [2]. Because chitin and chitin-based derivatives contain different ratios of UV chromophoric groups (two far regions), i.e., glucosamine and N-acetyl-glucosamine, their extinction coefficients do not fall to zero for the wavelengths shorter than about 225 nm. Because there is no indication that the residues of these two chromophoric groups interact inside the chitin chain and also get added to the total absorbance of these polymers at a specific wavelength [36], there are several applications reported of UV–vis spectroscopy, for instance, It can be used to determine the concentration of chitin in a sample by measuring the absorbance of chitin at specific UV wavelengths [34]. UV spectroscopy can be used to monitor changes in the surface chemistry of chitin nanoparticles by analyzing shifts in absorption peaks or the appearance of new peaks resulting from functionalization or coating with other materials. In situations where particle stability is essential, such as medication delivery, changes in the UV absorbance spectrum over time may signal aggregation or precipitation of the nanoparticles. Drug loading and drug release from the prepared drug loaded nano-formulation can be studied by UV–visible spectroscopy [1].

(4) **Dynamic Light Scattering (DLS)**

Dynamic Light Scattering (DLS) is based on the Brownian motion of solute particles or molecules in a solution, resulting from collisions with randomly moving solvent particles. As a laser beam passes through the sample, a photon detector captures changes in the scattered light caused by the particles' random motion. This technique measures zeta potential, polydispersity index (PDI), mean particle size, and particle size distribution. The PDI value, which indicates whether particles are polydisperse or monodisperse, is critical for forming stable formulations in an aqueous medium. According to [15, 40], a PDI value below 0.5 typically signifies monodispersity, while a value above 0.5 indicates polydispersity. Accurate DLS measurements depend on several factors: sample preparation, instrument calibration, and solvent selection.

According to Cummins et al. [12], Dynamic Light Scattering (DLS) was used to determine the particle size distribution of chitin nanoparticles. By monitoring changes in scattered light intensity due to Brownian motion, the Stokes–Einstein equation can be applied to calculate the diffusion coefficient (D), which is related to the particle's hydrodynamic radius (Rh) [12].

(5) **Microscopy**

The scanning electron microscope (SEM) plays a crucial role in studying the detailed shape and surface morphology of chitin-based nanoparticles, whether dispersed or agglomerated. It operates by directing a focused electron beam onto a sample to examine the morphological characteristics of prepared particles in nanoparticulate systems. To enhance contrast and improve the electrical conductivity of nanomaterials, ultrathin coatings containing metals such as gold, platinum, silver, or chromium are commonly applied [11]. During SEM imaging, a fine beam of high-energy electrons scans the nanomaterial sample, interacting with its surface and emitting various types of electron signals. These signals are detected, processed, and displayed as pixels on a monitor to generate three-dimensional images of the nanomaterials. The most frequent secondary electrons stimulated from nanomaterial surfaces are low-energy ones, followed by high-energy backscattered electrons and X-rays released from beneath the sample surface to reveal compositional information. SEM also yields data about the porosity, roughness and overall smoothness of the surface, which is useful in the interpretation of the results of other tests (e.g. in vivo reactions or dissolution behaviour). SEM has been extensively used to study the distribution of chitosan aggregates and the various types of chitosan/alginate nanoparticles [20].

(6) **Zeta Potential**

Zeta potential measurement, also known as the Zetasizer, determines the electric charge on particles, namely at the particle surface, following the principle of electrophoresis. An applied electric field causes the dispersed particles to migrate towards the oppositely charged electrode with a velocity directly proportional to their zeta potential. The laser Doppler antenna technique is generally employed to provide the migration velocity estimate. This parameter expresses the strength of electrostatic repulsive forces between charged particles.

Dispersion is prepared by choosing the suitable solvent, and the sample is tested on the zeta sizer. The resulting values marked of charged potential, either positive or negative, confirm that the particles present in the tested sample are susceptible to aggregation or stability of the system. [50], reported the determination of surface charge or zeta potential for carrageenan/chitin-based nanoparticles using laser Doppler anemometry.

5 Conclusion

This chapter illuminates the dynamic landscape of nanotechnology, specifically delving into the promising domain of chitin-based nanoparticles. Nanoparticles, ranging from nanotubes to dendrimers and lipid-based carriers, exhibit vast potential in reshaping drug delivery systems across electronics, energy, and biomedicine. Chitin, standing as the world's second most abundant polysaccharide, along with its derivative, chitosan, emerges as pivotal in the formulation of nanoparticles. The biodegradability and diverse applications of chitosan position it as a preferred choice for drug delivery systems, ensuring the safeguarding of bioactive compounds. The chapter thoroughly explores diverse methods for chitosan nanoparticle preparation, encompassing ionic gelation, coacervation, and emulsification-solvent evaporation. Significantly, it underscores the importance of characterization techniques like Fourier Transform Infrared Spectroscopy, X-ray diffraction Studies, UV–visible spectroscopy, Dynamic Light Scattering, microscopy, and Zeta Potential measurement. These methodologies unveil essential insights into the physicochemical properties, which are crucial for comprehending chitin-based nanoparticles' behavior and potential applications. Overall, the chapter emphasizes the transformative influence of nanotechnology, specifically in chitin-based nanoparticles, underscoring the imperative role of rigorous characterization to unlock their full potential across diverse applications.

References

1. AbdElhady MM (2012) Preparation and characterization of chitosan/zinc oxide nanoparticles for imparting antimicrobial and UV protection to cotton fabric. Int J Carbohydr Chem Article ID 840591. https://doi.org/10.1155/2012/840591
2. Alim-Al-Razy M, Bayazid GA, Rahman RU, Bosu R, Shamma SS (2020) Silver nanoparticle synthesis and UV-Vis spectroscopy are used to find particle size and measure resistance of colloidal solution. J Phys: Conf Ser 1706(1):012020. IOP Publishing
3. Alves NM, Mano JF (2008) Chitosan derivatives obtained by chemical modifications for biomedical and environmental applications. Int J Biol Macromol 43(5):401–414. https://doi.org/10.1016/j.ijbiomac.2008.09.007. PMID 18838086
4. Banerjee T, Mitra S, Kumar Singh A, Kumar Sharma R, Maitra A (2002) Preparation, characterization and biodistribution of ultrafine chitosan nanoparticles. Int J Pharm 243(1–2):93–105. https://doi.org/10.1016/S0378-5173(02)00267-3. PMID 12176298
5. Barclay TG, Day CM, Petrovsky N, Garg S (2019) Review of polysaccharide particle-based functional drug delivery. Carbohydr Polym 221:94–112. https://doi.org/10.1016/j.carbpol.2019.05.067. PMID 31227171
6. Begines B, Ortiz T, Pérez-Aranda M, Martínez G, Merinero M, Argüelles-Arias F et al (2020) Polymeric nanoparticles for drug delivery: recent developments and future prospects. Nanomaterials (Basel). 10(7):1403. https://doi.org/10.3390/nano10071403. PMID 32707641
7. Bodmeier R, Chen HG (1988) Preparation of biodegradable poly(±)-lactide microspheres using a spray-drying technique. J Pharm Pharmacol 40(11):754–757. https://doi.org/10.1111/j.2042-7158.1988.tb05166.x. PMID 2907552
8. Boruah S, Dutta P (2021) Fungus mediated biogenic synthesis and characterization of chitosan nanoparticles and its combine effect with Trichoderma asperellum against Fusarium oxysporum, Sclerotium rolfsii and Rhizoctonia solani. Indian Phytopathol 74(1):81–93. https://doi.org/10.1007/s42360-020-00289-w
9. Chandra R, Rustgi R (1998) Biodegradable polymers. Prog Polym Sci 23(7):1273–1335. https://doi.org/10.1016/S0079-6700(97)00039-7
10. Chawla A, Taylor KMG, Newton JM, Johnson MCR (1994) Production of spray dried salbutamol sulphate for use in dry powder aerosol formulations. Int J Pharm 108:233–240
11. Choudhary RC, Kumaraswamy RV, Kumari S, Pal A, Raliya R, Biswas P et al (2017) Synthesis, characterization, and application of chitosan nanomaterials loaded with zinc and copper for plant growth and protection. In: Prasad R, Kumar M, Kumar V (eds) Nanotechnology: an agricultural paradigm. Springer, Singapore, pp 227–247
12. Cummins HZ, Pike ER (1974) Photon correlation and light beating spectroscopy. Physics. https://doi.org/10.1007/978-1-4615-8906-8
13. Dmour I, Taha MO (2018) Natural and semisynthetic polymers in pharmaceutical nanotechnology. In: Grumezescu AM (ed) Organic materials as smart nanocarriers for drug delivery. Elsevier, William Andrew, pp 35–100. https://doi.org/10.1016/B978-0-12-813663-8.00002-6
14. Dmour I, Taha MO (2017) Novel nanoparticles based on chitosan-dicarboxylate conjugates via tandem ionotropic/covalent crosslinking with tripolyphosphate and subsequent evaluation as drug delivery vehicles. Int J Pharm 529(1–2):15–31. https://doi.org/10.1016/j.ijpharm.2017.06.061. PMID 28634140
15. Dutta P, Kumari A, Mahanta M (2023) Chitosan nanoparticle: synthesis, characterization, and use as plant health materials [internet]. Chitin and Chitosan - isolation, properties, and applications. IntechOpen. https://doi.org/10.5772/intechopen.106502
16. Edgar JYC, Wang H (2017) Introduction for design of nanoparticle-based drug delivery systems. Curr Pharm Des 23(14):2108–2112. https://doi.org/10.2174/13816128226661610 25154003. PMID 27784242
17. Essa D, Kondiah PPD, Choonara YE, Pillay V (2020) The design of poly(lactide-co-glycolide) nanocarriers for medical applications. Front Bioeng Biotechnol 8:48. https://doi.org/10.3389/fbioe.2020.00048. PMID 32117928

18. Farokhzad OC, Langer R (2009) Impact of nanotechnology on drug delivery. ACS Nano 3(1):16–20. https://doi.org/10.1021/nn900002m. PMID 19206243
19. Gander B, Wehrli E, Alder R, Merkle HP (1995) Quality improvement of spray-dried, protein-loaded d, l-PLA microspheres by appropriate polymer solvent selection. J Microencapsul 12(1):83–97. https://doi.org/10.3109/02652049509051129. PMID 7730960
20. Gazori T, Khoshayand MR, Azizi E, Yazdizade P, Nomani A, Haririan I (2009) Evaluation of alginate/chitosan nanoparticles as antisense delivery vector: formulation, optimization and in vitro characterization. Carbohydr Polym 77:599–606
21. Ghadi A, Mahjoub S, Tabandeh F, Talebnia F (2014) Synthesis and optimization of chitosan nanoparticles: potential applications in nanomedicine and biomedical engineering. Casp J Intern Med 5(3):156–161 PMID 25202443
22. Goycoolea FM, Lollo G, Remuñán-López C, Quaglia F, Alonso MJ (2009) Chitosan-alginate blended nanoparticles as carriers for the transmucosal delivery of macromolecules. Biomacromol 10(7):1736–1743. https://doi.org/10.1021/bm9001377. PMID 19545119
23. Gunay MS, Ozer AY, Chalon S (2016) Drug delivery systems for imaging and therapy of Parkinson's disease. Curr Neuropharmacol 14(4):376–391. https://doi.org/10.2174/1570159x14666151230124904. PMID 26714584
24. He YF, Zhang F, Saleh E, Vaithilingam J, Aboulkhair N, Begines B et al (2017) A tripropylene glycol diacrylate-based polymeric support ink for material jetting. Addit Manuf 16:153–161. https://doi.org/10.1016/j.addma.2017.06.001
25. Hoang NH, Le Thanh T, Sangpueak R, Treekoon J, Saengchan C, Thepbandit W et al (2022) Chitosan nanoparticles-based ionic gelation method: a promising candidate for plant disease management. Polymers (Basel) 14(4):662. https://doi.org/10.3390/polym14040662. PMID 35215574
26. Huang Q, Yu H, Ru Q (2010) Bioavailability and delivery of nutraceuticals using nanotechnology. J Food Sci 75(1):R50–R57. https://doi.org/10.1111/j.1750-3841.2009.01457.x. PMID 20492195
27. Hull AW (1919) A new method of chemical analysis. J Am Chem Soc 41:1168–1175
28. Iqbal J, Anwar F, Afridi S (2017) Targeted drug delivery systems and their therapeutic applications in cancer and immune pathological conditions. Infect Disord Drug Targets 17(3):149–159. https://doi.org/10.2174/1871526517666170606102623. PMID 28595539
29. Iwamoto T (2013) Clinical application of drug delivery systems in cancer chemotherapy: review of the efficacy and side effects of approved drugs. Biol Pharm Bull 36(5):715–718. https://doi.org/10.1248/bpb.b12-01102. PMID 23649331
30. Jafernik K, Ładniak A, Blicharska E, Czarnek K, Ekiert H, Wiącek AE et al (2023) Chitosan-based nanoparticles as effective drug delivery systems—a review. Molecules 28(4):1963. https://doi.org/10.3390/molecules28041963. PMID 36838951
31. Jiang Y, Wu J (2019) Recent development in chitosan nanocomposites for surface-based biosensor applications. Electrophoresis 40(16–17):2084–2097. https://doi.org/10.1002/elps.201900066. PMID 31081120
32. Joudeh N, Linke D (2022) Nanoparticle classification, physicochemical properties, characterization, and applications: a comprehensive review for biologists. J Nanobiotechnol 20(1):262. https://doi.org/10.1186/s12951-022-01477-8. PMID 35672712
33. Kafshgari MH, Khorram M, Mansouri M, Samimi A, Osfouri S (2012) Preparation of alginate and chitosan nanoparticles using a new reverse micellar system. Iran Polym J (Engl Ed) 21:99–107. https://doi.org/10.1007/s13726-011-0010-1
34. Kain D, Kumar S (2020) Synthesis and characterization of chitosan nanoparticles of Achillea millefolium L. and their activities. F1000 Res 9:1297
35. Kim JY, Lee JK, Lee TS, Park WH (2003) Synthesis of chitooligosaccharide derivative with quaternary ammonium group and its antimicrobial activity against Streptococcus mutans. Int J Biol Macromol 32(1–2):23–27. https://doi.org/10.1016/s0141-8130(03)00021-7. PMID 12719128
36. Liu D, Wei Y, Yao P, Jiang L (2006) Determination of the degree of acetylation of chitosan by UV spectrophotometry using dual standards. Carbohydr Res 341(6):782–785

37. Marco-Dufort B, Willi J, Vielba-Gomez F, Gatti F, Tibbitt MW (2021) Environment controls biomolecule release from dynamic covalent hydrogels. Biomacromol 22(1):146–157. https://doi.org/10.1021/acs.biomac.0c00895. PMID 32813504
38. Miola M, Verné E, Ciraldo FE, Cordero-Arias L, Boccaccini AR (2015) Electrophoretic deposition of chitosan/45S5 bioactive glass composite coatings doped with Zn and Sr. Front Front Bioeng Biotechnol. 3:159. https://doi.org/10.3389/fbioe.2015.00159. PMID 26539431
39. Mohammed MA, Syeda JTM, Wasan KM, Wasan EK (2017) An overview of chitosan nanoparticles and its application in non-parenteral drug delivery. Pharmaceutics 9(4):53. https://doi.org/10.3390/pharmaceutics9040053. PMID 29156634
40. Nagarajan E, Shanmugasundaram P, Ravichandiran V, Vijayalakshmi A, Senthilnathan B, Masilamani K (2015) Development and evaluation of chitosan based polymeric nanoparticles of an antiulcer drug lansoprazole. J Appl Pharm Sci. 5:20–25. https://doi.org/10.7324/JAPS.2015.50404
41. Naskar S, Sharma S, Kuotsu K (2019) Chitosan-based nanoparticles: an overview of biomedical applications and its preparation. J Drug Deliv Sci Technol. 49:66–81. https://doi.org/10.1016/j.jddst.2018.10.022
42. Nasrollahzadeh M, Sajadi SM, Sajjadi M, Issaabadi Z (2019) An introduction to nanotechnology. Interface Sci Technol. 28:1–27. https://doi.org/10.1016/B978-0-12-813586-0.00001-8
43. Niederhofer A, Müller BW (2004) A method for direct preparation of chitosan with low molecular weight from fungi. Eur J Pharm Biopharm 57(1):101–105. https://doi.org/10.1016/s0939-6411(03)00189-9. PMID 14729085
44. Nunes R, Sousa Â, Simaite A, Aido A, Buzgo M (2021) Sub-100 NM chitosan-triphosphate-DNA nanoparticles for delivery of DNA vaccines. Proceedings 78:12. https://doi.org/10.3390/IECP2020-08653
45. Patterson AL (1939) The Scherrer formula for x-ray particle size determination. Phys Rev 56:978–982
46. Pillai CKS, Paul W, Sharma CP (2009) Chitin and chitosan polymers: chemistry, solubility and fiber formation. Prog Polym Sci 34(7):641–678. https://doi.org/10.1016/j.progpolymsci.2009.04.001
47. Quiñones JP, Peniche H, Peniche C (2018) Chitosan based self-assembled nanoparticles in drug delivery. Polymers (Basel) 10(3):235. https://doi.org/10.3390/polym10030235. PMID30966270
48. Riva R, Ragelle H, des Rieux A, Duhem N, Jérôme C, Préat V (2011) 'Chitosan and chitosan derivatives in drug delivery and tissue engineering,' in chitosan for Biomaterials II. Adv Polym Sci 244:19–44. https://doi.org/10.1007/12_2011_137
49. Roco MC (2017) Overview: affirmation of nanotechnology between 2000 and 2030. In: Nanotechnology commercialization: manufacturing processes and products. Wiley, Hoboken, NJ, pp 1–23
50. Rodrigues S, da Costa AM, Grenha A (2012) Chitosan/carrageenan nanoparticles: effect of crosslinking with tripolyphosphate and charge ratios. Carbohydr Polym 89(1):282–289
51. Santos VP, Marques NSS, Maia PCSV, Lima MABd, Franco LO, Campos-Takaki GMd (2020) Seafood waste as attractive source of chitin and chitosan production and their applications. Int J Mol Sci 21(12):4290. https://doi.org/10.3390/ijms21124290. PMID 32560250
52. Savin CL, Popa M, Delaite C, Costuleanu M, Costin D, Peptu CA (2019) Chitosan grafted-poly (ethylene glycol) methacrylate nanoparticles as carrier for controlled release of bevacizumab. Mater Sci Eng C Mater Biol Appl 98:843–860. https://doi.org/10.1016/j.msec.2019.01.036. PMID 30813091
53. Shanmuganathan R, Edison TNJI, LewisOscar F, Kumar P, Shanmugam S, Pugazhendhi A (2019) Chitosan nanopolymers: an overview of drug delivery against cancer. Int J Biol Macromol 130:727–736. https://doi.org/10.1016/j.ijbiomac.2019.02.060. PMID 30771392
54. Shi C, Zhu Y, Ran X, Wang M, Su Y, Cheng T (2006) Therapeutic potential of chitosan and its derivatives in regenerative medicine. J Surg Res 133(2):185–192. https://doi.org/10.1016/j.jss.2005.12.013. PMID 16458923

55. Siddaramaiah K, Kumar P, Divya KH, Mhemavathi BT, Manjula DS (2006) Chitosan/HPMC polymer blends for developing transdermal drug delivery systems. J Macromol Sci A 43(3):601–607. https://doi.org/10.1080/10601320600575231. Google Scholar
56. Sim S, Wong NK (2021) Nanotechnology and its use in imaging and drug delivery (review) [review]. Biomed Rep 14(5):42. https://doi.org/10.3892/br.2021.1418. PMID 33728048
57. Singla AK, Chawla M (2001) Chitosan: some pharmaceutical and biological aspects - an update. J Pharm Pharmacol 53(8):1047–1067. https://doi.org/10.1211/0022357011776441. PMID 11518015
58. Swierczewska M, Han HS, Kim K, Park JH, Lee S (2016) Polysaccharide-based nanoparticles for theranostic nanomedicine. Adv Drug Deliv Rev 99(A):70–84. https://doi.org/10.1016/j.addr.2015.11.015. PMID 26639578
59. Van Bavel N, Issler T, Pang L, Anikovskiy M, Prenner EJ (2023) A simple method for synthesis of chitosan nanoparticles with ionic gelation and homogenization. Molecules 28(11):4328. https://doi.org/10.3390/molecules28114328. PMID 37298804
60. Wang Y, Wang EL, Wu ZM, Li H, Zhu Z, Zhu XS et al (2014) Synthesis of chitosan molecularly imprinted polymers for solid-phase extraction of methandrostenolone. Carbohydr Polym 101:517–523. https://doi.org/10.1016/j.carbpol.2013.09.078. PMID 24299807
61. Wilson B, Samanta MK, Santhi K, Kumar KP, Ramasamy M, Suresh B (2010) Chitosan nanoparticles as a new delivery system for the anti-Alzheimer drug tacrine. Nanomedicine 6(1):144–152. https://doi.org/10.1016/j.nano.2009.04.001
62. Yin J, Deng X, Zhang J, Lin J (2018) Current understanding of interactions between nanoparticles and ABC transporters in cancer cells. Curr Med Chem 25(42):5930–5944. https://doi.org/10.2174/0929867325666180314122200. PMID 29542407
63. Yusuf A, Almotairy ARZ, Henidi H, Alshehri OY, Aldughaim MS (2023) Nanoparticles as drug delivery systems: a review of the implication of nanoparticles' physicochemical properties on responses in biological systems. Polymers 15(7):1596. https://doi.org/10.3390/polym15071596. PMID 37050210
64. Zhang Y, Lim CT, Ramakrishna S, Huang ZM (2005) Recent development of polymer nanofibers for biomedical and biotechnological applications. J Mater Sci Mater Med 16(10):933–946. https://doi.org/10.1007/s10856-005-4428-x. PMID 16167102

Role of Agrochemical-Loaded Chitin Nanoparticles for Insect Pest Management

Meseret Zebeaman Birhanu, Rahel Gebeyehu Shiferaw, Rakesh Kumar Bachheti, and Ibrahim Nasser Ahmed

Abstract In recent years, chitin nanoparticles, especially the cationic form, have shown promising insecticidal activity on insect pests that threaten the world's food security. Though no extensive researches were done on the topic, all the reported findings agreed that the cationic nano chitin crystal alone or as a commercial insecticide carrier or by mixing with commercial insecticide has insecticidal activity and even reduces pesticide usage by up to 50%. The cationic nano chitin crystal also demonstrated excellent fungicidal activity against certain plant pathogens and enhanced plant immunity. This chapter presents the insecticidal role of chitin nanoparticles or crystals and their mechanism of action; the developed chitin nanoparticle syntheses and characterization techniques are also highlighted in the chapter.

Keywords Chitin · Nanoparticle · Insecticidal activity

1 Introduction

Insect pest management is an ongoing and nonstop worldwide struggle to secure sustainable global food security. This is because there are around 10,000 insect and mite pests. Only in 2020, 60 million Euros were spent on the purchase and global use of various pesticides to control insect pests [15]. The use of synthetic insecticides

M. Z. Birhanu · I. N. Ahmed (✉)
Nanotechnology Center of Excellence, Addis Ababa Science and Technology University, P.O. Box 16417, Addis Ababa, Ethiopia
e-mail: ibrahim.nasser@aastu.edu.et

Department of Industrial Chemistry, College of Natural and Applied Sciences, Addis Ababa Science and Technology University, P.O. Box 16417, Addis Ababa, Ethiopia

R. G. Shiferaw
Communicable and Non-communicable Diseases Research Directorate, Armauer Hansen Research Institute, P. O. Box 1005, Addis Ababa, Ethiopia

R. K. Bachheti
University Centre for Research and Development, Chandigarh University, Gharuan 140413, Punjab, India

R. K. Bachheti et al. (eds.), *Chitin-Based Nanoparticles for the Agriculture Sectors*, Smart Nanomaterials Technology, https://doi.org/10.1007/978-981-96-0920-8_4

such as organophosphates, carbamates, and pyrethroids has been recommended to control insect pests; however, due to long-time usage, the insects showed resistance to these insecticides. Nanomaterials are now an emerging new class of chemicals used in insect pest control. Besides the metal nanoparticles, chitin and chitosan nanoparticles loaded with synthetic insecticides are now new insecticide chemicals advancing to the agricultural sector. Chitin and its derivative chitosan, which is synthesized by deacetylation of chitin (Fig. 1), are abundant natural polymers next to cellulose. Chitin is widely available in waste streams of the seafood industry: crustacean shells and insects, the cell walls of fungi, and even in the beaks of octopuses [17].

Insecticide nanoparticles are not limited to metal and natural polymer nanoparticles. Rather, other synthetic chemical nanoparticles also showed insecticidal activity. A report indicated that calcium borate nanoparticle with mushroom structure has a comparative insecticidal activity with the commercial insecticidal called Dimilin (Diflubenzuron), which causes chitin synthesis inhibitor on the insect pest *Spodoptera littoralis* [2].

The general mechanism of action of the nanoparticle's role in insect pest management varies with the type of nanoparticles. For instance, nanoparticle-based RNA insecticides shut down insect gene expression and cause lethal insect pests [16]. In the same way, synthetic nanoparticle insecticides like calcium borate NPs inhibit chitin synthesis of insects [2], and metal nanoparticles such as silver and sulfur NPs mostly kill plant pathogens [1]. Chitin and chitosan NPs can have two featured mechanisms of action; the first is the NPs directly participate and enhance the host plant defense system by accumulating pathogen-related proteins, proteinase inhibitors, and Phytoalexins. The second is the NPs acting as insecticide carriers or drug delivery and hence enhancing the pesticide activity.

Therefore, in this chapter, the role of insecticide-loaded chitin nanoparticles (CNPs) for insect pest management will be discussed briefly, starting from CNPs synthesis to its insecticidal activity and with some conclusion and future perspective remarks.

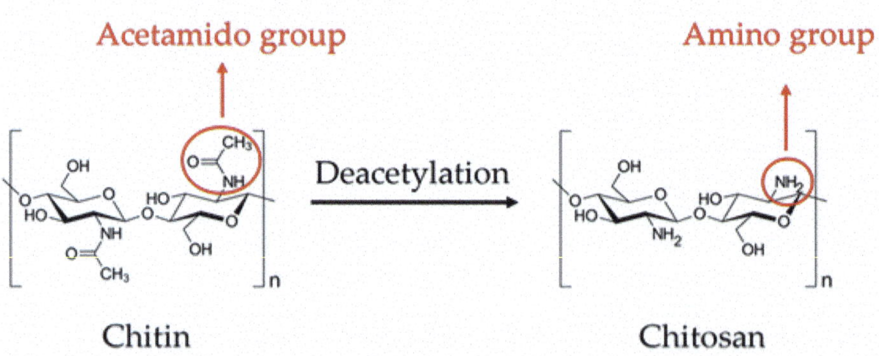

Fig. 1 Chemical structure of chitin and chitosan [17]

2 Chitin Nanoparticle Synthesis Methods

NPs can be manufactured either by following bottom-up or top-down methods (Fig. 2). While the top-down method uses millers to grind the macromolecule to desired nanoscale level, bottom-up method uses chemical reactions to build the nanomaterial from the atoms of the macromolecules [18].

For the purpose of controlling the size and charge type of the nanoparticles, the bottom-up approach has been employed; for instance, CNPs have been synthesized from alkali chitin solution using regeneration or neutralization reaction [19]. In this report, 7 nm CNPs with good chemical stability like dispersity or emulsifier, heat resistance, and high viscosity were attained. To characterize the CNPs, UV–Vis, FTIR, XRD, SEM, and TGA/DTA were utilized. Another study used a hydrolysis reaction to prepare the CNPs [7]. This work used a hydrolytic medium termed deep eutectic solvent obtained from a 1:1 molar ratio of ferric chloride hexahydrate and betaine chloride to prepare the CNPs. The prepared CNPs had an average diameter of 10 nm and length of 268 nm, a crystallinity of 89.2%, and an 88.5% yield. In the study, the prepared CNPs were sufficiently characterized utilizing the FTIR, TEM, XRD, and TGA/DTA. The FTIR broad bands at 3477 and 3264 cm^{-1} ascribed for the O–H and N–H stretching, respectively, whereas the typical characteristic peaks of chitin are located at 1664, 1623, and 1565 cm^{-1} for amides I and II (Fig. 3). The TEM described the CNPs have road like shape and the XRD depicts the 19.2, 20.6, and 23.2° 2θ attributed to the chitin nanoparticle crystalline planes of (110), (120), and (130), respectively.

The other Chitin nanocrystal or nanoparticle synthesis methods, as schematically described in Fig. 4, are acid hydrolysis and TEMPO oxidation [17]. Acid hydrolysis, the most used method, involves highly concentrated acid and ultrasound treatment to remove β-1-4-glycosidic bonds of amorphous chitin regions, as shown in Fig. 4. A study done by [5] used 3 M hydrochloric acid (HCl) to prepare CNPs at temperatures 80–110 °C for 1–6 h. Generally, the reaction is stopped by dilution with water, and the residual acid is removed by dialysis tubing, centrifuging, and washing. Finally, the CNPs are freeze-dried, and the CNPs powder is collected.

The other alternative method for CNPs synthesis is TEMPO (2,2,6,6-tetramethyl-piperidinyl-1-oxyl)-mediated oxidation, which selectively targets the

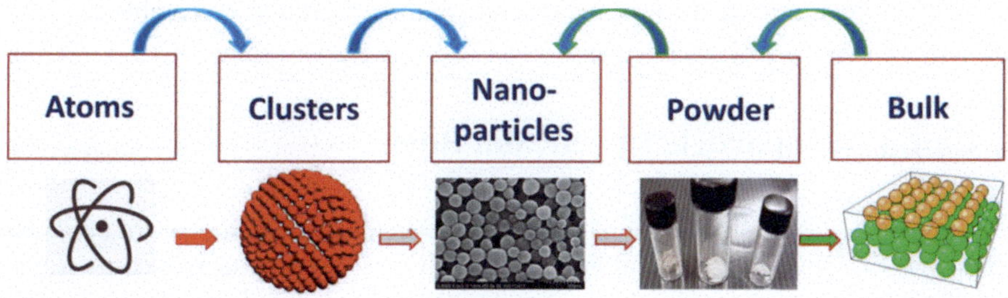

Fig. 2 General method of nanoparticle synthesis [1]

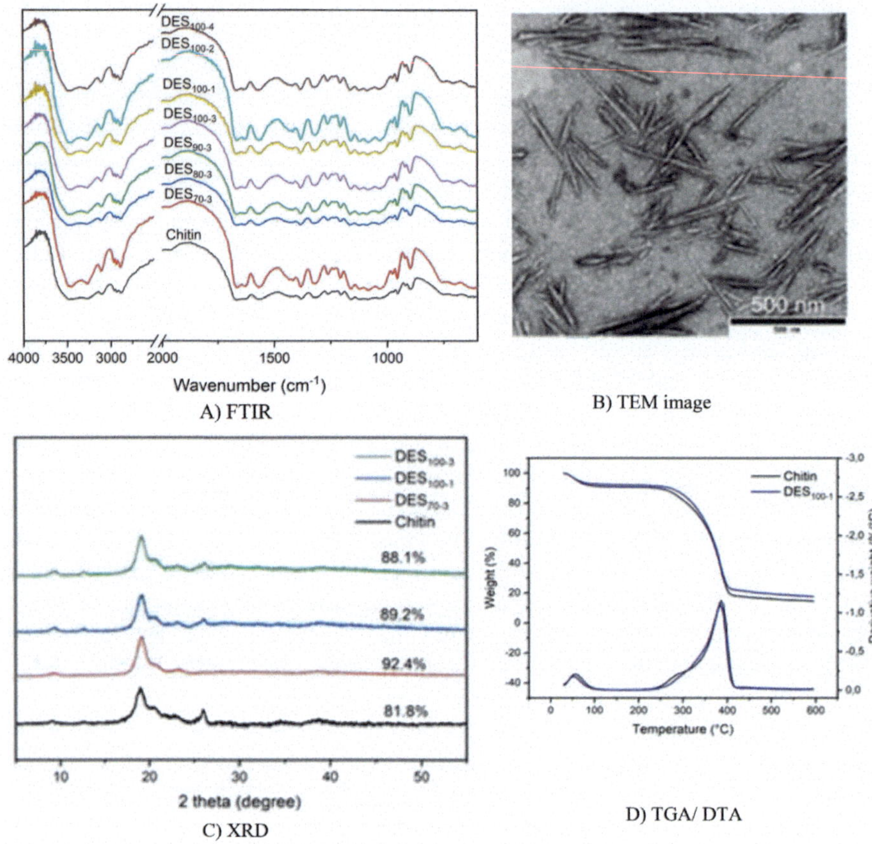

Fig. 3 FTIR (**a**), TEM (**b**), XRD (**c**), and TGA/DTA (**d**) results of Chitin and CNPs prepared by using Deep Eutectic Solvent (DES) at different reaction conditions [7]

alcoholic hydroxyl group (the 6th carbon of the chain) (Fig. 4). The reaction is mainly carried out at pH 10, with chitin dispersed in an aqueous solution of the oxidizing agents mostly NaClO (Sodium hypochlorite) and TEMPO under constant stirring [14]. Different studies by different scholars on the synthesis of chitin nanoparticles and their application other than pesticide application are tabulated in Table 1.

3 Chitin Nanoparticle Insecticidal Role

In recent years, the application of nanotechnology in the agricultural sector has been increasing, like food and medical applications. It was possible to reduce pesticide usage by 50% if cationic nano chitin crystal/whisker loaded with commercial pesticides and used to control white aphides insects [9]. The nano chitin crystals were

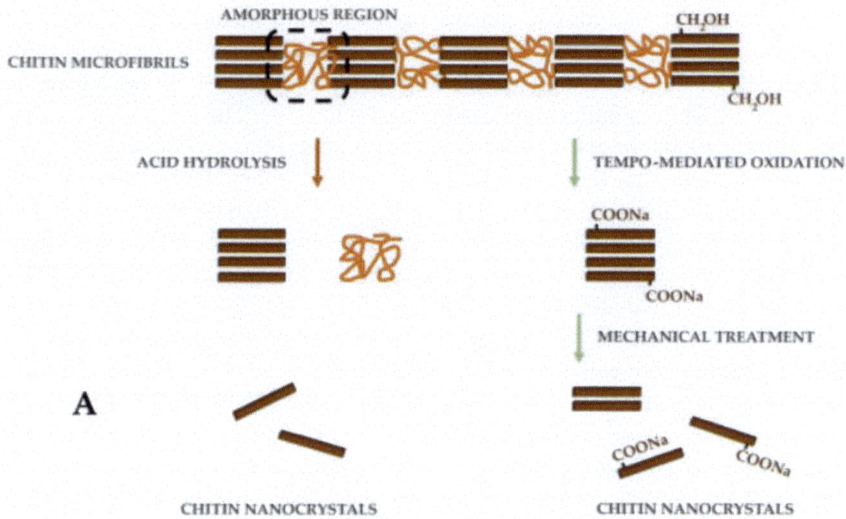

Fig. 4 Schematic diagram of acid hydrolysis (**a**) and TEMPO-oxidation (**b**) method for CNPs synthesis [17]

Table 1 Summary of different synthesis method and applications of chitin nanomaterial

Nanomaterial type	Method of synthesis	Application	References
α- Chitin nanoparticle	Acid hydrolysis	Adsorption of Methylene Blue, Bromophenol Blue, and Coomassie Brilliant Blue	[6]
Chitin nanoparticle	Sulfuric acid hydrolysis, TEMPO-oxidation, and ammonium persulfate methods	Chitin-silver hybrid nanoparticles as antibacterial and filler in food packaging material	[12]
Chitin nanoparticle	Low-energy deconstruction of partially deacetylated chitin	Oil/Water interfacial superstabilization	[3]
Chitin nanocrystal	Acid hydrolysis	Anisotropic adhesive	[11]
Chitin nanocrystal	Acid hydrolysis	Chitin nanocrystal hybrid with –Zn2+ to treat tobacco mosaic virus	[4]
Chitin nanocrystal	Acid hydrolysis	As reinforcing with thermoplastic starch film	[13]

synthesized by acidic hydrolysis, and the particles were mixed with commercial pesticides Omethoate (40% EC), Imidacloprid (10% WP), and Acetamiprid (40% WG). Spraying of the combined complex on the white aphides lay on a leaf; the mortality rate of white aphides increased from 80 to 95%. The entrance of mixed drug through the mouth of aphides was traced by adding Fluorescein isothiocyanate on the nano chitin crystal (Fig. 5). The safety of the combined complex was well affirmed because the Lethal Dos 50% (LD50) was very high, which indicates it does not harm unless it is orally taken in very high concentration. The LD50 they found is greater than 5000 mg/kg BW (body weight) in rats; acute dermal toxicity LD50 was more significant than 2000 mg/kg BW.

Another report indicated chitin nanoparticles have the potential to stabilize the biopesticide Abamectin which needs a short duration in conventional formulations that limit its application [8]. They prepared the CNPs by ultrasonication combined with low-concentration acidic hydrolysis and mixed it with Abamectin by conjugation in dispersion naturally. The CNPs significantly promoted the thermal stability and water solubility of Abamectin. Similarly, CNPs decreased photodegradation of Abamectin by 48% when exposed to UV light for 56 h. Finally, the Abamectin combined with CNPs significantly controls the larva growth and enhances mortality against 2 instar larvae of noctuid. More importantly, the two mixtures effectively reduced pesticide dosage in insect pest control. This finding indicates that Abamectin-CNPs can be used as a new bioactive delivery system in future pest management.

In agricultural use, chitin nanoparticles were not only limited to insect pest control management. Rather, it has plant pathogen protection. The mechanism either enhances the complex plant defense system or directly kills the pathogens. For example, the study by [20] showed that cationic nano chitin has enhanced the plant immune system against *Phytophthora* by enhancing phenylalanine ammonia-lyase activity and PR gene expression in the plant. Similarly, a study showed that the

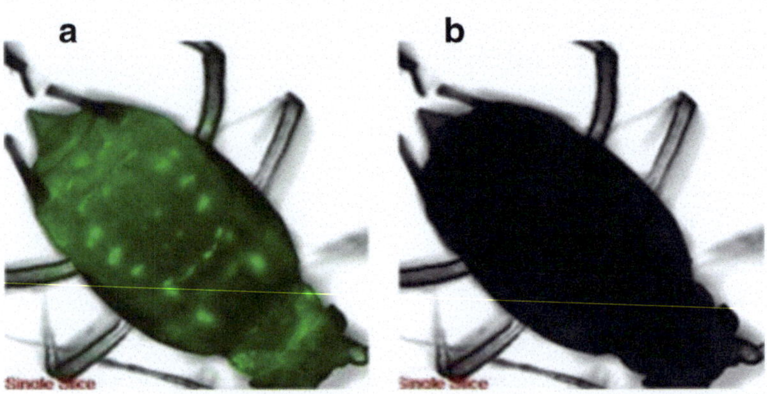

Fig. 5 Fluorescein isothiocyanate labeling on the nano chitin crystal; **b** control and **a** after taking of Fluorescein isothiocyanate labeled nano chitin crystal by white aphides insect [9]

cationic nano chitin crystal, which has a rod-like structure, was effective against plant fungi that cause Crown rot, a disease of wheat plants [10]. In the report, the cationic nano chitin crystal demonstrated significant inhibition of the mycelial growth and conidial production of the fungi. Moreover, the study noted that combining the cationic nano chitin crystal with commercial fungicide increased fungicide activity from 79.30 to 90.02%.

4 Conclusion and Future Perspective

Chitin nanomaterial application attracts many scholars because chitin is the second most abundant biomass. Chitin nanoparticle synthesis with bottom-up methods follows the following different methods: acid hydrolysis, TEMPO-oxidation, regeneration/neutralization, and hydrolysis using a special hydrolytic medium. So far, the cationic chitin nanoparticle and crystal are promising insecticidal alone, used as a carrier, or mixed with bio and synthetic pesticides. In addition to stabilizing biopesticides, it also reduces pesticide usage by up to 50% since it enhances commercial insecticides' insecticidal activity, which has many advantages, like reducing pesticide consumption and related problems. In addition to pesticide usage, chitin nanoparticles also have antifungal activity for plants, and similarly, they also enhance the plant's immunity systems. In conclusion, even though nanochitin has shown pesticide and antifungal activity, more research needs to be done. Therefore, more research should be done to be more confident and commercialize it on the market.

References

1. Ahmed AI, Lee YS (2015) Nanoparticles as alternative pesticides: concept, manufacturing and activities. Korean J Mycol 43(4):207–215
2. Ayoub HA, Khairy M, Rashwan FA, Abdel-Hafez HF (2021) Comparison of the insecticidal activity of calcium borate with mushroom-like structures and chitin synthesis inhibitor insecticide (dimilin) against Egyptian cotton leaf worm. J Entomol Zool Stud 9:119–124
3. Bai L, Huan S, Xiang W, Liu L, Yang Y, Nugroho RWN, Fan Y, Rojas OJ (2019) Self-assembled networks of short and long chitin nanoparticles for oil/water interfacial superstabilization. ACS Sustain Chem & Eng 7(7):6497–6511
4. Cao Z, Ma X, Lv D, Wang J, Shen Y, Peng S, Yang S, Huang J, Sun X (2023) Synthesis of chitin nanocrystals supported zn2+ with high activity against tobacco mosaic virus. Int J Biol Macromol 250:126168
5. Cheikh FB, Mabrouk AB, Magnin A, Putaux J-L, Boufi S (2021) Chitin nanocrystals as pickering stabilizer for o/w emulsions: effect of the oil chemical structure on the emulsion properties. Colloids Surf B: Biointerfaces 200:111604
6. Dhananasekaran S, Palanivel R, Pappu S (2016) Adsorption of methylene blue, bromophenol blue, and coomassie brilliant blue by α-chitin nanoparticles. J Adv Res 7(1):113–124
7. Hong S, Yuan Y, Zhang K, Lian H, Liimatainen H (2020) Efficient hydrolysis of chitin in a deep eutectic solvent synergism for production of chitin nanocrystals. Nanomaterials 10(5):869

8. Li Z, Su L, Wang H, An S, Yin X (2020) Physicochemical and biological properties of nanochitin—abamectin conjugate for noctuidae insect pest control. J Nanoparticle Res 22:1–16
9. Li Z, Wang H, An S, Yin XJ (2021) Nanochitin whisker enhances insecticidal activity of chemical pesticide for pest insect control and toxicity. J Nanobiotechnol 19:1–13
10. Liang R, Li X, Yuan W, Jin S, Hou S, Wang M, Wang H (2018) Antifungal activity of nanochitin whisker against crown rot diseases of wheat. J Agric Food Chem 66(38):9907–9913
11. Liu H, Feng Y, Cao X, Luo B, Liu M (2021) Chitin nanocrystals as an eco-friendly and strong anisotropic adhesive. ACS Appli Mater & Interfaces 13(9):11356–11368
12. Oun AA, Rhim J-W (2018) Effect of isolation methods of chitin nanocrystals on the properties of chitin-silver hybrid nanoparticles. Carbohydr Polym 197:349–358
13. Salaberria AM, Diaz RH, Labidi J, Fernandes SC (2015) Role of chitin nanocrystals and nanofibers on physical, mechanical and functional properties in thermoplastic starch films. Food Hydrocoll 46:93–102
14. Tran TH, Nguyen HL, Hwang DS, Lee JY, Cha HG, Koo JM, Hwang SY, Park J, Oh DX (2019) Five different chitin nanomaterials from identical source with different advantageous functions and performances. Carbohydr Polym 205:392–400
15. Vreysen MJ, Abd-Alla AM, Bourtzis K, Bouyer J, Caceres C, de Beer C, Oliveira Carvalho D, Maiga H, Mamai W, Nikolouli K, Yamada H, Pereira R (2021) The insect pest control laboratory of the joint fao/iaea programme: ten years (2010–2020) of research and development, achievements and challenges in support of the sterile insect technique. Insects 12(4):346
16. Yan S, Yin M-Z, Shen J (2022) Nanoparticle-based nontransformative rna insecticides for sustainable pest control: mechanisms, current status and challenges. Entomol Gen 43:21–30
17. Yanat M, Schroën K (2023) Advances in chitin-based nanoparticle use in biodegradable polymers: a review. Carbohydr Polym 312:120789
18. Zebeaman, M, Bachheti, R. K, Bachheti, A, Pandey, D, & Husen, A (2023) Green and cost-effective nanoparticles synthesis from some frequently used medicinal plants and their various applications. In: Secondary metabolites from medicinal plants. CRC Press, pp 287–304
19. Zewude DA, Noguchi T, Sato K, Izawa H, Ifuku S (2022) Production of chitin nanoparticles by bottom-up approach from alkaline chitin solution. Int J Biol Macromol 210:123–127
20. Zhou Y, Jing M, Levy A, Wang H, Jiang S, Dou D (2020) Molecular mechanism of nanochitin whisker elicits plant resistance against phytophthora and the receptors in plants. Int J Biol Macromol 165:2660–2667

Chitin-Based Nanoparticles for Plant Disease Management

Noureddine Chaachouay and Lahcen Zidane

Abstract The use of chitin-based nanoparticles in plant disease management has gained significant attention due to their unique properties and potential applications. Chitin, a natural polysaccharide found in the exoskeletons of crustaceans, insects, and fungal cell walls, is a promising material for developing nanoparticles. Different methods can synthesize chitin-based nanoparticles, such as mechanochemical, enzymatic hydrolysis, and ionic liquid-mediated synthesis. The nanoparticles exhibit antifungal and antibacterial properties, making them effective agents against various plant pathogens. Additionally, the biocompatibility and biodegradability of chitin-based nanoparticles make them environmentally friendly alternatives to traditional chemical pesticides. This chapter also discusses how these nanoparticles interact with plant pathogens and enhance plant defense mechanisms. Chitin-based nanoparticles present a promising and sustainable substitute for traditional chemical-based approaches to plant disease management, given their biodegradability, biocompatibility, and non-toxic nature. Consequently, these nanoparticles hold significant potential to transform the landscape of agricultural practices and plant disease management.

Keywords Agriculture · Chitin · Nanoparticles · Plant disease management · Synthesis

List of Abbreviations

CD Callose Deposition
CDPKs Calcium-Dependent Protein Kinases

N. Chaachouay (✉)
Agri-Food and Health Laboratory (AFHL), ENS of Berrechid, Hassan First University, Po Box 382, 26000 Settat, Morocco
e-mail: nour.chay@gmail.com

L. Zidane
Plant, Animal Productions and Agro-Industry Laboratory, Department of Biology, Faculty of Sciences, Ibn Tofail University, B.P. 133 14000, Kenitra, Morocco

R. K. Bachheti et al. (eds.), *Chitin-Based Nanoparticles for the Agriculture Sectors*, Smart Nanomaterials Technology, https://doi.org/10.1007/978-981-96-0920-8_5

CERK1	Chitin Elicitor Receptor Kinase 1
CS	Calcium Signaling
LSD	Lysergic Acid Diethylamide
MAPK	Mitogen-Activated Protein Kinase
NPs	Chitin Nanoparticles
PR	Pathogenesis-Related
PRRs	Pattern Recognition Receptors
ROS	Reactive Oxygen Species
SAR	Systemic Acquired Resistance
SAS	Salicylic Acid Signaling
SIPK	Salicylate-Induced Protein Kinase
ToMV	Tomato Mosaic Virus
UV	Ultraviolet
WIPK	Wound-Induced Protein Kinase

1 Introduction

Nanoparticles have dimensions on the nanometer scale, typically ranging from 1 to 100 nm [33]. These materials can be engineered and manipulated at the nanoscale, leading to unique properties and behaviors that differ from those of the same material in bulk [83]. Chitin-based nanoparticles are tiny particles derived from chitin, a natural polymer in the exoskeletons of crustaceans, insects, plants, fungi, and bacteria [25]. Chitin-based nanoparticles have many potential applications in biomedicine, biotechnology, and agriculture due to their biocompatibility, biodegradability, antimicrobial, and immunostimulatory properties. Some examples of chitin-based nanoparticles are chitin nanocrystals, chitin nanofibrils, and chitosan nanoparticles [75, 76]. Chitin-based nanoparticles have gained attention in plant disease management due to their unique properties and potential applications. Chitin is a natural polysaccharide in the exoskeletons of crustaceans, insects, and fungal cell walls [59]. Chitosan, a derivative of chitin, is obtained by deacetylating chitin and has been extensively studied for its antimicrobial and plant protection properties [78].

Nanoparticles derived from chitin or chitosan offer significant advantages in managing plant diseases. With inherent antimicrobial properties, these nanoparticles exhibit the capability to impede the growth of a range of pathogens, encompassing fungi, bacteria, and viruses, rendering them promising candidates for controlling plant diseases arising from these microorganisms [66]. Moreover, the biodegradability and environmental friendliness of chitin and chitosan contribute to sustainable agricultural practices, as these nanoparticles can disintegrate into harmless components, minimizing environmental impact when compared to specific synthetic alternatives [62]. Additionally, chitin-based nanoparticles have been identified as an inducer of plant defense mechanisms, fortifying resistance against pathogens, while chitin-based nanoparticles function as elicitors, prompting a systemic acquired resistance (response in plants and bolstering their ability to fend off diseases [48, 61]. This chapter highlights the significant role of chitin-based nanoparticles in revolutionizing approaches to plant disease control with a focus on sustainability and environmental impact.

2 Chitin

2.1 *History and Terminology*

The attribution of the discovery of chitin is ascribed to the French scientist Henri Braconnot in 1811 [10]. He was the first individual to elucidate the chemical substance derived from mushrooms, which exhibits insolubility in sulfuric acid. The term he gave it was "fungine" [22]. Auguste Odier, a French chemist, successfully extracted the molecule from beetle cuticles and named it chitin. The term "chitine" was taken from the Greek word "χιτών" ("chiton"), which means covering [19]. Subsequently, the compound's name was altered to chitin. Later, chitin has found applications in the fields of medicine and industry. Due to its non-toxic nature, biodegradability, and antibacterial capabilities may be used in many applications. For example, it has been used in the fabrication of biodegradable sutures. Additionally, it has been used as an ingredient in certain creams and cosmetics. The Swiss chemist Albert Hofmann (1906–2008), who is well known for his discovery of lysergic acid diethylamide (LSD), revealed the structure of chitin in 1929 [18, 22].

2.2 *Chitin Molecular Structure*

Chitin is a colorless, rigid, non-stretchable, nitrogen-containing polysaccharide (R. A. [58]). It is the second most prevalent biopolymer after cellulose. It is present in the outer skeletons of crustaceans, insects, some fish scales, and even in the cellular walls of fungi, plants, and bacteria (Fig. 1). Chitosan is a hydrophilic polysaccharide obtained from chitin by deacetylation using alkalis [24]. While chitin and chitosan may be used to create films for food packing, their primary purpose is to serve as an edible covering that prolongs the freshness of fresh fruits and vegetables [87].

Chitin has a similar structural role to keratin, a protein found in human hair and nails. However, it distinguishes itself from keratin due to its polysaccharide structural component. In addition, chitin exists in three distinct polymeric forms in nature: α, β, and γ. The form is ubiquitously present and is typically obtained from the exoskeletons of crabs and prawns. It exhibits an antiparallel arrangement [9]. On the other hand, the state is mostly derived from mollusks and displays a parallel arrangement distinct from the form. The α and β forms of chitin exhibit economic

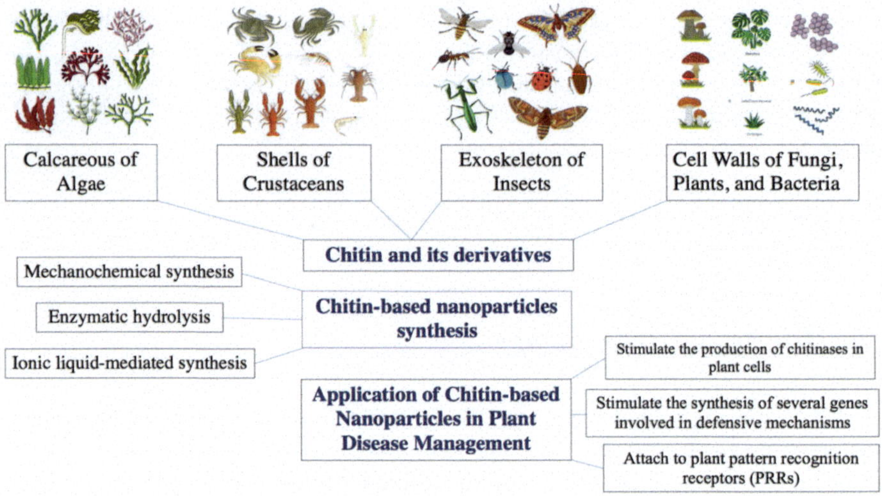

Fig. 1 Chitin Sources, synthesis, and its applications in plant disease management

characteristics, and they vary in the structure and polarity of the linear chains inside the chitin sheets [66]. In the case of γ, chitin consists of two parallel strands and two antiparallel strands. The β form cannot be transformed into the alpha form, but γ chitin may be converted into α chitin using lithium thiocyanate (Fig. 2).

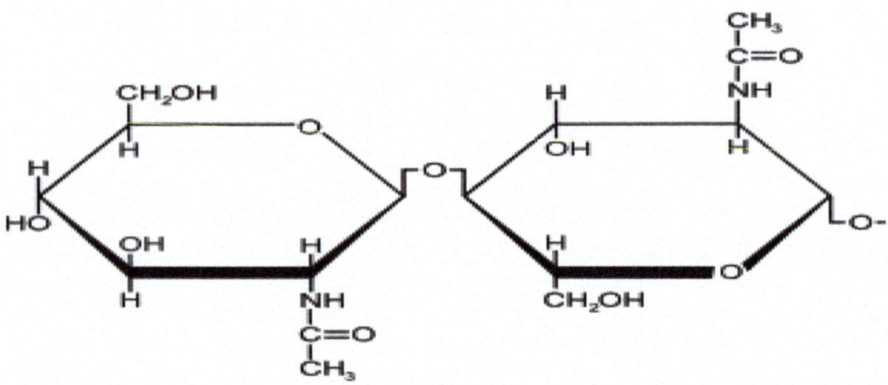

Fig. 2 Schematic illustrating the distinctions among the three polymorphs of chitin

2.3 Chitin Molecular Extraction

The primary sources of raw material for chitin manufacturing are the exoskeletons of different crustaceans, namely crabs and shrimps. Chitin is present in crustaceans, particularly shellfish, as a component of an intricate structure that combines with proteins [59]. This structure serves as a foundation for the deposition of calcium carbonate, forming a sturdy shell [32]. The process of isolating chitin from shellfish involves removing the shell's two main components, proteins via deproteinization and inorganic calcium carbonate through demineralization, along with tiny quantities of colors and lipids that are typically eliminated during the preceding processes [55]. Decolorization is sometimes used to eliminate any remaining pigments in certain instances and produce chitin in its pure form, yet a universally accepted approach has yet to be established [63]. Chemical or enzymatic solutions may be used for both deproteinization and demineralization. There may be some advantage in reversing the sequence of the two previously described stages, particularly when considering enzymatic therapy [88]. Microbial fermentation is used, whereby deproteinization and demineralization procedures are carried out concurrently.

2.4 Chitin Molecular Properties

Chitin is insoluble in water owing to intermolecular solid hydrogen bonding. It exhibits enhanced resistance to degradation and increased capacity to establish robust connections with other molecules [78]. However, specific water-soluble derivatives derived from chitin exist, such as chitosan or carboxymethyl, which can be extracted from chitin [9]. The primary benefits of these compounds lie in their exceptional flexibility, allowing them to be shaped into diverse forms, including fibers, hydrogels, beads, sponges, and membranes [68]. Chitin has a crystalline structure and possesses much greater tensile strength than most synthetic materials. The substantial hydrogen bonding inside the chitin chains accounts for the elevated tensile strength of chitin [87]. Chitin is abundantly found in almost all fungi, providing proof of its vast distribution in the fungus kingdom and the Animalia kingdom's phyla Arthropoda and Mollusca.

3 Chitin-Based Nanoparticles

Chitin-based nanoparticles are nanoscale compounds made from chitin, a natural polysaccharide. Chitin is present in the exoskeletons of crustaceans (e.g., prawns and crabs), insects, and fungus cell walls [86]. Nanoparticles are often generated by disintegrating or altering chitin using several methods to get nanoscale formations with distinctive characteristics. Furthermore, chitosan, derived from chitin by a process of partial deacetylation, is often used in producing these nanoparticles [49].

3.1 Chitin-Based Nanoparticles Synthesis

The process of synthesizing chitin-based nanoparticles entails the transformation of chitin or its derivative, chitosan, into particles on the nanoscale. Multiple techniques may be used to accomplish this, and the course selection relies on the nanoparticles' desired characteristics and planned uses [25]. Below are many prevalent techniques used in the production of chitin-based nanoparticles (Fig. 1):

- **Mechanochemical synthesis** is a process that involves the grinding or milling of chitin sources together with different additives or agents to generate nanocrystals or nanofibers. Investigation outlines a technique for producing carboxylate chitosan nanocrystals from chitin found in crab shells [57]. This is achieved by using sodium hydroxide and acetic acid. The nanocrystals produced have a large surface area and may be excellent substrates for gold-catalyzed processes [5].
- **Enzymatic hydrolysis** is a process that involves the degradation of chitin-bound proteins via the use of enzymes or microbes. Investigations examine the current advancements in creating and improving nanoscale chitin and chitosan products from different sources via enzymatic hydrolysis [46]. This technology can maintain the inherent structure and characteristics of chitin and chitosan while enabling the production of nanoparticles of various shapes and sizes [50].
- **Ionic liquid-mediated synthesis** refers to using ionic liquids as solvents or catalysts to dissolve or modify chitin sources. A study examines new work using ionic liquids to produce chitin from fungal and insect cuticles as alternate sources [57]. An advantage of this approach is its ability to surpass the constraints associated with traditional solvents, including elevated temperature, toxicity, and environmental repercussions.

The selection of the synthesis method is contingent upon the individual application requirements, and fine-tuning of the circumstances is often required to get the desired characteristics of chitin-based nanoparticles. Scientists are actively investigating and creating novel techniques to improve the effectiveness and adaptability of chitin-based nanoparticle production.

3.2 Chitin Nanoparticles Examples

Chitin and its derivatives possess distinctive characteristics that make them essential in nanotechnology and nanomaterial research. Several examples of chitin nanoparticles include:

- **Chitin nanocrystals and nanofibrils**: These are obtained from the exoskeletons of crustaceans and insects. These materials provide appealing mechanical and biological characteristics, including a high Young's modulus, biocompatibility, antibacterial capabilities, and immunogenicity. They may serve as additives in composites composed of biodegradable polymers for medicinal purposes [70].
- **Chitosan nanocrystals**: These are derived from chitin by enzymatic hydrolysis. They possess characteristics comparable to chitin nanocrystals; however, they have a reduced level of crystallinity and an elevated level of acetylation. Additionally, they may serve as additives in composites composed of biodegradable polymers for medicinal purposes [26].
- **Chitin nanofibers**: These are generated using either mechanical or chemical means. Due to their elevated aspect ratio and extensive surface area, they are well-suited for various applications, including drug delivery systems, wound dressings, tissue engineering scaffolds, and biosensors [35].

4 Role of Chitin in Plant Defense Mechanisms

4.1 Natural Presence of Chitin in Plant Cell Walls

Chitin is a naturally occurring polysaccharide that is widely present in various organisms. It can be found in arthropods, marine invertebrate exoskeletons, fungal cell walls, insect exoskeletons, crustaceans, and mollusks. In terms of abundance, it is the second most prevalent polysaccharide in nature, following cellulose [44]. Chitin possesses a well-structured crystalline arrangement and exhibits limited chemical reactivity. It is not soluble in water or organic solvents. Chitin may also be a signaling molecule to trigger plant defensive responses against fungal diseases and rhizobacterial symbionts [23]. Chitin can stimulate the production of chitinases in plant cells. These chitinases are responsible for breaking down the cell wall of fungi and, therefore, restricting their invasion. Chitin may also stimulate the synthesis of several defense-related genes, including those responsible for lignin biosynthesis, pathogenesis-related proteins, and phytoalexins. Chitin does not exist in plant cell walls, but it may interact with other cell wall elements to modify its structure and function [38]. Chitin can attach itself to cellulose microfibrils and create connections that enhance the stiffness and durability of the cell wall [7]. In addition, chitin can interact with hemicellulose and pectin polymers, forming complexes that impact the cell wall's water retention and permeability.

4.2 Chitin as an Elicitor, Triggering Plant Defense Responses

Within plant biology, chitin has been identified as a powerful stimulant of plant defense mechanisms. Elicitors are chemical compounds that may trigger a defensive response in plants, priming them to withstand prospective assaults from pathogens. The interplay between chitin and plants exemplifies the activation of the intrinsic immune system in plants [45]. Chitin can stimulate the production of chitinases in plant cells. These chitinases then break down the cell walls of fungi and restrict their ability to invade. Chitinases are enzymes that catalyze the hydrolysis of the β-1,4 glycosidic linkages present in chitin, resulting in the formation of chitin oligosaccharides and chitosan oligosaccharides [34]. These compounds may also function as elicitors that stimulate the plant's immune system, triggering ion fluxes, enhancing chitinase activity, synthesizing phytoalexins, generating reactive oxygen species (ROS), and promoting the expression of defense genes.

Chitin may also stimulate the synthesis of several genes involved in defensive mechanisms, including those responsible for lignin formation, pathogenesis-related proteins, and phytoalexins. Phytoalexins are antimicrobial substances that build up in plant tissues when pathogens attack them [16, 65]. Lignin is an intricate polymer that enhances the structure of the cell wall and imparts robustness and resilience against both living and non-living pressures. Pathogenesis-related proteins are a category of proteins that are triggered by pathogen invasion and exhibit many roles, including antifungal properties, alteration of cell wall structure, and transmission of signals [73].

Moreover, chitin can attach to plant pattern recognition receptors (PRRs) either on the plasma membrane or inside the cytoplasm of plant cells. PRRs are proteins that detect conserved chemical patterns of microorganisms, including chitin, and activate signal transduction pathways that trigger plant defensive responses [51, 67]. For instance, chitin can attach itself to the chitin elicitor receptor kinase 1 (CERK1) in *Arabidopsis thaliana*. This attachment triggers a series of events known as a mitogen-activated protein kinase (MAPK) cascade, which leads to the activation of defense genes [17, 56]. Gaining insight into the molecular pathways that drive plant defensive responses triggered by chitin is essential for devising methods to bolster plant resistance to pathogens. Scientists are now investigating ways to use this information to enhance crop defense mechanisms and decrease the need for chemical pesticides.

4.3 Biochemical Pathways Involved in Plant Defense Activated by Chitin

Chitin, a prevalent constituent of fungal cell walls, is an inherent polysaccharide that can stimulate plant defense mechanisms [13]. Plants induce the production of antimicrobial compounds, the activation of diverse biochemical pathways, and cell wall reinforcement in response to chitin perception [6]. When chitin is present, it stimulates the following biochemical pathways involved in plant defense:

- **Salicylic Acid Signaling (SAS)**: Although the ethylene and jasmonic acid pathways are frequently linked to resistance against necrotrophic pathogens, this pathway is more often associated with resistance against biotrophic pathogens [29]. Additionally, the activation of the SA pathway by chitin perception can result in the production of defense genes and the development of systemic acquired resistance [40].

- **Calcium Signaling (CS)**: The detection of chitin by plant receptors initiates the entry of calcium ions (Ca^{2+}) into the plant cell. Increased concentrations of cytosolic calcium function as a secondary messenger and are vital in delivering the signal induced by chitin. Calcium-dependent protein kinases (CDPKs) are often stimulated, triggering subsequent signaling processes [79].
- **Activation of Ethylene and Jasmonic Acid Pathways**: The chitin signaling route often triggers the activation of the ethylene and jasmonic acid signaling pathways. Phytohormones have crucial functions in controlling defensive responses against infections. They regulate the expression of defense-related genes, particularly those involved in producing antimicrobial substances [77].
- **Pathogenesis-Related (PR) Protein Synthesis**: Chitin perception triggers the production and buildup of PR proteins. These proteins possess antibacterial characteristics and play a role in the plant's immune response against pathogens. Chitin-induced defense responses often result in the increased expression of genes that encode PR proteins [37, 65].
- **Mitogen-Activated Protein Kinase (MAPK) Cascades**: Chitin identification induces the activation of MAPK cascades, signaling modules consisting of protein kinases [82]. These cascades consist of a series of phosphorylation events that activate transcription factors responsible for controlling the expression of genes associated with defense. The activation of chitin triggers the activation of many MAPKs, including MPK3, MPK4, and MPK6 in *Arabidopsis thaliana*, and salicylate-induced protein kinase (SIPK) and wound-induced protein kinase (WIPK) in *Nicotiana tabacum* [84, 85]. These MAPKs regulate the expression of genes related to the production of PR proteins, phytoalexins, and lignin.
- **Systemic Acquired Resistance (SAR)**: Chitin-induced defensive responses can expand beyond the initially contaminated region [14]. The plant can trigger SAR, a process in which signaling molecules such as salicylic acid move throughout the plant to prepare distant regions for more robust defense against possible future assaults by pathogens [28].
- **Production of Reactive Oxygen Species (ROS)**: Chitin perception stimulates the prompt generation of ROS in plant cells. ROS, such as superoxide radicals and hydrogen peroxide, are signaling molecules that trigger subsequent defensive responses. Additionally, they possess direct antibacterial properties [42].
- **Callose Deposition (CD)**: The detection of chitin leads to the formation of callose, a polymer made of β-1,3-glucan, in the cell wall of plants [27]. Callose functions as a physical barrier that limits the dissemination of infections. Chitin-induced signaling pathways often govern callose production [80].

5 Antifungal Properties of Chitin-Based Nanoparticles

Chitin nanoparticles (NPs) have shown favorable antifungal characteristics against a range of fungal infections, including *Penicillium chrysogenum*, *Candida spp*, *Aspergillus niger*, and *Fusarium oxysporum* [39]. These fungi can potentially induce severe diseases and illnesses in people, animals, and plants.

The antifungal properties of chitin nanoparticles include several processes, including the inhibition of fungal enzymes, the breakdown of fungal cell membranes and metabolic pathways, the generation of oxidative stress and death, and the modification of fungal gene expression and signaling pathways [69]. Chitin nanoparticles, for instance, can attach to ergosterol found in the cell membrane of fungi. This attachment leads to the disruption of the membrane, resulting in the leaking of the contents of the cytoplasm [36]. Chitin nanoparticles may further impede the function of fungal enzymes, including chitin synthase, glucan synthase, and ergosterol synthase, and disrupt the production of vital constituents of the fungal cell wall [15]. In addition, chitin nanoparticles can produce ROS and trigger oxidative stress and death in fungal cells. Chitin nanoparticles may also impact the expression and functionality of fungal genes and proteins associated with the cell cycle, stress response, virulence, and drug resistance [21].

Chitin nanoparticles may also augment the plant's defensive mechanisms against fungal diseases by promoting the synthesis of phytohormones and antimicrobial peptides. Phytohormones are plant hormones that control several aspects of plant growth and development, including cell division, differentiation, aging, and stress resistance [1]. Antimicrobial peptides are diminutive compounds that possess antibacterial and antifungal properties and can regulate the immune system of plants. Chitin nanoparticles can stimulate the production and communication of plant hormones, including salicylic acid, jasmonic acid, and ethylene [4]. They also trigger the activation of antimicrobial peptides inside plant cells, including defensins, thionins, and chitinases. These chemicals can augment the plants' resilience and safeguard them against fungal infections [60].

6 Enhancement of Plant Resistance

Chitin-based nanoparticles may augment plant immunity to illnesses by functioning as innate elicitors, which are compounds that stimulate plant defense mechanisms against infections [53]. These effects are achieved by their binding to particular receptors on plant cells and subsequent activation of signaling cascades. As a result of this activation, antimicrobial chemicals such as phytoalexins, phenolics, and terpenoids are produced [41]. In addition, they stimulate the activation of genes associated with plant defense mechanisms, including glucanases, pathogenesis-related proteins, peroxidases, and chitinases [2]. Additionally, these nanoparticles augment the functionality of enzymes responsible for plant defense, including superoxide dismutase, catalase, and ascorbate peroxidase, efficiently eliminating reactive oxygen species and averting oxidative harm [43]. In addition, nanoparticles made from chitin can regulate the levels of plant hormones, including salicylic acid, jasmonic acid, and ethylene. These hormones are essential for controlling several elements of plant immunology [3]. Finally, they enhance structural and biochemical defenses in plant cells, including increasing cell wall thickness, promoting lignification, facilitating cuticle development, and promoting callose deposition [39]. These improvements together function to inhibit the ingress and dissemination of infections.

Chitin-based nanoparticles may bolster plant resistance against a range of illnesses produced by bacteria (Bacterial wilt of tomato caused by *Ralstonia solanacearum*), fungus (Fusarium wilt of banana caused by *Fusarium oxysporum f. sp cubense.*), viruses (Tomato mosaic virus (ToMV) infection), and nematodes (Root-knot nematode disease caused by *Meloidogyne incognita*) by the utilization of these methods [52, 74]. Furthermore, using chitin-based nanoparticles might enhance plant growth, productivity, and overall quality by mitigating the adverse effects of abiotic stressors, including drought, salt, temperature fluctuations, and heavy metal toxicity. Chitin-based nanoparticles are regarded as auspicious instruments for promoting sustainable agriculture and safeguarding the environment [53, 69].

7 Targeted Delivery and Formulation

Optimizing the transportation of chitin-based nanoparticles to plants is crucial to maximize their influence on plant growth, well-being, and ability to fight diseases. Multiple approaches may be investigated for precise distribution, such as tailored nanoparticles with surface changes that increase their attraction to plant tissues (Gupta & [30]). Enhancing the functionality of chitin nanoparticles by attaching specific ligands that may interact with receptors on plant cells may improve their absorption and movement inside the plant [12]. In addition, enclosing chitin nanoparticles in biocompatible materials helps safeguard them from environmental deterioration, guaranteeing their stability while being transported to specific locations inside the plant [81]. Knowledge about the physiological and biochemical processes during plant-nanoparticle interactions is essential for developing efficient delivery methods.

Integrating chitin nanoparticles into various agricultural inputs provides versatility in the application techniques. Chitin nanoparticles may improve nutrient availability, stimulate root growth, and boost soil health when used as a supplement [39, 72]. Seed coatings provide a direct pathway for introducing nanoparticles to germination seeds, therefore impacting the initial growth of plants [31]. Foliar sprays offer a rapid and effective method for directly administering chitin nanoparticles to the aboveground sections of plants, thereby addressing concerns related to insect resistance and tolerance to stress [71]. Adapting the application strategy to suit the unique objectives of the agricultural system is essential, and using a mix of these ways may be the most effective way to enhance overall plant health. The durability of chitin-based nanoparticles is crucial for their effective use in agriculture. The stability of nanoparticle formulations may be influenced by factors such as pH, temperature, and exposure to UV radiation [20]. Researchers should prioritize developing stable formulations that maintain their effectiveness in practical field circumstances.

Furthermore, factors such as cost-effectiveness, scalability, and environmental impact are crucial in determining the actual implementation of these formulations. Collaboration among scientists, agronomists, and industry professionals is essential to connect laboratory research with practical applications. This will ensure that chitin-based nanoparticle technologies make a beneficial contribution to sustainable agricultural practices.

8 Synergistic Effects with Bioactive Compounds

Incorporating chitin-based nanoparticles with fungicides or other bioactive substances in agricultural research has great potential for sustainable crop protection [72]. Chitin, an organic compound obtained from the exoskeletons of crustaceans, has garnered interest because of its compatibility with living organisms, capacity to break down naturally, and intrinsic ability to combat fungal growth [52]. When integrated into nanoparticles, chitin improves the durability and transportation of bioactive substances, such as fungicides or plant extracts, to specific disease-causing agents. Several research studies have investigated this novel technique to create agricultural practices that are both more effective and ecologically sustainable [75, 76, 86, 87].

The investigations have emphasized the significant benefits of combining chitin-based nanoparticles with traditional fungicides or other bioactive agents, resulting in greater effectiveness via synergistic effects. The synergistic interactions often lead to enhanced disease management, heightened plant resilience, and optimized utilization of resources. Moreover, using nanoparticles made from chitin can decrease the dependence on conventional agrochemicals, therefore tackling problems associated with environmental consequences and pesticide resistance [54]. This integrated approach aligns with the increasing need for sustainable agriculture, providing a viable way to improve crop protection while reducing the environmental impact of chemical inputs [72, 75, 76]. Continued research is necessary to understand the mechanisms that drive these synergistic effects and their potential application in various crops and environmental conditions. This knowledge is essential for successfully adopting chitin-based nanoparticle technologies in modern agriculture.

9 Challenges and Future Directions

The utilization of chitin-based nanoparticles in diverse domains, notably agriculture, encounters several obstacles and restrictions that need meticulous deliberation. A significant challenge is the efficient manufacturing of these nanoparticles on a big scale. Although chitin is plentiful in the natural world, the logistical difficulties of collecting and converting it into nanoparticles on a large scale for broad agricultural use may be significant. Moreover, the manufacturing procedure may include intricate and resource-intensive techniques, resulting in escalated expenses. Therefore, the successful adoption of chitin-based nanoparticles in agriculture heavily relies on overcoming the fundamental challenge of discovering manufacturing techniques that are both scalable and cost-effective [8].

The utilization of chitin-based nanoparticles raises a notable apprehension about its environmental repercussions. Chemical treatments used in extracting and processing chitin may pose significant ecological risks [47]. Assessing the environmental consequences at every stage of chitin-based nanoparticle synthesis and use is crucial to guarantee sustainability. Furthermore, the correct handling and disposal of these nanoparticles must be meticulously regulated to avoid any inadvertent impacts on ecosystems [11]. Considering regulatory issues is a vital feature of using chitin-based nanoparticles in agriculture. Due to its novelty, regulatory authorities may have restricted regulations on using these nanoparticles [64]. It is crucial to establish a solid and transparent regulatory framework to promote the use of chitin-based nanoparticles while also ensuring their safety for the environment and human health.

To enhance the utilization of chitin-based nanoparticles in agriculture, forthcoming investigations and advancements should prioritize the resolution of these obstacles. This encompasses investigating cutting-edge and eco-friendly manufacturing techniques, evaluating and reducing ecological footprints, and actively collaborating with regulatory authorities to set unambiguous protocols for their implementation. In addition, research may focus on gaining a deeper comprehension of the interactions between chitin-based nanoparticles and crops, enabling customized applications that optimize their advantages while reducing any possible hazards. Effective collaboration among academics, industry partners, and regulatory agencies is essential for fully harnessing the promise of chitin-based nanoparticles in sustainable agriculture.

10 Conclusion

Chitin-based nanoparticles show great potential as a nanomaterial for controlling plant diseases due to their many benefits compared to traditional chemical pesticides. Chitin-based nanoparticles may be manufactured using many techniques, including ionic gelation, emulsion, and precipitation. Chitin-based nanoparticles may trigger plant defensive responses, generating reactive oxygen species, phytoalexins, and pathogenesis-related proteins. They can regulate the expression of genes associated with defense. Chitin-based nanoparticles have shown efficacy against several plant pathogens, including fungi, bacteria, and viruses, resulting in decreased occurrence and intensity of plant diseases. Chitin-based nanoparticles may enhance plant development, increase yield, and improve quality.

Additionally, they can alleviate the negative impacts of abiotic stressors, such as drought, salt, and heavy metals. Chitin-based nanoparticles provide a sustainable and environmentally benign alternative to chemical pesticides. They do not contribute to environmental degradation, soil contamination, or developing resistance to plant diseases. Moreover, the regulatory framework for using chitin-based nanoparticles in agriculture necessitates progress and enhancement to establish unambiguous rules and benchmarks. Researchers, industry partners, and regulatory organizations must work together to create a framework that guarantees the safety and effectiveness of these nanoparticles in managing plant diseases. While facing these issues, promoting continuous research and development projects is equally crucial. Future efforts should prioritize enhancing manufacturing methods, reducing environmental effects, and acquiring a more comprehensive understanding of the intricate relationships between chitin-based nanoparticles and crops.

Declaration of Competing Interest The authors affirm that they do not own any identifiable conflicting financial interests or personal ties that may have influenced the findings presented in this study.

References

1. Ali S, Baloch AM (2020) Overview of sustainable plant growth and differentiation and the role of hormones in controlling growth and development of plants under various stresses. Recent Pat Food Nutr Agric 11(2):105–114
2. Aoun M (2017) A review of host defense mechanisms during fungal pathogenesis and how these are overcome in susceptible plants. Int J Bot 13(2):82–102
3. Arif Y, Siddiqui H, Hayat S (2021) Role of chitosan nanoparticles in regulation of plant physiology under abiotic stress. In: Faizan M, Hayat S, Yu F (eds) Sustainable agriculture reviews, vol 53. Springer International Publishing, pp 399–413. https://doi.org/10.1007/978-3-030-86876-5_16
4. Asif M, Pervez A, Ahmad R (2019) Role of melatonin and plant-growth-promoting rhizobacteria in the growth and development of plants. CLEAN–Soil Air Water 47(6):1800459
5. Asrahwi MA, Rosman N, Shahri NNM, Santos JH, Kusrini E, Thongratkaew S, Faungnawakij K, Hassan S, Mahadi AH, Usman A (2023) Solid-state mechanochemical synthesis of chitosan from mud crab (Scylla serrata) chitin. Carbohydr Res 534:108971
6. Bacete L, Mélida H, Miedes E, Molina A (2018) Plant cell wall-mediated immunity: cell wall changes trigger disease resistance responses. Plant J 93(4):614–636. https://doi.org/10.1111/tpj.13807
7. Badreddine I, Lafitte C, Heux L, Skandalis N, Spanou Z, Martinez Y, Esquerré-Tugayé M-T, Bulone V, Dumas B, Bottin A (2008) Cell Wall chitosaccharides are essential components and exposed patterns of the phytopathogenic oomycete *Aphanomyces euteiches*. Eukaryot Cell 7(11):1980–1993. https://doi.org/10.1128/EC.00091-08
8. Bai L, Liu L, Esquivel M, Tardy BL, Huan S, Niu X, Liu S, Yang G, Fan Y, Rojas OJ (2022) Nanochitin: chemistry, structure, assembly, and applications. Chem Rev 122(13):11604–11674. https://doi.org/10.1021/acs.chemrev.2c00125
9. Bastiaens L, Soetemans L, D'Hondt E, Elst K (2019) Sources of chitin and chitosan and their isolation. In: Broek LAM, Boeriu CG (eds) Chitin and chitosan, 1st edn. Wiley, pp 1–34. https://doi.org/10.1002/9781119450467.ch1
10. Berezina N (2016) Production and application of chitin. Phys Scie Rev 1(9). https://doi.org/10.1515/psr-2016-0048
11. Besha AT, Liu Y, Bekele DN, Dong Z, Naidu R, Gebremariam GN (2020) Sustainability and environmental ethics for the application of engineered nanoparticles. Environ Sci Policy 103:85–98

12. Bhatia S (2016) Nanoparticle types, classification, characterization, fabrication methods, and drug delivery applications. In: Bhatia S (ed) Natural polymer drug delivery systems. Springer International Publishing, pp 33–93. https://doi.org/10.1007/978-3-319-41129-3_2
13. Brown HE, Esher SK, Alspaugh JA (2019) Chitin: a "hidden figure" in the fungal cell wall. In: Latgé J-P (ed) The fungal cell wall, vol 425. Springer International Publishing, pp 83–111. https://doi.org/10.1007/82_2019_184
14. Busam G, Kassemeyer H-H, Matern U (1997) Differential expression of chitinases in Vitis vinifera L. responding to systemic acquired resistance activators or fungal challenge. Plant Physiol 115(3):1029–1038
15. Carrillo-Munoz AJ, Giusiano G, Ezkurra PA, Quindós G (2006) Antifungal agents: mode of action in yeast cells. Rev Esp Quimioter 19(2):130–139
16. Chaachouay N, Azeroual A, Bencherki B, Douira A, Zidane L (2023) Various metabolites and or bioactive compounds from vegetables, and their use nanoparticles synthesis, and applications. In: Husen A (eds) Nanomaterials from agricultural and horticultural products. Smart Nanomaterials Technology. Springer, Singapore. https://doi.org/10.1007/978-981-99-3435-5_10
17. Chaachouay N, Douira A, Zidane L (2021) COVID-19, prevention and treatment with herbal medicine in the herbal markets of Salé Prefecture, North-Western Morocco. Eur J Integr Med 42:101285. https://doi.org/10.1016/j.eujim.2021.101285
18. Chaachouay N, Zidane L (2024) Role of carbon nanotubes for herbicide detection and remediation. In: Bachheti A, Bachheti RK, Husen A (eds) Carbon-based nanomaterials. Smart Nanomaterials Technology. Springer, Singapore. https://doi.org/10.1007/978-981-97-0240-4_8
19. Chaachouay N, Douira A, Zidane L (2022) Herbal medicine used in the treatment of human diseases in the Rif, Northern Morocco. Arab J Sci Eng 47:131–153. https://doi.org/10.1007/s13369-021-05501-1
20. Chaachouay N, Azeroual A, Benkhnigue O, Zidane L (2023) Alkaloids: A Suitable Precursor for Nanomaterials Synthesis, and Their Various Applications. In: Husen, A. (eds) Secondary Metabolites Based Green Synthesis of Nanomaterials and Their Applications. Smart Nanomaterials Technology. Springer, Singapore. https://doi.org/10.1007/978-981-99-0927-8_2
21. Dhillon GS, Kaur S, Brar SK, Verma M (2013) Green synthesis approach: extraction of chitosan from fungus mycelia. Crit Rev Biotechnol 33(4):379–403. https://doi.org/10.3109/07388551.2012.717217
22. Dongre R (2018) Chitin-chitosan: myriad functionalities in science and technology. BoD–books on demand. https://books.google.com/books?hl=fr&lr=&id=EXeQDwAAQBAJ&oi=fnd&pg=IA2&dq=The+attribution+of+the+discovery+of+chitin+is+ascribed+to+the+French+scientist+Henri+Braconnot+in+1811.+&ots=v01wecSrGL&sig=zbvwlVIzIcimYVS8dGC7RsFcV4o
23. El Gueddari NE, Rauchhaus U, Moerschbacher BM, Deising HB (2002) Developmentally regulated conversion of surface-exposed chitin to chitosan in plant pathogenic fungi cell walls. New Phytol 156(1):103–112. https://doi.org/10.1046/j.1469-8137.2002.00487.x
24. El Knidri H, Belaabed R, Addaou A, Laajeb A, Lahsini A (2018) Extraction, chemical modification, and characterization of chitin and chitosan. Int J Biol Macromol 120:1181–1189
25. Elieh-Ali-Komi D, Hamblin MR (2016) Chitin and chitosan: production and application of versatile biomedical nanomaterials. Int J Adv Res 4(3):411
26. Fan Y, Saito T, Isogai A (2008) Chitin nanocrystals prepared by TEMPO-mediated oxidation of α-chitin. Biomacromol 9(1):192–198. https://doi.org/10.1021/bm700966g
27. Fesel PH, Zuccaro A (2016) β-glucan: Crucial component of the fungal cell wall and elusive MAMP in plants. Fungal Genet Biol 90:53–60
28. Fu ZQ, Dong X (2013) Systemic acquired resistance: turning local infection into global defense. Annu Rev Plant Biol 64(1):839–863. https://doi.org/10.1146/annurev-arplant-042811-105606
29. Ghozlan MH, Eman E-A, Tokgöz S, Lakshman DK, Mitra A (2020) Plant defense against necrotrophic pathogens. Am J Plant Sci 11(12):2122–2138

30. Gupta AK, Gupta M (2005) Synthesis and surface engineering of iron oxide nanoparticles for biomedical applications. Biomaterials 26(18):3995–4021
31. Hajong M, Devi NO, Debbarma M, Majumder D (2019) Nanotechnology: an emerging tool for management of biotic stresses in plants. In: Prasad R (ed) Plant nanobionics. Springer International Publishing, p 299-335. https://doi.org/10.1007/978-3-030-16379-2_11
32. Hayes M, Carney B, Slater J, Brück W (2008) Mining marine shellfish wastes for bioactive molecules: chitin and chitosan ndash; part A: extraction methods. Biotechnol J 3(7):871–877. https://doi.org/10.1002/biot.200700197
33. Hochella MF, Lower SK, Maurice PA, Penn RL, Sahai N, Sparks DL, Twining BS (2008) Nanominerals, mineral nanoparticles, and earth systems. Science 319(5870):1631–1635. https://doi.org/10.1126/science.1141134
34. Hoell IA, Vaaje-Kolstad G, Eijsink VGH (2010) Structure and function of enzymes acting on chitin and chitosan. Biotechnol Genet Eng Rev 27(1):331–366. https://doi.org/10.1080/02648725.2010.10648156
35. Ifuku S, Saimoto H (2012) Chitin nanofibers: preparations, modifications, and applications. Nanoscale 4(11):3308–3318
36. Ing LY, Zin NM, Sarwar A, Katas H (2012) Antifungal activity of chitosan nanoparticles and correlation with their physical properties. Int J Biomater 1:e632698. https://doi.org/10.1155/2012/632698
37. Jain D, Khurana JP (2018) Role of pathogenesis-related (PR) proteins in plant defense mechanism. In: Singh A, Singh IK (eds) Molecular aspects of plant-pathogen interaction. Springer Singapore, pp 265-281. https://doi.org/10.1007/978-981-10-7371-7_12
38. Kaku H, Nishizawa Y, Ishii-Minami N, Akimoto-Tomiyama C, Dohmae N, Takio K, Minami E, Shibuya N (2006) Plant cells recognize chitin fragments for defense signaling through a plasma membrane receptor. Proc Natl Acad Sci 103(29):11086–11091. https://doi.org/10.1073/pnas.0508882103
39. Kalagatur NK, Nirmal Ghosh OS, Sundararaj N, Mudili V (2018) Antifungal activity of chitosan nanoparticles encapsulated with Cymbopogon martinii essential oil on plant pathogenic fungi Fusarium graminearum. Front Pharmacol 9:610
40. Kamle M, Borah R, Bora H, Jaiswal AK, Singh RK, Kumar P (2020) Systemic acquired resistance (SAR) and induced systemic resistance (ISR): role and mechanism of action against phytopathogens. In: Hesham AE-L, Upadhyay RS, Sharma GD, Manoharachary C, Gupta VK (eds) Fungal biotechnology and bioengineering. Springer International Publishing, pp 457-470. https://doi.org/10.1007/978-3-030-41870-0_20
41. Kaur S, Samota MK, Choudhary M, Choudhary M, Pandey AK, Sharma A, Thakur J (2022) How do plants defend themselves against pathogens, biochemical mechanisms, and genetic interventions? Physiol Mol Biol Plants 28(2):485–504. https://doi.org/10.1007/s12298-022-01146-y
42. Kawasaki T, Yamada K, Yoshimura S, Yamaguchi K (2017) Chitin receptor-mediated activation of MAP kinases and ROS production in rice and Arabidopsis. Plant Signal Behav 12(9):e1361076. https://doi.org/10.1080/15592324.2017.1361076
43. Khan MN, Mobin M, Abbas ZK, AlMutairi KA, Siddiqui ZH (2017) Role of nanomaterials in plants under challenging environments. Plant Physiol Biochem 110:194–209
44. Kubicek CP, Seidl V, Seiboth B (2014) Plant cell wall and chitin degradation. In: Borkovich KA, Ebbole DJ (eds) Cellular and molecular biology of filamentous fungi. ASM Press, pp 396-413. https://doi.org/10.1128/9781555816636.ch27
45. La Camera S, Gouzerh G, Dhondt S, Hoffmann L, Fritig B, Legrand M, Heitz T (2004) Metabolic reprogramming in plant innate immunity: the contributions of phenylpropanoid and oxylipin pathways. Immunol Rev 198(1):267–284. https://doi.org/10.1111/j.0105-2896.2004.0129.x
46. Li J, Huang W-C, Gao L, Sun J, Liu Z, Mao X (2019) Efficient enzymatic hydrolysis of ionic liquid pretreated chitin and its dissolution mechanism. Carbohydr Polym 211:329–335
47. Lopes C, Antelo LT, Franco-Uría A, Alonso AA, Pérez-Martín R (2018) Chitin production from crustacean biomass: sustainability assessment of chemical and enzymatic processes. J Clean Prod 172:4140–4151

48. López M, Miranda E, Ramos C, García H, Neira-Carrillo A (2020) Activation of early defense signals in seedlings of nicotiana benthamiana treated with chitin nanoparticles. Plants 9(5):607
49. Lv J, Lv X, Ma M, Oh D-H, Jiang Z, Fu X (2023) Chitin and chitin-based biomaterials: a review of advances in processing and food applications. Carbohydr Polym 299:120142
50. Lv J, Zhang Y, Jin Y, Oh D-H, Fu X (2024) Chitin nanofibers prepared by enzymatic hydrolysis: characterization and application for pickering emulsions. Int J Biol Macromol 254:127662
51. Macho AP, Zipfel C (2014) Plant PRRs and the activation of innate immune signaling. Mol Cell 54(2):263–272
52. Mahmoud MG, El Kady EM, Asker MS (2019) Chitin, chitosan, and glucan, properties and applications. World J Agric Soil Sci 3(1):1–19
53. Malerba M, Cerana R (2019) Recent applications of chitin-and chitosan-based polymers in plants. Polymers 11(5):839
54. Maluin FN, Hussein MZ (2020) Chitosan-based agronanochemicals as a sustainable alternative in crop protection. Molecules 25(7):1611
55. Mohammed MH, Williams PA, Tverezovskaya O (2013) Extraction of chitin from prawn shells and conversion to low molecular mass chitosan. Food Hydrocoll 31(2):166–171
56. Muhr M (2022) Characterization of populus x canescens LysM receptor-like kinases CERK1-1 and CERK1-2 and their role in chitin signaling. https://ediss.uni-goettingen.de/handle/11858/13943
57. Muzzarelli RA (2011) Biomedical exploitation of chitin and chitosan via mechano-chemical disassembly, electrospinning, dissolution in imidazolium ionic liquids, and supercritical drying. Mar Drugs 9(9):1510–1533
58. Muzzarelli RA (2013) Chitin. Elsevier. https://books.google.com/books?hl=fr&lr=&id=ZA7LBAAAQBAJ&oi=fnd&pg=PP1&dq=chitin&ots=NwqNY5JCHm&sig=P9BGxAHIJSlj-3t2SySfk7XWKH4
59. Muzzarelli RAA (2011) Chitin nanostructures in living organisms. In: Gupta NS (Éd.), Chitin, vol 34). Springer Netherlands, pp 1-34. https://doi.org/10.1007/978-90-481-9684-5_1
60. Nadeem SM, Ahmad M, Zahir ZA, Kharal MA (2016) Role of phytohormones in stress tolerance of plants. In: Hakeem KR, Akhtar MS (eds) Plant, soil and microbes. Springer International Publishing, pp 385-421. https://doi.org/10.1007/978-3-319-29573-2_17
61. Nisar N, Tsuzuki T, Lowe A, Rossiter JT, Javaid A, Powell G, Waseem R, Al-Mijalli SH, Iqbal M (2021) Chitin nanofibers trigger membrane-bound defense signaling and induce elicitor activity in plants. Int J Biol Macromol 178:253–262
62. Oyervides-Muñoz E, Oyervides-Muñoz MA, Garcia-Lobato MA (2023) Chitin and chitosan nanocomposites: from the synthesis to the application. In: Avalos Belmontes F, González FJ, López-Manchado MÁ (eds) Green-based nanocomposite materials and applications. Springer International Publishing, pp 101-118. https://doi.org/10.1007/978-3-031-18428-4_6
63. Percot A, Viton C, Domard A (2003) Optimization of chitin extraction from shrimp shells. Biomacromol 4(1):12–18. https://doi.org/10.1021/bm025602k
64. Preston CJ, Sheinin MY, Sproat DJ, Swarup VP (2010) The novelty of nano and the regulatory challenge of newness. NanoEthics 4(1):13–26. https://doi.org/10.1007/s11569-010-0083-x
65. Pusztahelyi T (2018) Chitin and chitin-related compounds in plant–fungal interactions. Mycology 9(3):189–201. https://doi.org/10.1080/21501203.2018.1473299
66. Rameshthangam P, Solairaj D, Arunachalam G, Ramasamy P (2018) Chitin and chitinases: biomedical and environmental applications of chitin and its derivatives. J Enzym 1(1):20–43
67. Ranf S (2017) Sensing of molecular patterns through cell surface immune receptors. Curr Opin Plant Biol 38:68–77
68. Rinaudo M (2006) Chitin and chitosan: properties and applications. Prog Polym Sci 31(7):603–632
69. Sahraee S, Milani JM, Ghanbarzadeh B, Hamishehkar H (2017) Physicochemical and antifungal properties of bio-nanocomposite film based on gelatin-chitin nanoparticles. Int J Biol Macromol 97:373–381. https://doi.org/10.1016/j.ijbiomac.2016.12.066
70. Salaberria AM, Labidi J, Fernandes SC (2014) Chitin nanocrystals and nanofibers as nanosized fillers into thermoplastic starch-based biocomposites processed by melt-mixing. Chem Eng J 256:356–364

71. Sarkar MM, Sarkar A, Roy S (2022) Interventions of nanotechnology for the growth and stress tolerance in crop plants. In: Roy S, Mathur P, Chakraborty AP, Saha SP (eds) Plant stress: challenges and management in the new decade. Springer International Publishing, pp 421-443. https://doi.org/10.1007/978-3-030-95365-2_26
72. Shahrajabian MH, Chaski C, Polyzos N, Tzortzakis N, Petropoulos SA (2021) Sustainable agriculture systems in vegetable production using chitin and chitosan as plant biostimulants. Biomolecules 11(6):819
73. Sharma B, Sharma S, Jain P (2021) Leveraging advances in chemistry to design biodegradable polymeric implants using chitosan and other biomaterials. Int J Biol Macromol 169:414–427
74. Shrestha R, Thenissery A, Khupse R, Rajashekara G (2023) Strategies for preparing chitosan derivatives for antimicrobial, drug delivery, and agricultural applications: a review. Molecules 28(22):7659
75. Singh RP, Handa R, Manchanda G (2021) Nanoparticles in sustainable agriculture: an emerging opportunity. J Control Release 329:1234–1248
76. Singh R, Upadhyay SK, Singh M, Sharma I, Sharma P, Kamboj P, Saini A, Voraha R, Sharma AK, Upadhyay TK (2021) Chitin, chitinases and chitin derivatives in biopharmaceutical, agricultural and environmental perspective. Biointerface Res Appl Chem 11(3):9985–10005
77. Son GH, Wan J, Kim HJ, Nguyen XC, Chung WS, Hong JC, Stacey G (2012) Ethylene-responsive element-binding factor 5, ERF5, Is involved in chitin-induced innate immunity response. Mol Plant-Microbe Interact® 25(1):48–60. https://doi.org/10.1094/MPMI-06-11-0165
78. Synowiecki J, Al-Khateeb NA (2003) Production, properties, and some new applications of chitin and its derivatives. Crit Rev Food Sci Nutr 43(2):145–171. https://doi.org/10.1080/10408690390826473
79. Tian W, Wang C, Gao Q, Li L, Luan S (2020) Calcium spikes, waves, and oscillations in plant development and biotic interactions. Nat Plants 6(7):750–759
80. Trillas MI, Cotxarrera L, Casanova E, Cortadellas N (2000) Ultrastructural changes and localization of chitin and callose in compatible and incompatible interactions between carnation callus and Fusarium oxysporum. Physiol Mol Plant Pathol 56(3):107–116
81. Vega-Vásquez P, Mosier NS, Irudayaraj J (2020) Nanoscale drug delivery systems: from medicine to agriculture. Front Bioeng Biotechnol 8:79
82. Wan J, Zhang S, Stacey G (2004) Activation of a mitogen-activated protein kinase pathway in *Arabidopsis* by chitin. Mol Plant Pathol 5(2):125–135. https://doi.org/10.1111/j.1364-3703.2004.00215.x
83. Xu L, Ma W, Wang L, Xu C, Kuang H, Kotov NA (2013) Nanoparticle assemblies: dimensional transformation of nanomaterials and scalability. Chem Soc Rev 42(7):3114–3126
84. Yamada K, Yamaguchi K, Shirakawa T, Nakagami H, Mine A, Ishikawa K, Fujiwara M, Narusaka M, Narusaka Y, Ichimura K, Kobayashi Y, Matsui H, Nomura Y, Nomoto M, Tada Y, Fukao Y, Fukamizo T, Tsuda K, Shirasu K, Shibuya, N, Kawasaki T (2016) The *Arabidopsis* CERK 1-associated kinase PBL 27 connects chitin perception to MAPK activation. EMBO J 35(22):2468-2483. https://doi.org/10.15252/embj.201694248
85. Yamada K, Yamaguchi K, Yoshimura S, Terauchi A, Kawasaki T (2017) Conservation of chitin-induced MAPK signaling pathways in rice and Arabidopsis. Plant Cell Physiol 58(6):993–1002
86. Yanat M, Schroën K (2023) Advances in chitin-based nanoparticle use in biodegradable polymers: a review. Carbohydr Polym 312:120789
87. Younes I, Rinaudo M (2015) Chitin and chitosan preparation from marine sources. Structure, properties and applications. Mar Drugs 13(3):1133–1174
88. Zainol Abidin NA, Kormin F, Zainol Abidin NA, Mohamed Anuar NAF, Abu Bakar MF (2020) The potential of insects as alternative sources of chitin: an overview on the chemical method of extraction from various sources. Int J Mol Sci 21(14):4978

Chitin-Based Nanoparticles for Soil Remediation

Marie Yayinie, Limenew Abate Worku, Rakesh Kumar Bachheti, and Archana Bachheti

Abstract The majority of important material resources and the basis for a vast range of living things are available in the soil. However, the amount of toxic and hazardous materials and chemicals released into the environment has dramatically increased due to rapid industrial activity and urbanization development. In this regard, soil becomes a default sink for those detrimental anthropogenic pollutants such as heavy metals, pesticides, herbicides, and many toxic organic compounds. Nano-based remediation technologies are a new approach to mitigate the pollution crisis of the soil environment. This book chapter aims to review the properties, synthesis, application, and uses of chitin-based nanomaterials for soil remediation from various pollutants. Chitin-based nanomaterials are effective in many agricultural processes like, crop defense, pathogen resistance, plant growth, soil conditioning agents, and seed coatings. However, like most of the nano remediation technologies, the application of chitin-based nanomaterials for soil remediation is confined to lab spaces and needs more commerce. Thus, the field's future possibilities lie in commercializing nano remediation techniques.

Keywords Pollution · Chitin · Nano particles · Phytoremediation · Soil remediation

M. Yayinie (✉) · L. A. Worku
Department of Chemistry, College of Natural and Computational Science, Debre Tabor University, P. O. Box. 272, Debre Tabor, Ethiopia
e-mail: marieyayinie@gmail.com

R. K. Bachheti
Department of Industrial Chemistry, Addis Ababa Science and Technology University, P.O. Box: 16417, Addis Ababa, Ethiopia

A. Bachheti
Department of Environment Science, Graphic Era University, Dehradun 248002, Uttarakhand, India

R. K. Bachheti et al. (eds.), *Chitin-Based Nanoparticles for the Agriculture Sectors*, Smart Nanomaterials Technology, https://doi.org/10.1007/978-981-96-0920-8_6

1 Introduction

1.1 Overview of Soil Contamination

Soil is vital in offering numerous ecosystem services essential for life on Earth. The most valuable material resources and the foundation for numerous animals and plants are in the soil [17, 50]. But soil, which naturally absorbs a variety of contaminants, is becoming alarmingly polluted. The term "soil pollution" refers to the contaminating effects of natural and human activities on soil that may harm living things. The release of toxic and dangerous substances and materials into the environment has dramatically increased due to rapid industrial activity and urbanization development. This leads to a swift decline in the quality of our soils and the priceless advantages it provides. The toxic and harmful chemical contamination of soil has become one of today's world's most significant environmental challenges [17, 23, 57].

Research has been conducted around the globe to estimate the degree of soil pollution. For example, a national soil assessment conducted in China between April 2005 and December 2013 revealed that 16.1% of all soils under investigation had organic or inorganic pollution contamination. Every year, the country's contaminated arable soils lose over 10 million tons of crop production [48]. Comparable studies in the globe estimated that over 10 million soil fields were polluted, including 100,000 sites in the United States of America, 80,000 sites in the European Union, and 50,000 sites in Australia [32]. Thus, soil contamination is a global problem that has drawn the attention of academicians and politicians as one of today's biggest environmental issues [19].

Soil typically consists of inorganic mineral matter, which is the consequence of numerous chemical and mechanical processes that have broken down the different types of rocks that make up the Earth's crust, including volcanic, sedimentary, and metamorphic rocks; organic stuff, or humus, primarily waste products and leftovers from plants and animals; Soil water, soil oxygen, and the living things that greatly enhance its fertility [3].

When foreign objects enter the soil, their presence alters its makeup, resulting in both chemical and physical changes. Furthermore, because complex chemical compounds leak from damaged or ruptured subsurface storage tanks, the composition and natural properties of the soil might be altered [2, 7].

1.2 Need for Innovative and Sustainable Soil Remediation Techniques

These days, soil contamination poses a serious threat to human health and the health of the ecosystem worldwide. Its direct contribution to water and air pollution demonstrates the impact on the ecosystem. For example, it causes acid rain by releasing large amounts of ammonia into the atmosphere. The way that contaminated soil impacts human health varies widely depending on the type of pollutant and might occur through direct contact with the soil or via inhalation pathway of attack and vulnerability of the exposed population. Moreover, contaminated soil loses its ability to produce, which causes a shortage of food. To avoid these threats, various soil remediation techniques on diverse chemistry, biology, agroecology, and other related topics have been introduced occasionally [6, 17].

Contaminated soils have traditionally been cleaned using physicochemical processes such as excavation, separation, extraction, incineration, vitrification, washing, oxidation, reduction, and conveyance to a landfill under control. However, their high costs and environmentally disruptive nature limit their adoption [1, 31, 34]. Although applying them can often successfully remove or immobilize the intended contaminants, the method may not necessarily enhance the remediated soil's ecological quality. Conversely, the soil biota is often degraded when these traditional physicochemical methods are applied [4, 31].

As a result of these limitations, numerous biological and environmentally friendly remediation techniques, often referred to as Gentle Remediation Options (GROs), have emerged. As shown in Fig. 1, these techniques fall under phytoremediation, bioremediation, and vermiremediation, each with its own sub-divisions. These methods usually restore soil functionality, lower the total bioavailable concentration of soil contaminants, and sometimes produce renewable resources for the biobased economy. Additionally, they are less intrusive and more considerate of the soil ecosystem and its biota [11, 27, 29].

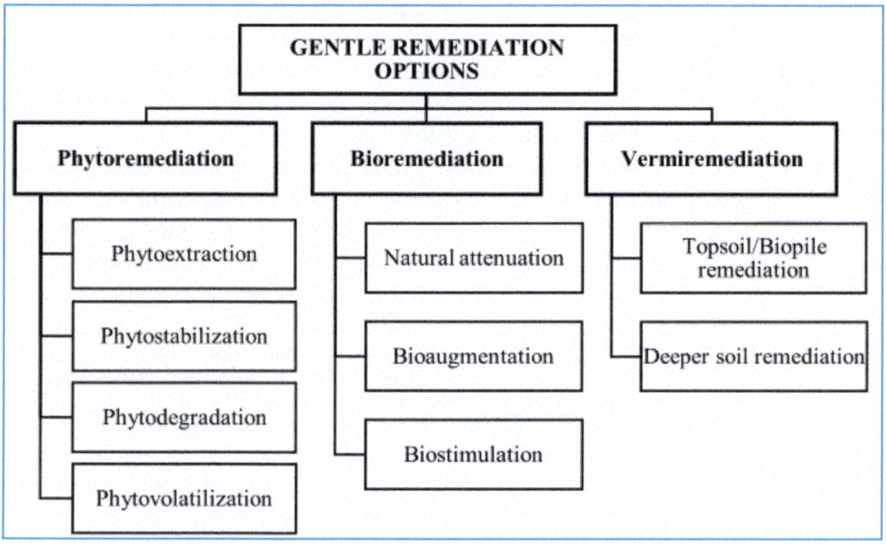

Fig. 1 Remediation options (adopted from [31])

Using plants and related microbes to lessen the number of pollutants in the environment and their harmful consequences is known as phytoremediation. Bioremediation involves using microorganisms, primarily bacteria and fungi, to clean up contaminated soils as a sustainable alternative for remediating contaminated soils. Vermiremediation, on the other hand, is the process of using earthworms to remove pollutants from soil. The world's soil is occasionally becoming more contaminated, which lowers crop yield and causes hunger. On the other hand, stringent laws that control fertilizer consumption, pesticide application, and materials that control plant growth have created a market for organic substitutes. Therefore, the current global focus is either developing a novel soil remediation methodology or refining and upgrading the current remediation approach [16, 31].

1.3 Introduction to Chitin-Based Nanoparticles as a Promising Solution

Chitin is a ubiquitous biopolymer in many types of algae, bottom-feeding crustaceans, microorganisms, plants, fungi, and insect exoskeletons. Chitin is a (1–4)-linked N-acetyl-ß–D-glucosamine polysaccharide that gives species a hard exterior, which is easily treated using straightforward chemical methods without endangering the species. Chemically, chitin is the natural carbohydrate polymer of N-acetyl D-glucosamine named poly-β-(1–4)-N-acetyl-D-glucosamine. This biopolymer has many applications, as depicted in Fig. 2. For instance, in medicine, it has an application in tissue regeneration, dressing of wound, and the treatment of cancer cases. In the environment, chitin can be used to remove dyes and remediate organic and inorganic contaminants, and in agricultural sectors, it is functional in pathogen resistance, crop growth as well as crop defense [13, 39, 47].

Fruit retardants, soil conditioners, fertilizers, seed coatings, and agents for controlling plant diseases can all be effectively achieved with chitin and its derivatives.

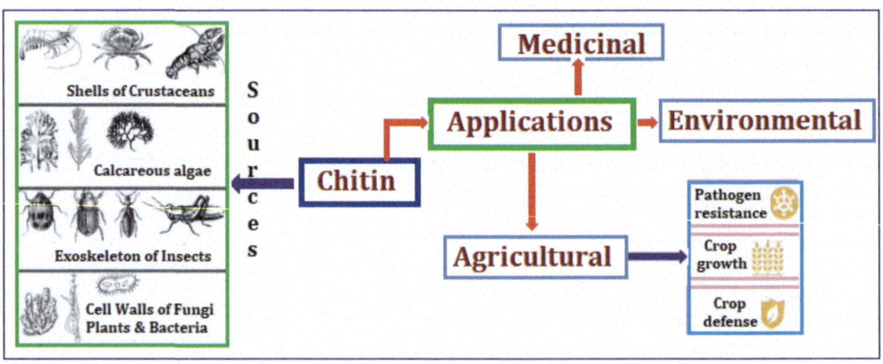

Fig. 2 Sources of chitin and its applications adapted from [13]

They are recognized as growth regulators, growth promoters, anti-stress agents, and inducers of the synthesis of secondary metabolites in plants. They also increase or induce natural defense systems in plants [22, 38]. When the growth effects of organic fertilizer on crops were compared to those of man-made fertilizers, it was discovered that the usage of artificial fertilizers does have a damaging effect on the soil and is harmful to the environment. Conversely, chitin's fertilizing function results from the polymer's biodegradation in the soil into ammonia, which is usable form of nitrogen nutrient. Furthermore, once ammonia is present in the soil, it encourages the development of particular microbes and enzymes, such as those that break down the polymer chitin into its oligomers and monomers, ultimately producing ammonia. Chitinases found in bacteria, like exochitinases and endochitinases are responsible for the biodegradation of chitin [13, 54]. Another advantage of chitin is its carbon-to-nitrogen ratio, which determines how quickly it breaks down and releases its useful form for the plant. Regarding the NPK ratio, chitin does not contain phosphorus or potassium, where the nitrogen ratio becomes highest, and it is an additional important fertilizer characteristic [13, 24, 35]. Because the structure of chitin lacks free amino groups, it lacks "direct" antibacterial capabilities yet nevertheless exhibits indirect suppression of microorganisms. The production of oligomeric fragment byproducts during chitin breakdown leads to the indirect suppression of the pathogens. These fragments serve as elicitors to activate the plant's defense mechanisms against the invading pathogen. Here, oligomeric fragments are identified as one of the molecular patterns linked with microbes, which in turn sets off various plant defensive mechanisms. Additionally, a combination of chitin and urea has been demonstrated to be a successful nematicide when irrigated onto various plants [37, 42, 47].

It has been found that chitin made from crushed shrimp waste, controls the pathogen *Streptomyces scabies*, which causes scab disease on potato tubers by reducing the risk of infection for scab-vulnerable potato cultivars. Deacetylating chitin oligomers yields chitosan, a substance with strong antibacterial properties that has been demonstrated to regulate plant viral infections. Different chain lengths of chitin and chitosan have been practically tested and noted on agricultural plants, with a variety of positive agronomical responses that can be seen in four categories. (1) Direct antibiosis against pests and pathogens of crops; (2). Augmentation of useful microbes, both in plant defense and growth; (3) Stimulation of plant defense responses against biotic stress; (4) Up-regulation of plant growth and tolerance to abiotic tensions [28, 36, 49]. The different agricultural input chemicals, such as pesticides and fertilizers introduced to improve productivity, greatly impact the soil environment. In finding more eco-friendly systems to control plant stresses, chitin and its deacetylated derivative, chitosan, seems to be an auspicious solution to solve environmental problems and maximize crop production in agricultural areas.

2 Chitin-Based Nanoparticles: Synthesis and Properties

2.1 Methods of Synthesizing Chitin-Based Nanoparticles

Chitin is the second most common polysaccharide on Earth, next to cellulose. Chitin can be obtained from a wide range of diverse organisms except in higher plants and vertebrate animals [49]. From an economic perspective, the primary industrial source utilized for synthesizing or extracting this biopolymer is fishing industry waste. There are now several ways to extract pure chitin from its sources, including chemical and biological techniques that employ chemicals and microbes, enzymes as well as combinations of them to produce an even more effective product. The objective remains the same regardless of the method used: remove proteins, minerals, lipids, and colors until the substance of interest remains, as illustrated in Fig. 3. The source material underwent a preliminary treatment using deionized water and sodium hypochlorite baths to eliminate certain impurities before the chemical or biological approach was applied [9].

Chemical methods: In this scheme of chitin preparation, the source material is deproteinized by the action of basic solutions, commonly with NaOH, and demineralized by the treatment of inorganic acids, commonly HCl acid. Treatment with organic solvent is also applied to eliminate lipids and colour. For the elimination of protein content (deproteinization) at about 95 °C present in the raw material KOH

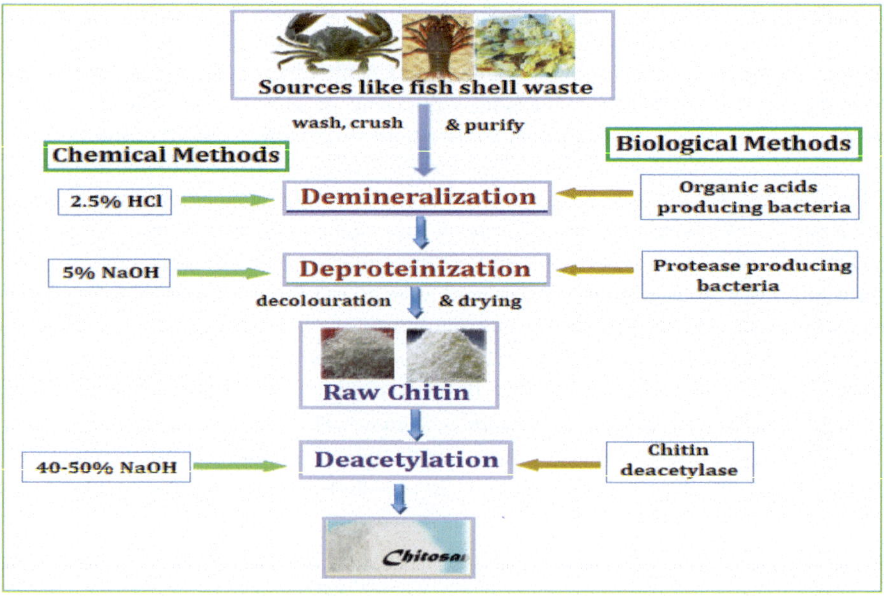

Fig. 3 Chemical and biological methods of chitin and chitosan production adapted from [14, 39, 52]

can also be used as a second choice of alkali in place of the NaOH. For removing the inorganic fillers such as carbonates and phosphates of calcium (demineralization step), it is also possible to use HNO_3 and H_2SO_4 suspicions, and strong organic acids like the HCOOH and CH_3COOH. The shortcomings that appeared in this approach might be possibility for the denaturation of the chitin, the method might lead to the creation of monomers to trimers (lacks tetramers to heptane's), and the organic solvents utilized may be mutagenic, poisonous, corrosive, or degradative, rendering the resultant chitin unfit for use as medicine. Additionally, the waste handling expense may increase [9, 39].

Biological methods: This novel approach uses fungi, proteolytic microbes, or pure enzymes to synthesize chitin from the ingredients. It does not denature the chitin and produces oligomers with the ideal polymerization level for various applications. A selective enzymatic extraction technique can obtain higher molecular weight chitin. This method has environmental benefits, as it eliminates hazards, reactive reagents, and economic advantages by minimizing the cost of carbon foundations (glucose, lactose, sucrose, or dextrose). Additionally, it has been demonstrated that techniques utilizing biological processes are more effective in obtaining chitin in greater purity levels with significantly lower molecular weight. This method's drawbacks are the time required for fermentation and the expensive price of some enzymes [9, 39].

The major steps to extract chitin by this approach are:

1. Preparation of inoculums: This will be done from lactobacillus cell in 5 ml MRS broth and incubation at 30 °C for about 24 h.
2. Fermentation: In this step, well-grind shellfish waste will be mixed properly with 10% of any carbon sources and 10% prepared culture inoculums.
   A firm cake is produced after 180 h of incubation time, filtering, and hot air oven drying.
3. Deproteinization: This is a protein separation step. Proteases released into the fermentation medium are responsible for this intense protein separation process. Protease-producing microorganisms or the addition of exo-proteases are further methods for deproteinization.
4. Calcium carbonate separation: This stage of demineralization is facilitated by bacteria that produce lactic acid and convert it into an additional carbon source [39].

2.2 *Structural and Chemical Properties of Chitin-Based Nanoparticles*

Chitin is a natural amino polysaccharide polymer with an inimitable structure, properties, highly sophisticated functions, and a wide range of applications. Structurally, chitin is very similar to cellulose. The acetamide group in chitin is located at the C-2 position, whereas in cellulose, a hydroxyl group occupies the same position (Fig. 4a). Chitin the β-(1–4)-2-acetamido-2-deoxy-β-D-glucose contains β-D-glucosamine molecules and N-acetyl groups of monomers [9, 33]. Naturally, chitin

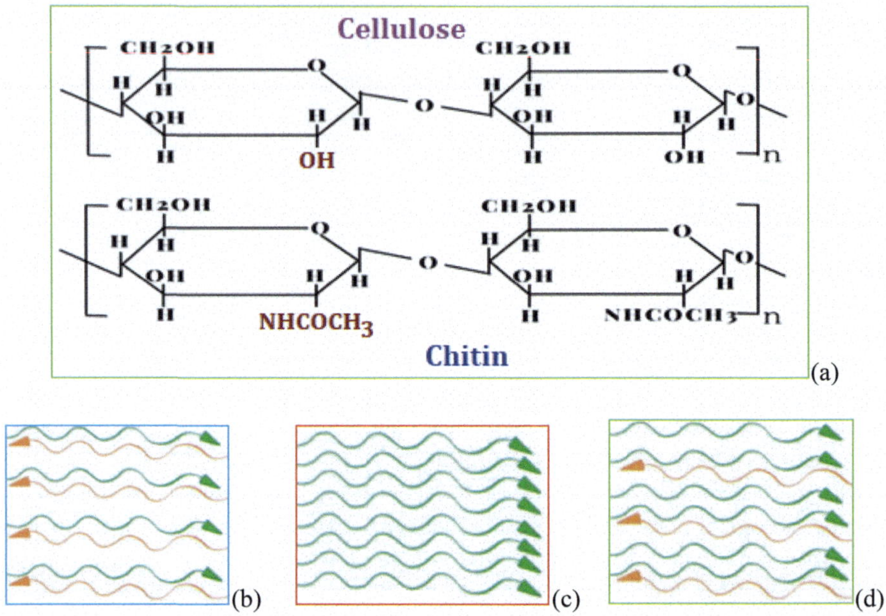

Fig. 4 Molecular assembly of **a** cellulose and chitin molecule **b** α-chitin **c** ß-chitin and **c** γ-chitin adopted from [13]

$(C_8H_{13}O_5N)_n$ molecules assemble into larger packages in at least three different crystal allomorphs distinguished as α, ß, and γ, differing each other in the orientation of hydrogen bond chains.

The α-chitin, commonly found in the shells of crustaceans like shrimp and crabs, does have antiparallel chains allowing for strong intermolecular bonding. β-chitin, usually available in squid pens, contains sequences affiliated in a parallel fashion, possessing fragile intermolecular interactions. While in the γ-chitin (typically found in insects and the abdomen of the Loligo squid), the orientation of two consecutive chains are parallel. However, the third chain goes in an antiparallel direction as obtained through XRD analysis and illustrated in Fig. 4b–d [13, 20].

Chitin is non-toxic, excellent in its biocompatibility, and possesses an admirable biodegradation with an ecologically safe nature. Because of the strong hydrogen connections between the hydroxy (OH) and acetamide ($NHCOCH_3$) groups, it is often insoluble in water and common organic solvents despite being incredibly stable due to the beta bonds between carbons 1 and 4. Mineral acids, hot concentrated aqueous thiocyanate solutions, and irritating organic solvents like hexafluoro isopropanol alcohol and formic acid are among the few solvents that can dissolve chitin. However, ß-chitin has been identified to have a relatively better reactivity, solubility, and affinity for certain solvents than α-chitin. In its application, chitin is versatile, like controlled drug release capsule production, food and textile industries, agriculture, wastewater treatment, and medicine for humans and animals. Furthermore, cast films of chitin nanofibers (CNFs) have been used in making transparent papers and as sutures, which make up the majority of the material implants found in human bodies [9, 13, 33].

Chitin has three classes of surface morphologies; the first one is hard and rough surface without pores or nanofibers, the second class is solely composed of nanofibers and the third category is a surface holding both pores and nanofibers. The size of chitin nanofibers (CNFs) is approximately 5 nm wide and has lengths ranging from hundreds of nanometers (nm) to a few micrometers (μm) [26, 55]. Chemically, chitin is an interesting polysaccharide possessing two functional groups that may be appropriately altered to impart desired characteristics and unique biological roles. As with cellulose it undergoes many reactions such as etherification, esterification, cross-linking, acetylation, and quaternization. Moreover, it can react with alkanes and alkenones and undergo alkylation, metal chelation, and graft copolymerization reactions [33, 41].

3 Mechanisms of Soil Contamination

3.1 Types of Soil Contaminants

The presence of xenobiotics-chemicals created by humans or other changes to the natural soil environment leads to soil deterioration as a component of land degradation. The paths where pollutants enter the soil matrix vary. The most frequent materials or chemicals that contaminate soil are (1) heavy metals like Pb, Cd, As, Cr, Sn, and Hg; (2) organic pollutants like microplastics and petroleum hydrocarbons; (3) agricultural input chemicals like weed killers, herbicides, and pesticides; and other substances like acid rain [2, 18]. The ultimate sources of these pollutants are:

Industrial wastes: these are discharged from various industrial origins such as oil refineries, tanneries, sugar factories, pulp and paper mills, steel, distilleries, drugs, glass, cement, coal, and mineral mining industries, acid rain due to pollutant gas inputs by various industries into the atmosphere, petroleum, and many other engineering industries [43].

Urban wastes: These classes of wastes are commonly named anthropogenic wastes and comprise both commercial as well as domestic trash entailing dried sludge and sewage. To mention some of them: debris and bunkum things like plastics, fibres, paper, rubbers, glasses, fuel residues, metallic cans, containers, seepages from a landfill, solid waste seepages, percolation of contaminated water into the soil, abandoned vehicles and many other discarded manufactured as well as domestic products [18, 40].

Agricultural practices: In advancing agrotechnology, humans use huge quantities of agricultural inputs like fertilizers and weedkillers to maximize crop production. The cumulative effect of these chemicals pollutes the soil. Furthermore, farm wastes like manure, various debris from the crops and the slurry, and soil sedimentation through erosion that contains mostly mineral chemicals are informed to cause soil pollution [15].

Radioactive pollutants: Improper disposal of wastes from radioactive/nuclear stations, research laboratory rooms, or production factories that might put nuclear dust into the soil environment can severely contaminate the soil. Radionuclides of Radium, Thorium, Uranium, isotopes of Potassium (K-40) and Carbon (C-14), Strontium-90, Iodine-129, and Cesium-137; the explosion of hydrogen weapons; nuclear reactors, as well as radiation wastes can be taken as common pollutants of this category [51]. Furthermore, natural phenomena such as volcanic eruptions and forest fires also contribute to soil pollution by releasing many toxic substances into the environment [2, 45].

3.2 Impact of Soil Contamination on the Environment and Human Health

After the pollutants have entered the soil matrix, cleanup becomes very difficult and expensive. Moreover, they can pose significant health risks to various organisms, including humans, and affect the whole ecosystem. The health effects of soil contamination exposure differ significantly according to the kind of pollutant, the route of exposure, and the population's susceptibility. For instance, long-term exposure to solvents, petroleum, heavy metals like lead, as well as many pesticide and herbicide formulations, can result in mutagenicity, carcinogenesis, congenital abnormalities, and other long-term health impacts. Furthermore, research has indicated that heavy metals like arsenic and cadmium diminish crop productivity and significantly hinder crop quality and food security by diminishing soil fertility. Different pollutants might also reduce nitrogen fixation, increasing erodibility of the soil, creating imbalance in soil fauna and flora [17, 57].

Because of their propensity to bind to proteins, nucleic acids, and minute metabolites in living things, heavy metal ions like Hg^{2+}, Cd^{2+}, Pb^{2+}, Ni^{2+}, and Cu^{2+} are well recognized to cause severe health concerns in an organism. Humans exposed to Cd^{2+} and Zn^{2+}metals are exposed to the demolition of red blood cells and testicular tissue, high blood pressure, as well as kidney impairment. In certain biological enzymes, Cd^{2+} can replace Zn^{2+}, changing their stereo-structure and reducing their catalytic activity [5, 21].

4 Chitin-Based Nanoparticles for Heavy Metal Remediation

Heavy metals are those elements whose density surpasses 5 g/cm^3 with toxic or poisonous nature even at low concentrations, and they are categorized as persistent inorganic pollutants. The mechanisms involved during remediation techniques are commonly degradation, immobilization, isolation, or extraction, which are applied through chemical, physical, or biological processes. However, unlike organic pollutants, it is difficult to remove heavy metals from the soil by a simple physical process because they do not degrade or cannot be easily mineralized. Rather, the heavy metals possess a bioaccumulation nature in an organism's body across the food chain. Thus, it can only be transformed into a less toxic form or immobilized to reduce its bioavailability in the environment. In this regard, various mechanisms have been used for the elimination of metals from a fresh natural environment like soil and water, such as ion electrodialysis, sedimentation, nanofiltration, coagulation/flocculation, solid-phase extraction, chelation, complexation and adsorption [8, 53].

Chitin as a biopolymer, especially in its derivative form, chitosan, can be used to eliminate heavy metals. These days, adsorption by chitin is one of the many methods mentioned that is receiving greater scientific interest due to its availability of the biomaterial, high efficiency, cheap cost, and ease of handling. The lesser the nanomaterial size, the better the efficiency of removing metal. Chitin as a nanostructured adsorbent has considerably superior efficiencies and quicker adsorption rates owing to its extremely elevated surface area, improved active sites, multiple functional groups, suitable pore size and volume, and strong sorption properties. Besides, its adsorption ability, mechanical firmness, compatibility, ease of accessibility and regeneration, cost-effectiveness, environmental friendliness, and simple processing approach make chitin and its derivative a choice for metal removal. In addition to the particle size of the adsorbent, it is also important to consider that adsorption was found to be dependent on the initial pH, temperature, contact time, initial concentration of the metal, availability of other ions, and biomass dose [5, 53].

Furthermore, in vitro tests of chitin-based and its composite nanoparticles for antibacterial and environmental applications were successful.

For instance, Cu^{2+} is effectively removed from the soil, and its bioaccumulation in wheat plants is decreased by carboxylated graphene oxide/chitosan/cellulose composite beads. Clay soils polluted with Cesium can be cleaned using magnetic microgels based on chitosan. The soil matrix is cleansed of Pb^{2+} and Cu^{2+} by the nano-fungal chitosan. It is noteworthy that a composite consisting of chitosan and water hyacinth plant (*Eichhornia crasspes*) exhibits a high capacity for absorbing Cu^{2+}, Pb^{2+}, and Cd^{2+} ions [36].

A test done on removing Fe^{3+} by chitin nanoparticle showed that a particle size of 0.21 mm attains advanced adsorption efficiency than the size of 0.50 and 0.71 mm chitin nanoparticle under the same condition. The kinetic study showed that equilibrium was achieved after 8 min, and an inhibition of adsorption was observed in the presence of chloride ions. In the mentioned test, increasing the chloride ion concentration from 0 to 100 mg per liter decreases the removal percentage of iron from 64.5 to 27.65%. The possible explanation for the decrease in removal is that the greater the chloride ion concentration, the higher the formation of soluble metal chloro-complexes, and the less free Fe^{3+} ion is available. Furthermore, an increase in temperature from 20 to 50 °C increases the removal of Fe^{3+} due to the widening of the pore size, activation of the adsorbent surface and increasing the mobility of the metal ion [25].

5 Chitin-Based Nanoparticles for Organic Pollutant Remediation

Different pollutants can poison soil, which is one of the natural environments. Various factories like dye production, rubber, cosmetics, and printing industries put their effluent solutions, commonly containing water-soluble methylene blue dye, as waste flowing over the surface of the water, thereby adsorbing and contaminating the soil. Methylene blue is an organic molecule that can cause many health issues like shock, vomiting, jaundice, and cancer. It is very poisonous and resistant to treatment. Many organic molecules like phenols, o-chlorophenol, various petroleum fractionations, oils, and many other organic compounds might have a chance of entering the soil and polluting it.

Furthermore, the organic drugs can mix with the soil system and pollute it from pharmaceutical areas. The biopolymer Chitin as is or in adapted form have been extensively implemented for the amputation of pollutants. The derivative of chitin, named chitosan, and its different composites and modified beads exhibit a promising potential for adsorbing various organic pollutants from a waste solution. Adsorbents, sorbent beads, and composites prepared and derived from chitin and chitosan are low cost, efficient, and mostly eco-friendly. Studies showed that methylene blue and methylene orange dyes as organic pollutants are effectively adsorbed on chitin and grafted chitosan. Azo dye removal techniques also employ chitosan magnetic composites and beads that have been chemically crosslinked [12, 30].

Moreover, the cleanup of oil spilled into the environment has shown promise when using chitin and chitosan in flake and powder form. Oil spill cleanup has also used thiol-modified superhydrophobic and super oleophilic sponges made from chitosan. Both entrapment matrices and chitosan hydrogels have been explored for immobilizing biomass that breaks down hydrocarbons in the event of an oil spill [14]. Remarkably, waste from mushroom stems, including chitin and chitosan, can be used directly to adsorbing drugs like paracetamol and 17 α-ethynyl estradiol in soil and surface water. This reduces expenses and is environmentally friendly. Similarly, compounds containing chitin and chitosan can be utilized to contrast hydrocarbons and synthetic colour pollution; these materials are frequently used in conjunction with microorganisms and plants [36].

Three commonly used antibiotics—ciprofloxacin, erythromycin, and amoxicillin—were studied for adsorption and removal using a novel adsorbent called the composite magnetic-activated carbon/chitosan (MACC) system. By taking samples of antibiotic solutions with an initial concentration of 15, 60, and 60 ppm for ciprofloxacin, erythromycin, and amoxicillin, correspondingly, 1 mg of MACC adsorbent was added, for the first 30-min adsorption took place rapidly then after it decreases slowly and attain equilibrium at about 120 min. The maximum sorption capacity of the material using the Langmuir isotherm model for the different antibiotics, as reported by the author, was (90.01 mg/g for ciprofloxacin, 178.57 mg/g for erythromycin and 526.31 mg/g for amoxicillin indicating an excellent adsorption efficiency [12].

6 Chitin-Based Nanoparticles in Pesticide and Herbicide Remediation

Various agricultural inputs such as fertilizers, pesticides and herbicides are extensively used to maximize yields and quality. However, their persistence and hazardous nature associated with many agricultural input chemicals (like atrazine, known and most widely used pesticides) present a severe environmental and human health risk. Their large scale and excessive use intensify this situation. They might have a chance of undergoing soil adsorption–desorption, degradation, crop absorption, migration, as well as transformation processes and polluting the soil [56]. Therefore, it is crucial to create effective strategies for their removal from the environment, and from this standpoint, bio-adsorption is acknowledged as a viable and cost-effective solution. Moreover, the efficient removal of impurities permits the repurposing of increasingly scarce and valuable resources, like cultivable soil and freshwater [36].

Effective herbicide formulations with reduced environmental and human health risks have been produced by efficiently using nanoencapsulation for targeted delivery. The use of chitin- and chitosan-based adsorbents in these areas is well-tested due to their low cost of production, wide availability, biocompatible and biodegradable nature, high resistance to mechanical and antimicrobial attack, and consequent lack of generation of potentially hazardous secondary end products. Furthermore, chitin and chitosan's chemical properties make integrating with ions, molecules, and materials simple to create complex structures. Formulations containing chitin and chitosan can, therefore, be employed as a carrier to increase the effectiveness of herbicides by minimizing their hazardous buildup in the environment [44].

For example, Paraquat is a nonselective contact weed killer widely applied in cotton, tobacco, and rice and is highly toxic to aquatic organisms like algae and fish. It may cause severe lung and renal injuries in mammals. This herbicide has a half-life of 1.3–13 years in soil and binds tightly to it. Fruits and vegetables have been shown to have residual paraquat in amounts up to 70 mg/kg. However, paraquat loaded into chitosan nanoparticles (using chitosan as an herbicide carrier) shows a decreased soil sorption and penetration, cytotoxicity, and mutagenicity nature accompanied by either enhancing or at least preserving their herbicide activity than applying the paraquat alone. The encapsulation was done using the ionic gelification method, combining a paraquat solution with a chitosan solution [10]. The carrier helps to maximize the contact area of the herbicide and plant thus shortening the absorption time of the toxin by the soil. Similarly, chitosan-glyphosate formulations show less phytotoxicity and better herbicidal action than using pure glyphosate salt [36, 44, 46].

A study was conducted by Rashidipour M. and his colleagues to observe the effect of the carrier nanoparticle prepared from chitosan on the sorption activity of the soil, cytotoxicity as well as mutagenicity of the paraquat herbicide when using the paraquat alone and encapsulated with the nanoparticle by controlling many parameters. After a lot of vivid scientific experimental procedure, their finding presents that; (1) by soil column experiments: the free paraquat-treated columns displayed a comparatively constant herbicide concentration at various depths. On the other hand, when a nanoparticle-paraquat (NP:PQ) formulation is used, paraquat exhibits less infiltration deep into the soil because of a larger particle size and more retention of them by the colloids in the latter situation. (2) In their study of the soil sorption assay done in clay soil at 25 °C, their observation on using pure paraquat is that it swiftly and effectively binds to the soil compared to the formulation with the chitosan biopolymer. The basic reason mentioned here is that the toxin's absorption period shortens, and its contact area rises at the plant level. Their finding also exhibited that encapsulation of paraquat lowers its mutagenesis and its cytotoxicity nature compared to its free and commercial forms and magnifying its herbicidal activity [44].

7 Environmental Impact and Safety

Fruit retardants, antitranspirants, soil conditioners, fertilizers, seed coatings, and plant disease management agents can all benefit from chitin and its derivatives. They either strengthen or activate plants' innate defensive systems. They are acknowledged as elicitors of the production of secondary metabolites, growth stimulants, anti-stress agents, and plant growth regulators [47]. Made from the deacetylated form of chitin found in crab shells, chitosan is a linear polymer. This substance is inexpensive, readily available, harmless, and biodegradable. Plants can readily absorb it through their stems and leaves. Furthermore, chitin and chitin-based nanoparticle materials like chitosan are eco-friendly to the environment [36, 44].

8 Conclusion and Future Prospects

It is necessary to address the significant environmental impact of human activity with better, more economical, and environmentally friendly strategies. Nano-based remediation technologies are promising tools to mitigate contamination crises like soil pollution. Chitin-based nanomaterials have become our choice for this sustainable soil remediation purpose for two basic reasons. In one way, it is very abundant and cost-effective and can be exploited in large amounts. In another view, chitin can be easily transformed into chitosan, another very useful biopolymer in soil remediation, by simple deacetylation processes either biologically or chemically. Conversely, specific, and unspecific enzymes produced by certain soil-dwelling bacteria can readily breakdown chitin and chitosan. The products of own degeneration cannot poison humans. Therefore, their application as biostimulants of stress responses, organic agricultural agents, and bioremediation agents will become increasingly pertinent. Derivatives of chitin and chitosan-based nanoparticles have generally shown promise in various agricultural applications, including as fertilizers, soil conditioners, plant disease preventives, growth regulators for plants, and remediation of contaminated soil. They are also highly intriguing instruments for developing more environmentally friendly and sustainable food production methods. However, it is important to remember that most research on the effectiveness of using these materials and their preparations has been done in labs or greenhouses thus far, with large-scale field testing still needing to be present needing to be present. Furthermore, these biopolymer molecules must complex certain nutrients essential to plant growth. However, their effects on the various environments and humans or impacts on other living organisms must be studied as a prospect.

References

1. Adetitun D, Akinmayowa V, Atolani O, Olayemi A (2018) Biodegradation of jet fuel by three gram negative bacilli isolated from kerosene contaminated soil. Pollution 4(2):291–303
2. Al-Taai SHH (2021) Soil pollution-causes and effects. IOP Conf Ser: Earth Environ Sci 790(1):012009. https://doi.org/10.1088/1755-1315/790/1/012009. IOP Publishing
3. Al-Yasiri K (2019) Pesticides and their impact on contamination of agricultural soils located between Al-Kifl stream and Awfi river. Babylon University. J Hum Sci 5:91–96
4. Ali H, Khan E, Sajad MA (2013) Phytoremediation of heavy metals—concepts and applications. Chemosphere 91(7):869–881
5. Anastopoulos I, Bhatnagar A, Bikiaris DN, Kyzas GZ (2017) Chitin adsorbents for toxic metals: a review. Int J Mol Sci 18(1):114. https://doi.org/10.3390/ijms18010114
6. Azubuike CC, Chikere CB, Okpokwasili GC (2016) Bioremediation techniques–classification based on site of application: principles, advantages, limitations and prospects. World J Microbiol Biotechnol 32:1–18
7. Badran A (1988) Environmental pollution, its sources and its types. J Sci Technol Lett 4:7–8
8. Boulaiche W, Hamdi B, Trari M (2019) Removal of heavy metals by chitin: equilibrium, kinetic and thermodynamic studies. Appl Water Sci 9:1–10. https://doi.org/10.1007/s13201-019-0926-8

9. Broquá J, Zanin BG, Flach AM, Mallmann C, Taborda FGD, Machado LEL, Alves SML, Silva MM, Dias RJSP (2019) Methods of chitin production a short review. Am J Biomed Sci & Res 3(4):307–314. https://doi.org/10.34297/AJBSR.2019.03.000682
10. Calvo P, Remuñan-López C, Vila-Jato JL, Alonso MJ (1997) Chitosan and chitosan/ethylene oxide-propylene oxide block copolymer nanoparticles as novel carriers for proteins and vaccines. Pharm Res 14:1431–1436
11. Cundy AB, Bardos RP, Puschenreiter M, Mench M, Bert V, Friesl-Hanl W, Müller I, Li XN, Weyens N, Witters N, Vangronsveld J (2016) Brownfields to green fields: Realising wider benefits from practical contaminant phytomanagement strategies. J Environ Manag 184:67–77
12. Danalıoğlu ST, Bayazit ŞS, Kuyumcu ÖK, Salam MA (2017) Efficient removal of antibiotics by a novel magnetic adsorbent: Magnetic activated carbon/chitosan (MACC) nanocomposite. J Mol Liq 240:589–596
13. Dave U, Somanader E, Baharlouei P, Pham L, Rahman MA (2021) Applications of chitin in medical, environmental, and agricultural industries. J Mar Sci Eng 9(11):1173. https://doi.org/10.3390/jmse9111173
14. Escudero-Oñate C, Martínez-Francés E (2018) A review of chitosan-based materials for the removal of organic pollution from water and bioaugmentation. Chitin-Chitosan-Myriad Funct Sci Technol 1:72–87. https://doi.org/10.5772/intechopen.76540
15. Fadlallah S (2001) Environmental pollution and its impact on agricultural economic development. Assiut J Environ Stud 20:77–89
16. Fingerman M, Nagabhushanam R (2016) Bioremediation of aquatic and terrestrial ecosystems. Science Publishers Inc., Enfield CRC Press
17. Gao J, Faheem M, Yu X (2022) Global research on contaminated soil remediation: a bibliometric network analysis. Land 11(9):1581. https://doi.org/10.3390/land11091581
18. Gautam K, Sharma P, Dwivedi S, Singh A, Gaur VK, Varjani S, Srivastava JK, Pandey A, Chang JS, Ngo HH (2023) A review on control and abatement of soil pollution by heavy metals: emphasis on artificial intelligence in recovery of contaminated soil. Environ Res 225:115592. https://doi.org/10.1016/j.envres.2023.115592
19. Gooden J, 't Sas-Rolfes M (2020) A review of critical perspectives on private land conservation in academic literature. Ambio 49(5):1019–1034
20. Hajji S, Younes I, Ghorbel-Bellaaj O, Hajji R, Rinaudo M, Nasri M, Jellouli K (2014) Structural differences between chitin and chitosan extracted from three different marine sources. Int J Biol Macromol 65:298–306
21. Haripriyan U, Gopinath KP, Arun J (2022) Chitosan based nano adsorbents and its types for heavy metal removal: a mini review. Mater Lett 312:131670. https://doi.org/10.1016/j.matlet.2022.131670
22. Hassan O, Chang T (2017) Chitosan for eco-friendly control of plant disease. Asian J Plant Pathol 11(2):53–70. https://doi.org/10.3923/ajppaj.2017.53.70
23. Jie C, Jing-Zhang C, Man-Zhi T, Zi-tong G (2002) Soil degradation: a global problem endangering sustainable development. J Geog Sci 12:243–252. https://doi.org/10.1007/bf02837480
24. Kaplan L, Tlustoš P, Szakova J, Najmanova J, Brendova K (2016) The effect of NPK fertilizer with different nitrogen solubility on growth, nutrient uptake and use by chrysanthemum. J Plant Nutr 39(7):993–1000. https://doi.org/10.1080/01904167.2015.1106559
25. Karthikeyan G, Andal NM, Anbalagan K (2005) Adsorption studies of iron (III) on chitin. J Chem Sci 117:663–672
26. Kaya M, Baran T (2015) Description of a new surface morphology for chitin extracted from wings of cockroach (Periplaneta americana). Int J Biol Macromol 75:7–12
27. Kidd P, Mench M, Álvarez-López V, Bert V, Dimitriou I, Friesl-Hanl W, Herzig R, Olga Janssen J, Kolbas A, Müller I, Neu S (2015) Agronomic practices for improving gentle remediation of trace element-contaminated soils. Int J Phytorem 17(11):1005–1037
28. Kulikov SN, Chirkov SN, Il'Ina AV, Lopatin SA, Varlamov VP (2006) Effect of the molecular weight of chitosan on its antiviral activity in plants. Appl Biochem Microbiol 42:200–203

29. Kumpiene J, Bert V, Dimitriou I, Eriksson J, Friesl-Hanl W, Galazka R, Herzig R, Janssen J, Kidd P, Mench M, Müller I (2014) Selecting chemical and ecotoxicological test batteries for risk assessment of trace element-contaminated soils (phyto) managed by gentle remediation options (GRO). Sci Total Environ 496:510–522
30. Labidi A, Salaberria AM, Fernandes SC, Labidi J, Abderrabba M (2019) Functional chitosan derivative and chitin as decolorization materials for methylene blue and methyl orange from aqueous solution. Materials 12(3):361–379
31. Lacalle RG, Becerril JM, Garbisu C (2020) Biological methods of polluted soil remediation for an effective economically-optimal recovery of soil health and ecosystem services. J Environ Sci Public Heal 4(2):112–133. https://doi.org/10.26502/jesph.96120089
32. Lee HG, Kim HK, Noh HJ, Byun YJ, Chung HM, Kim JI (2021) Source identification and assessment of heavy metal contamination in urban soils based on cluster analysis and multiple pollution indices. J Soils Sediments 21:1947–1961
33. Lee S, Hao LT, Park J, Oh DX, Hwang DS (2023) Nanochitin and nanochitosan: chitin nanostructure engineering with multiscale properties for biomedical and environmental applications. Adv Mater 35(4):2203325. https://doi.org/10.1002/adma.202203325
34. Makombe N, Gwisai RD (2018.) Soil remediation practices for hydrocarbon and heavy metal reclamation in mining polluted soils, Hindawi. Sci World J. https://doi.org/10.1155/2018/5130430
35. Malerba M, Cerana R (2019) Recent applications of chitin-and chitosan-based polymers in plants. Polymers 11(5):839. https://doi.org/10.3390/polym11050839
36. Malerba M, Cerana R (2020) Chitin-and chitosan-based derivatives in plant protection against biotic and abiotic stresses and in recovery of contaminated soil and water. Polysaccharides 1(1):21–30. https://doi.org/10.3390/polysaccharides1010003
37. Okada M, Matsumura M, Ito Y, Shibuya N (2002) High-affinity binding proteins for N-acetylchitooligosaccharide elicitor in the plasma membranes from wheat, barley and carrot cells: conserved presence and correlation with the responsiveness to the elicitor. Plant Cell Physiol 43(5):505–512. https://doi.org/10.1093/pcp/pcf060
38. Orzali L, Corsi B, Forni C, Riccioni L (2017) Chitosan in agriculture: a new challenge for managing plant disease. In: Biological activities and application of marine polysaccharides, pp 17–36. https://doi.org/10.5772/66840
39. Pal J, Verma HO, Munka VK, Maurya SK, Roy D, Jitendra Kumar J (2014) Biological method of chitin extraction from shrimp waste an eco-friendly low cost technology and its advanced application. Int J Fisher Aquat Stud 1(6):104–107
40. Peña A (2022) A comprehensive review of recent research concerning the role of low molecular weight organic acids on the fate of organic pollutants in soil. J Hazard Mater 434:128875. https://doi.org/10.1016/j.jhazmat.2022.128875
41. Pillai CK, Paul W, Sharma CP (2009) Chitin and chitosan polymers: chemistry, solubility and fiber formation. Prog Polym Sci 34(7):641–678. https://doi.org/10.1016/j.progpolymsci.2009.04.001
42. Pusztahelyi T (2018) Chitin and chitin-related compounds in plant–fungal interactions. Mycology 9(3):189–201. https://doi.org/10.1080/21501203.2018.1473299
43. Rashid S (2017) Effects of the economic problems of environmental pollution in Iraq and methods for its treatment. Al-Mustansiriya J Arab Int Stud 14:106–112
44. Rashidipour M, Maleki A, Kordi S, Birjandi M, Pajouhi N, Mohammadi E, Heydari R, Rezaee R, Rasoulian B, Davari B (2019) Pectin/chitosan/tripolyphosphate nanoparticles: efficient carriers for reducing soil sorption, cytotoxicity, and mutagenicity of paraquat and enhancing its herbicide activity. J Agric Food Chem 67(20):5736–5745. https://doi.org/10.1021/acs.jafc.9b01106
45. Rodríguez-Eugenio N, McLaughlin M, Pennock D (2018) Soil pollution: a hidden reality. a hidden reality. Food and Agriculture Organization of United States Leadell Pennock, Rome University of Saskatchewan
46. Rychter P (2019) Chitosan/glyphosate formulation as a potential, environmental friendly herbicide with prolonged activity. J Environ Sci Health B 54(8):681–692

47. Shamshina JL, Kelly A, Oldham T, Rogers RD (2019) Agricultural uses of chitin polymers. Environ Chem Lett 18(1):53–60. https://doi.org/10.1007/s10311-019-00934-5
48. Shao H, Liye C, Gang X, Kun Y, Lihua Z, Junna S (2011) Progress in phytoremediating heavy-metal contaminated soils. In: Sherameti I, Varma A (eds) Detoxification of heavy metals. Springer, Berlin, pp 73–90
49. Sharp RG (2013) A review of the applications of chitin and its derivatives in agriculture to modify plant-microbial interactions and improve crop yields. Agronomy 3(4):757–793. https://doi.org/10.3390/agronomy3040757
50. Singh SP, Singh MK (2020) Soil pollution and human health. In: Plant responses to soil pollution. Singapore, pp 205–220
51. Smičiklas I, Šljivić-Ivanović M (2016) Radioactive contamination of the soil: assessments of pollutants mobility with implication to remediation strategies. In: Soil contamination–current consequences and further solutions. InTech [Internet], Rijeka, pp 253–276
52. Tharanathan RN, Kittur FS (2003) Chitin-the undisputed biomolecule of great potential. Crit Rev Food Sci Nutr 43(1):61–87. https://doi.org/10.1080/10408690390826455
53. Vasilachi IC, Gavrilescu M (2021) Methods for the remediation of soils polluted with heavy metals. Bull Polytechnol Inst Iaşi 67(71):23–40
54. Velásquez CL, Pirela MR (2016) Biochemical aspects of the chitin fungicidal activity in agricultural uses. In: Chitosan in the preservation of agricultural commodities, pp 279–298. https://doi.org/10.1016/B978-0-12-802735-6.00010-0
55. Wijesena RN, Tissera ND, Rathnayaka VWSG, de Silva RM, de Silva KN (2020) Colloidal stability of chitin nanofibers in aqueous systems: effect of pH, ionic strength, temperature & concentration. Carbohyd Polym 235:116024
56. Yu H, Liu Y, Shu X, Fang H, Sun X, Pan Y, Ma L (2020) Equilibrium, kinetic and thermodynamic studies on the adsorption of atrazine in soils of the water fluctuation zone in the three-gorges reservoir. Environ Sci Eur 32(1):1–10. https://doi.org/10.1186/s12302-020-00303-y
57. Zhang X, Gustave W, He L, Yang X (2023) Soil pollution, risk assessment and remediation. Front Media SA. https://doi.org/10.3389/fenvs.2023.1252139

Role of Chitin-Based Nanoparticles for Crop Nutrition and Fertilization

Priti Pal, Akhilesh Kumar Singh, Maharudra Pratap Singh, Om Prakash, Anand Prakash, Brijesh Pandey, Vishnu Agarwal, Laxuman Sharma, and Rahul Gupta

Abstract The escalating global demand for food production necessitates innovative solutions to enhance crop nutrition and fertilization while mitigating environmental impacts. This chapter explores the pivotal role of chitin-based nanoparticles in revolutionizing agricultural practices. Derived from the abundant biopolymer chitin, chitosan-based nanoparticles exhibit unique characteristics that make them promising candidates for improving nutrient delivery, controlled release, and fertilizer efficiency. These nanoparticles facilitate targeted delivery by encapsulating essential nutrients, reducing environmental pollution and ensuring sustained nutrient supply to plants. Additionally, their inherent antifungal properties contribute to soil health, offering a natural solution to combat soil-borne diseases. This abstract encapsulates the multifaceted contributions of chitin-based nanoparticles to sustainable agriculture, with the potential to reshape the future of crop nutrition. Recent studies have demonstrated the efficacy of chitin-based nanoparticles in enhancing crop nutrition and fertilization. For instance, researchers encapsulated nitrogen, phosphorus, and potassium within chitosan nanoparticles and observed a significant improvement in crop nutrient uptake. The controlled release provided by these nanoparticles enhanced fertilizer efficiency and minimized nutrient leaching, addressing concerns related to environmental impact. Furthermore, field trials revealed reduced soil-borne

P. Pal
Environmental Engineering, Shri Ramswaroop Memorial College of Engineering and Management, Tewariganj, Faizabad, Road, Lucknow 226028, India

A. K. Singh (✉) · M. P. Singh · O. Prakash · A. Prakash · B. Pandey
Departments of Biotechnology, School of Life Sciences, Mahatma Gandhi Central University, Motihari 845401, Bihar, India
e-mail: akhileshsingh@mgcub.ac.in

V. Agarwal
Department of Biotechnology, Motilal Nehru National Institute of Technology Allahabad, Prayagraj 211004, India

L. Sharma
Department of Horticulture, Sikkim University, Gangtok 737102, India

R. Gupta
Amity Institute of Biotechnology, Amity University Uttar Pradesh, Lucknow Campus, Lucknow 226028, India

R. K. Bachheti et al. (eds.), *Chitin-Based Nanoparticles for the Agriculture Sectors*, Smart Nanomaterials Technology, https://doi.org/10.1007/978-981-96-0920-8_7

diseases, showcasing the potential of chitin-based nanoparticles in promoting soil health. This example highlights the practical application of chitin-based nanoparticles in agriculture and their positive implications for crop yield and environmental sustainability.

Keywords Chitin-based nanoparticles · Crop nutrition · Fertilization · Nutrient delivery · Controlled release · Fertilizer efficiency · Soil health · Sustainable agriculture

1 Introduction

Agricultural practices are at the forefront of global challenges, grappling with the intricate balance between increasing food demand, environmental sustainability, and the need for innovative solutions. The imperative to enhance crop nutrition and fertilization has never been more pressing, given the surging global population and the imperative to ensure food security. This chapter explores a ground-breaking avenue within the realm of nanotechnology—specifically, the transformative role of chitin-based nanoparticles in revolutionizing crop nutrition and fertilization [48].

Chitin is the world's second most plentiful natural polysaccharide, surpassed in quantity solely by cellulose. Constituting a polymer made up of β-1,4-linked N-acetyl-D-glucosamine, chitin exhibits three distinct polymorphic forms based on the alignment of its polymer chains. The first, α-chitin, denotes a configuration where the chains lie antiparallel. In contrast, β-chitin refers to chitin with parallel chains, while γ-chitin represents a combination of both α- and β-chitin, featuring some chains arranged parallelly and others antiparallelly [58]. Chitin, a naturally occurring biopolymer, is abundantly found in the exoskeletons of crustaceans, insects and the cell walls of fungi. Its derivative, chitosan, has garnered significant attention for its biocompatibility, biodegradability, and non-toxicity. Despite its insolubility in water and many organic solvents, attributed to a high degree of crystallinity and robust hydrogen bonds between its chains, chitin has discovered applications across diverse fields such as textiles, paper manufacturing, medicine, and wastewater treatment [46]. Harnessing the unique properties of chitosan, researchers have ventured into the realm of nanotechnology to develop chitin-based nanoparticles. These nanoparticles exhibit a high surface area, making them ideal carriers for a myriad of applications, particularly in agriculture [30–35, 85]. The relevance of this research becomes apparent when considering the challenges faced by modern agriculture. Conventional fertilization practices often lead to nutrient leaching, environmental pollution, and inefficient nutrient uptake by plants. Additionally, soil-borne diseases pose a constant threat to crop health. In this context, chitin-based nanoparticles emerge as a promising solution, offering a multifaceted approach to address these challenges and optimize crop nutrition for sustainable agriculture [78].

To illustrate the practical implications of chitin-based nanoparticles in crop nutrition and fertilization, let us delve into a recent study conducted in a rice paddy field in Southeast Asia. Rice, a staple food for over half of the world's population, presents unique challenges in terms of nutrient management. Nitrogen, an essential nutrient for rice growth, is commonly supplied through conventional urea fertilizers. However, the inefficiency of urea application often results in nitrogen leaching, contributing to environmental pollution and reducing the overall effectiveness of fertilization millet [76]. In this study, researchers incorporated chitosan-based nanoparticles into the urea fertilizer formulation. These nanoparticles served as carriers for nitrogen, encapsulating the nutrient and protecting it from environmental factors. The unique surface properties of chitosan facilitated the controlled release of nitrogen, addressing the challenge of rapid nutrient leaching associated with conventional urea application. The results were compelling. The rice plants treated with chitin-based nanoparticles exhibited enhanced nutrient uptake, leading to higher yields than those treated with conventional urea alone. The controlled release mechanism not only optimized nutrient utilization by the rice plants but also reduced the environmental impact associated with nitrogen leaching into groundwater [17]. This example underscores the practical significance of chitin-based nanoparticles in enhancing fertilizer efficiency and mitigating the environmental consequences of conventional fertilization practices. Additionally, the practical utility of chitosan nanoparticles extends into agriculture, where they have been effectively employed for the controlled release of agrochemicals at specific sites [24]. Their application has proven beneficial in enhancing stability and decreasing the required dosages of pesticides and fertilizers, thereby promoting sustainable agricultural practices [65]. Noteworthy research findings include reports of increased grain yield for barley and finger millet [76].

Moreover, chitosan nanoparticles have demonstrated efficacy in wastewater treatment by serving as sorbents for toxic pollutants like heavy metals and dyes [38]. Leveraging their high surface area, chitosan nanoparticles exhibit a superior sorption capacity compared to conventional sorbents [90]. For those interested in a more comprehensive understanding of the applications of chitosan nanoparticles in agriculture and wastewater treatment, recent reviews by [107] offer detailed insights and information. As we navigate through the intricacies of chitin-based nanoparticles in crop nutrition and fertilization, it is essential to understand the underlying mechanisms driving their effectiveness. The subsequent sections of this chapter will delve into how chitin-based nanoparticles contribute to improved nutrient delivery, controlled release, enhanced fertilizer efficiency, and overall soil health. By comprehensively exploring these aspects, we aim to unravel the full spectrum of possibilities that chitin-based nanoparticles offer in sustainable agriculture.

2 Chitin-Based Nanoparticles: A Brief Overview

Chitin, a polysaccharide, is a naturally occurring biopolymer abundantly found in the exoskeletons of crustaceans, insects, and the cell walls of fungi. Its derivative, chitosan, has garnered significant attention for its unique properties that lend themselves to a myriad of applications, particularly in the realm of nanotechnology. The transition from chitosan to chitin-based nanoparticles represents a crucial advancement in harnessing the potential of this biopolymer for various purposes, including its transformative role in agriculture [40] shown in Fig. 1. Chitin-based nanoparticles are engineered structures with a particle size typically in the nanometer range. Reducing chitosan to nanoscale dimensions imparts distinct physical and chemical properties that set these nanoparticles apart from their bulk counterparts. The high surface area-to-volume ratio of nanoparticles is a pivotal factor in their efficacy, as it enhances their reactivity and interaction with other substances [43]. Chitin nanomaterials, exemplified by chitin nanocrystals (ChNCs) or chitin nanowhiskers (ChNWs) ranging from 100 to 800 nm in length and 6 to 60 nm in width, as well as chitin nanofibers (ChNFs) with lengths in the micrometer range and widths spanning 10 to 100 nm, can be extracted [74]. The specific characteristics of these chitin nanomaterials, including their form and dimensions, are influenced by factors such as the chitin source and the extraction method. The breakdown of natural chitin into ChNCs or ChNFs leads to nanomaterials with an exceptionally high aspect ratio. Nanomaterials with a high aspect ratio, such as these chitin variants, confer distinct advantages in enhancing transparency, reactivity, mechanical strength, and other material properties [103]. For example, incorporating ChNCs with a high aspect ratio into natural rubber (NR) latex to produce NR/ChNCs composites has proven to significantly improve the resulting rubber composites' mechanical properties and thermal stability [51]. Chitin nanomaterials can be isolated through two main approaches: the "top-down" and "bottom-up" methods. In the top-down approach, purified chitin is broken down into nanomaterials using processes such as acid hydrolysis, TEMPO (2,2,6,6-Tetramethylpiperidine 1-oxyl) or APS (ammonium persulfate)-mediated oxidation, mechanical treatment, and partial deacetylation [74]. Conversely, in the bottom-up approach, chitin nanomaterials are regenerated from chitin solutions or gels using techniques like electrospinning and self-assembly [74].

One of the defining features of chitin-based nanoparticles is their biocompatibility. This characteristic makes them suitable for applications across diverse fields, from medicine to agriculture. In agriculture, the biocompatibility of these nanoparticles ensures minimal adverse effects on the environment, crops, and the surrounding ecosystem. Moreover, chitosan, and by extension chitin-based nanoparticles, exhibit biodegradability. This property is crucial in sustainable agriculture, as it ensures that the nanoparticles break down over time, leaving no persistent residues in the soil or water. The biodegradability of chitin-based nanoparticles aligns with the global shift toward environmentally friendly and sustainable farming practices. This is particularly pertinent when considering their use in the cultivation of food crops, where the potential for nanoparticle residues in the harvested produce raises concerns about food safety [53]. Various methods are employed to achieve the desired nanoscale dimensions in synthesizing chitin-based nanoparticles. These methods include physical techniques such as ball milling and chemical methods like acid hydrolysis. Each approach has advantages and limitations, influencing particle size distribution, stability, and scalability. The choice of synthesis method depends on the intended application and the specific characteristics required for optimal performance [8].

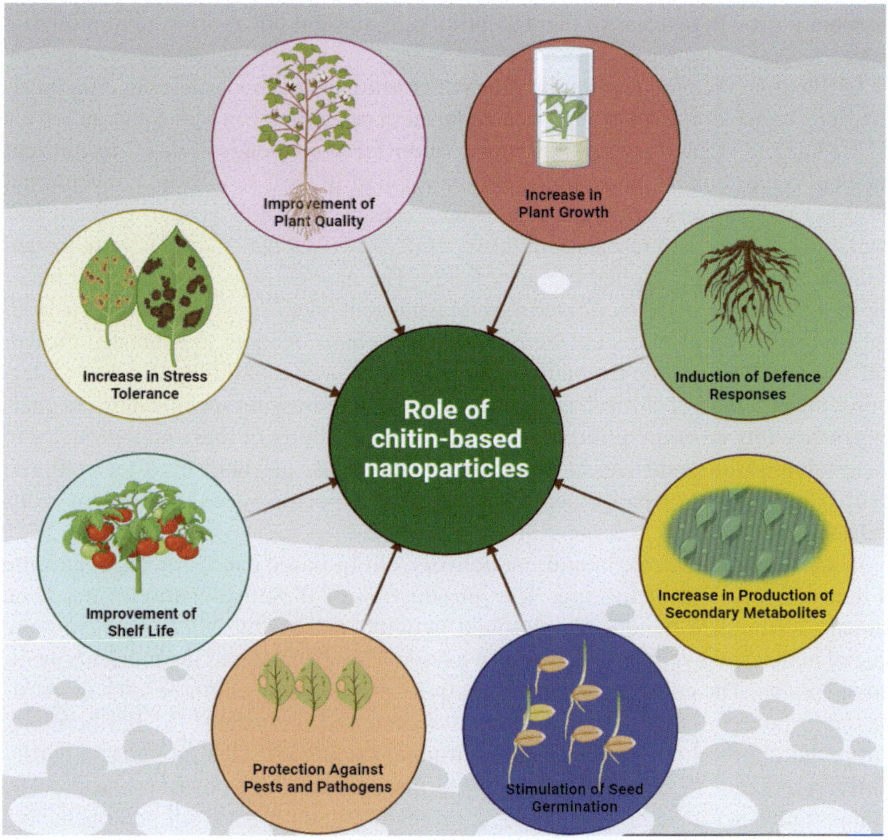

Fig. 1 Utilizing chitin-based nanoparticles in enhancing plant growth and safeguarding crops

In recent years, researchers have focused on exploring the potential of chitin-based nanoparticles in improving crop nutrition and fertilization. The high surface area of these nanoparticles makes them excellent carriers for various nutrients, both macro and micronutrients. For instance, nitrogen, phosphorus, potassium, iron, zinc, and copper can be encapsulated within chitin-based nanoparticles, offering a versatile platform for plant nutrient delivery [28]. The encapsulation of nutrients within chitin-based nanoparticles serves multiple purposes. Firstly, it protects the nutrients from environmental factors that can degrade their efficacy. This protection is particularly relevant in regions with adverse soil conditions, where factors such as high rainfall or alkaline soils can compromise the stability of conventional fertilizers. The nanoparticles act as a shield, ensuring the nutrients reach the plant in a readily available form for uptake. Secondly, chitin-based nanoparticles enable the controlled release of nutrients [39]. Unlike conventional fertilizers that may release nutrients rapidly, leading to potential leaching and inefficient uptake, chitin-based nanoparticles offer a more gradual and sustained release. This controlled release aligns with the natural growth patterns of plants, optimizing nutrient utilization and minimizing waste.

To illustrate the practical implications of chitin-based nanoparticles in crop nutrition, let's consider a scenario in a maize farming community in Sub-Saharan Africa [11]. Maize, a staple crop in the region, often faces challenges related to nutrient deficiency, particularly phosphorus. When applied in their raw form, conventional phosphorus fertilizers often undergo rapid fixation in the soil, rendering them less available to plants. In a research initiative, chitosan-based nanoparticles were loaded with phosphorus and applied to maize fields. The nanoparticles protected the phosphorus from fixation and facilitated a gradual nutrient release over an extended period. This improved phosphorus uptake by the maize plants, leading to enhanced growth and ultimately increasing the maize yield [11]. The success of this application underscores the potential of chitin-based nanoparticles in addressing specific nutrient challenges faced by crops in different regions. The adaptability of these nanoparticles to encapsulate various nutrients and their controlled release mechanisms offer a tailored approach to nutrient management, catering to the specific needs of different crops and soils [37].

In addition to their role in nutrient delivery, chitin-based nanoparticles contribute to the overall health of the soil. The introduction of these nanoparticles has been shown to stimulate beneficial microbial activity in the soil. Microorganisms are crucial in nutrient cycling, breaking down organic matter and making nutrients available to plants. The enhanced microbial activity promoted by chitin-based nanoparticles contributes to a more balanced and sustainable soil ecosystem [16]. Furthermore, chitin and chitosan possess inherent antifungal properties. This property is particularly significant in combating soil-borne diseases that can devastate crops. The nanoparticles act as natural fungicides, protecting plant roots from pathogens and contributing to crop disease resistance. This dual functionality—improved nutrient delivery and natural disease resistance—positions chitin-based nanoparticles as holistic solutions for promoting crop health and productivity [47]. As we delve into the intricacies of chitin-based nanoparticles, we must acknowledge the challenges and considerations associated with their application in agriculture. One notable concern is the scalability of nanoparticle production. While laboratory-scale synthesis methods have been well-established, scaling up production to meet the demands of large-scale agriculture poses challenges related to cost, efficiency, and environmental impact. Additionally, the potential ecological impacts of chitin-based nanoparticles require thorough investigation. Understanding their fate in the environment, including their persistence and interactions with other components in the ecosystem, is crucial for ensuring their safe and responsible use. Research efforts must also address regulatory aspects to establish guidelines for incorporating chitin-based nanoparticles in agriculture [55].

3 Nanoparticles as Nutrient Carriers

One of the key attributes that distinguishes chitin-based nanoparticles in agriculture is their role as exceptional carriers for essential nutrients. These nanoparticles' high surface area and biocompatibility make them ideal candidates for encapsulating a wide range of nutrients, including both macronutrients and micronutrients. This section delves into the significance of chitin-based nanoparticles as nutrient carriers and illustrates their practical applications through a pertinent example. The encapsulation of nutrients within chitin-based nanoparticles is a strategic approach to address challenges associated with conventional nutrient delivery in agriculture. Nutrient loss through leaching and volatilization is common, particularly with water-soluble fertilizers [89]. Chitin-based nanoparticles offer a protective shield, safeguarding nutrients from environmental factors and ensuring their availability to plants over an extended period. To illustrate the efficacy of chitin-based nanoparticles as nutrient carriers, let's consider a scenario in tomato cultivation [12].

Tomatoes are a widely cultivated crop, and their optimal growth requires a balanced supply of nutrients, including nitrogen, phosphorus, and potassium. Conventional fertilization methods often involve the application of water-soluble fertilizers, which may result in nutrient runoff, especially in regions with high rainfall. In a research study, chitosan-based nanoparticles were synthesized and loaded with nitrogen, phosphorus, and potassium. These nanoparticles were then applied to the soil around tomato plants [12]. The encapsulation of nutrients within the nanoparticles served multiple purposes in this context. Firstly, the nanoparticles protected the nutrients from leaching caused by heavy rainfall. Unlike traditional fertilizers that may be washed away by rainwater, the nanoparticles provided a physical barrier, preventing nutrient loss and ensuring a more efficient nutrient supply to the tomato plants. Secondly, the controlled release properties of chitin-based nanoparticles played a crucial role. The nanoparticles gradually released the encapsulated nutrients, matching the uptake patterns of the tomato plants during their growth stages [58]. This controlled release optimized nutrient utilization and reduced the frequency of fertilizer applications, resulting in resource efficiency and cost-effectiveness for the farmer. The study's results demonstrated a significant improvement in tomato plant growth and yield compared to conventional fertilization methods. The plants treated with chitin-based nanoparticles exhibited enhanced nutrient uptake, manifested in healthier foliage, increased fruit production, and improved overall crop quality.

Furthermore, the controlled release of nutrients contributed to a more sustainable and environmentally friendly approach to fertilization. This example showcases the practical implications of using chitin-based nanoparticles as nutrient carriers in agriculture. The encapsulation of nutrients not only protects them from environmental factors but also enables a tailored and sustained release, aligning with the specific nutritional needs of the crop [4]. The versatility of chitin-based nanoparticles to encapsulate various nutrients allows for a customized approach to address nutrient deficiencies in different crops and soil conditions. The application of chitin-based nanoparticles as nutrient carriers extends beyond traditional fertilization practices. Chitin-based nanoparticles can be introduced to enhance nutrient availability in hydroponic systems, where plants grow in nutrient-rich water without soil. The nanoparticles can encapsulate essential nutrients in a form easily assimilated by plants in water-based environments, contributing to the efficiency and sustainability of hydroponic agriculture.

Furthermore, using chitin-based nanoparticles as nutrient carriers has implications for precision agriculture [41]. In precision agriculture, where technology is leveraged to optimize field-level management, the controlled release of nutrients facilitated by nanoparticles aligns with targeted and efficient resource utilization principles. By tailoring nutrient delivery to the specific needs of plants, farmers can enhance crop yields while minimizing the environmental impact of excessive fertilizer use. Despite the promising benefits, challenges associated with adopting chitin-based nanoparticles as nutrient carriers still need to be addressed. Scaling up production to meet the demands of large-scale agriculture, ensuring cost-effectiveness, and addressing potential ecological impacts are critical considerations that require further research and development efforts [19].

4 Controlled Release of Nutrients

One of the distinctive features that make chitin-based nanoparticles invaluable in agriculture is their ability to facilitate the controlled release of nutrients. The controlled release mechanism addresses a critical challenge in conventional fertilization methods, where rapid nutrient release can lead to inefficient utilization, nutrient leaching, and environmental pollution. This section explores the significance of controlled nutrient release enabled by chitin-based nanoparticles, delving into the underlying mechanisms and showcasing practical applications through a relevant example. The controlled release of nutrients refers to the gradual and sustained delivery of essential elements to plants over an extended period [77]. This approach aligns with plants' natural growth patterns and nutrient uptake kinetics, optimizing resource utilization and minimizing the negative environmental impacts associated with conventional fertilization. Chitin-based nanoparticles, owing to their unique physicochemical properties, excel in providing a controlled release of nutrients. The nanoparticles act as carriers, encapsulating nutrients within their structure and releasing them gradually based on environmental conditions, plant demand, and the specific characteristics of the nanoparticles [23]. This controlled release offers several advantages in the context of sustainable agriculture. To illustrate the significance of controlled nutrient release, let's consider an example in the cultivation of wheat, a staple crop with high global significance. Wheat cultivation often relies on applying nitrogen-based fertilizers to ensure optimal growth and yield. However, the rapid release of nitrogen from conventional fertilizers can result in nutrient loss through leaching and volatilization [104]. In a research study, chitosan-based nanoparticles were formulated to encapsulate nitrogen, a key nutrient for wheat. These nanoparticles were then incorporated into the soil during the wheat planting phase. The controlled release properties of the nanoparticles played a pivotal role in this scenario. As the wheat plants grew, they exhibited varying stages of nutrient demand corresponding to their growth phases. The chitin-based nanoparticles responded to these demands by gradually releasing nitrogen, synchronizing with the plants' requirements. During periods of heightened growth, such as tillering and flowering, the controlled release ensured an increased nitrogen supply, promoting optimal development [93]. The results of the study were noteworthy. The wheat plants treated with chitin-based nanoparticles showed improved nitrogen utilization efficiency, evidenced by enhanced biomass production and grain yield. Moreover, the controlled release mitigated nitrogen leaching, reducing the soil's environmental footprint associated with excess nitrogen. This example underscores the practical implications of chitin-based nanoparticles in providing controlled nutrient release, optimizing plant growth, and mitigating environmental concerns [105]. The mechanisms behind the controlled release of nutrients facilitated by chitin-based nanoparticles are multifaceted. The nanoparticle structure allows for a gradual diffusion of nutrients, influenced by particle size, surface charge, and interactions with the surrounding environment. Additionally, environmental factors, including soil moisture, temperature, and microbial activity, can influence the release kinetics of nutrients from the nanoparticles [100]. The inherent properties of chitin and chitosan also influence the controlled release. These biopolymers possess a unique capacity to interact with nutrients through electrostatic interactions and chemical bonding. The encapsulation of nutrients within the chitin-based nanoparticles shields them from external factors that could compromise their stability, allowing for a more controlled and sustained release [60]. The practical implications of controlled nutrient release extend beyond traditional crop cultivation. In orchards and vineyards, where perennial crops have distinct growth stages, chitin-based nanoparticles can be strategically timed to coincide with increased nutrient demand. This precision in nutrient delivery enhances resource utilization efficiency, optimizing crop quality and yield [69]. Moreover, the controlled release provided by chitin-based nanoparticles aligns with the principles of organic farming. Organic agriculture emphasizes minimizing external inputs, promoting soil health, and reducing environmental impact. Organic farmers can enhance crop availability by utilizing chitin-based nanoparticles to provide controlled nutrient release while adhering to sustainable and environmentally conscious practices. While the advantages of controlled nutrient release facilitated by chitin-based nanoparticles are evident, challenges remain in optimizing and standardizing this approach [14]. Fine-tuning the release kinetics to match specific crop requirements, soil types, and environmental conditions requires ongoing research. Additionally, the scalability of nanoparticle production and considerations related to cost-effectiveness and regulatory approval demand attention for broader implementation.

5 Enhanced Fertilizer Efficiency

In the realm of modern agriculture, where the challenge of feeding a growing global population intersects with the imperative of sustainable resource management, the efficiency of fertilizer use plays a pivotal role. Conventional fertilization methods often need to improve in optimizing nutrient utilization, leading to issues such as nutrient runoff, leaching, and environmental pollution. With their unique characteristics, Chitin-based nanoparticles offer a transformative solution by significantly enhancing fertilizer efficiency [87]. This section explores the nuances of how these nanoparticles contribute to improved fertilizer efficiency and illustrates their practical impact through a relevant example. Fertilizers encompass a range of substances, both organic and inorganic, containing essential nutrients like nitrogen, potassium, and phosphorous, crucial for plant growth and immunity [81]. Chitin plays a noteworthy role in natural fertilizers, serving as an organic nitrogen-type fertilizer (C). Derived from sources like shellfish waste, comprising approximately 75% of total weight and containing 20–58% chitin, it is utilized in fertilizer preparation [99]. The residues of these shells, primarily composed of chitin and calcium carbonate, can be directly applied to the soil, functioning as a fertilizer or soil conditioner. The chitin-rich shell residues, abundant in nitrogen, phosphorous, and micronutrients, are effective fertilizers for crops and plants. The slow degradation of chitin in the soil, facilitated by soil organisms, releases nitrogen over an extended period, enhancing the soil's nitrogen concentration. Chitin's biodegradation produces ammonia-derived compounds with inherent fertilizer effects, fostering microbial activity and nutrient availability in the soil [13]. Incorporating chitin into the soil stimulates the growth of soil microbes, including chitin-degrading bacteria and fungi, resulting in increased nitrogen and calcium content. Additionally, when combined with micronutrients like selenium, chitin has demonstrated a reduction of nitrogen and phosphate fertilizer usage, offering potential benefits for sustainable agriculture practices [106]. Studies have indicated the upregulation of N metabolism and transport-related genes, affirming chitin's positive effects on soil microbes and suggesting its potential for sustainable agriculture [106]. The dual benefits of chitin, providing both calcium and nitrogen to the soil, underscore its significance in enhancing soil fertility. The nitrogen content in chitin-rich shell residue is approximately 6.9% [73]. Moreover, shrimp meal fertilizer, a precursor to chitin purification, contains essential nutrients like potassium, phosphorus, and nitrogen, further supporting the use of shellfish shells as early-period fertilizers [81]. Notably, applying chitin nanofibers in combination with protein/$CaCO_3$ has been observed to significantly enhance the growth of tomatoes in hydroponic systems [2]. Additionally, combining chitin with selenium has shown promise in promoting phosphorus accumulation in leaves and reducing phosphorus leaching from the soil [95].

Beyond traditional chitin, nano chitin emerges as a promising player in agricultural fertilizers, exhibiting substantial potential for positively influencing plant growth and soil fertility. A noteworthy study by [102] highlighted the significant benefits of nano chitin application. In their research, winter wheat varieties, specifically multi-spike wheat (MSW) and large-spike wheat (LSW) experienced remarkable growth enhancements when treated with 0.006 g/kg nano chitin. MSW and LSW exhibited growth increases of 23.0 and 33.4%, respectively. Moreover, nano chitin treatment resulted in elevated grain protein, iron, and zinc contents in wheat, showcasing increments of 5.0, 10.3, and 22.1% for MSW, and 33.4, 32.0, and 27.0% for LSW, respectively. Similarly, research conducted by [10] delved into the positive effects of chitin nanowhiskers (ChNWs) on agricultural productivity. Applying ChNWs at a concentration of 6 mg/kg demonstrated enhanced nitrogen metabolism, contributing to a 27.56% increase in winter wheat growth and a 13.26% rise in crude protein concentration. These findings underscore the potential of nano chitin, particularly in the form of ChNWs, as a valuable agent for promoting agricultural yield and nutrient content in crops, as shown in Fig. 2.

Fertilizer efficiency refers to the ability of a fertilizer to deliver nutrients to plants in a form that is readily accessible and utilized, minimizing waste and environmental impact. When applied in their raw form, conventional fertilizers can be prone to losses through various pathways, including leaching into groundwater, volatilization into the atmosphere, and fixation in the soil, making the nutrients less available to plants. Chitin-based nanoparticles address these challenges by acting as carriers that encapsulate nutrients, protecting them from external factors and facilitating their controlled release [14]. This encapsulation shields the nutrients and allows for a more targeted and efficient delivery to plants, thereby enhancing fertilizer efficiency. To illustrate the impact of chitin-based nanoparticles on fertilizer efficiency, let's

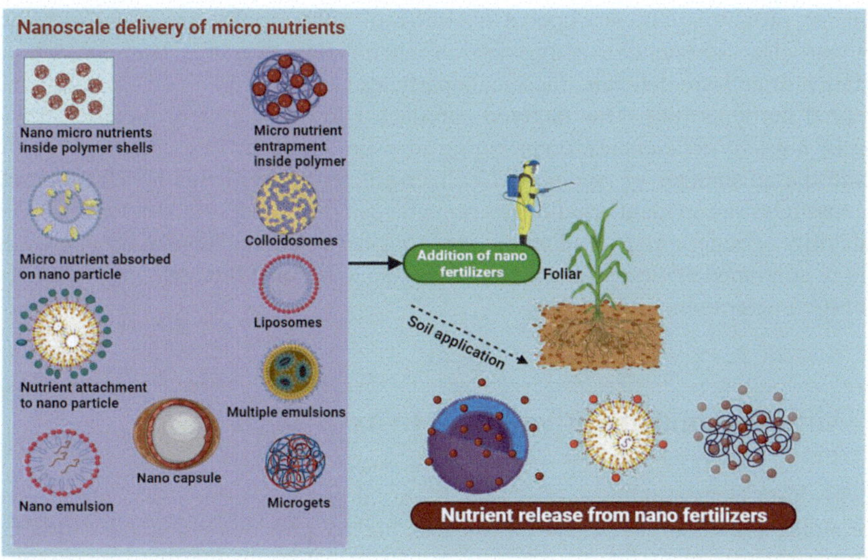

Fig. 2 Exploring recent progress and future prospects in chitosan-based delivery systems for plants

consider the cultivation of soybeans, a globally significant crop valued for its protein-rich seeds. Like many legumes, soybeans have a high demand for nitrogen, a critical nutrient for protein synthesis. When applied in excess, conventional nitrogen fertilizers often result in significant nitrogen losses, impacting both the environment and the economic viability of farming operations [63]. Chitosan-based nanoparticles were formulated to encapsulate nitrogen in a research study focused on soybean cultivation. These nanoparticles were then incorporated into the soil during the soybean planting phase. The encapsulation of nitrogen within the nanoparticles served to protect the nutrient from leaching and volatilization, common issues associated with conventional nitrogen fertilization [63].

The results of the study demonstrated notable improvements in fertilizer efficiency. The soybean plants treated with chitin-based nanoparticles exhibited enhanced nitrogen uptake, increasing the seeds' protein content. The controlled release of nitrogen from the nanoparticles ensured a sustained nutrient supply, aligning with the soybeans' growth stages and optimizing nutrient utilization. Additionally, reducing nitrogen losses through leaching and volatilization contributed to a more environmentally sustainable approach to soybean cultivation. The efficiency gains achieved in this example highlight the transformative potential of chitin-based nanoparticles in fertilizer management. Beyond nitrogen, these nanoparticles can be tailored to encapsulate various nutrients, including phosphorus and potassium, addressing the specific needs of different crops and soil conditions. The enhanced fertilizer efficiency facilitated by chitin-based nanoparticles can be attributed to several key factors [88]. The encapsulation of nutrients within the nanoparticles protects them from environmental factors that can compromise their efficacy. This protection ensures that nutrients remain available in forms that are easily accessible to plants, promoting efficient uptake and utilization. The application of chitin-based nanoparticles also extends to addressing nutrient imbalances in soils. In regions where certain nutrients are deficient, these nanoparticles can be tailored to encapsulate the required nutrients [86]. This targeted approach allows for a more precise correction of nutrient deficiencies, contributing to overall soil fertility and crop health. While the advantages of enhanced fertilizer efficiency facilitated by chitin-based nanoparticles are evident, challenges remain regarding widespread adoption. The scalability of nanoparticle production, cost considerations, and regulatory approvals require continued attention for these innovations to become integral components of mainstream agricultural practices.

6 Antimicrobial and Eliciting Activity

While chitin lacks significant direct antibacterial properties due to the absence of free amino groups in its structure [80], it indirectly provides protective action by activating the immune defense mechanism of plants through eliciting activity [97].

This eliciting activity involves the activation of chitinase enzymes, which are chitin-degrading enzymes that break down chitin into oligomeric fragments [69]. Consequently, chitinolytic enzymes are released when a pathogen invades a plant, breaking down chitin into oligomeric fragments. The conversion of chitin to its corresponding oligomers by chitinolytic enzymes identifies them as microbe-associated molecular patterns (MAMPs), triggering various defense responses in plants [69]. These defense responses include generating and activating phytoalexins, phenolics, terpenes, and reactive oxygen species (ROS) [83]. Additionally, physical barriers, such as the deposition of callose and lignin and the formation of tyloses, are developed as part of the defense mechanisms [83]. The enhanced expression of defense-related genes supports the impact of chitin supplementation. For example, in strawberry plants supplemented with chitin, the defense-related genes FaPR1 and FaWRKY1 exhibited increased expression, protecting against fungal infections [15]. Similarly, [62] reported the upregulation of defense-related genes, including mitogen-activated protein kinase (MPK3) and pathogenesis-related gene (PR2), in cabbage and strawberry plants treated with chitin nanofibrils at concentrations ranging from 1 to 10 mg/ml. These findings underscore the beneficial effects of chitin application in reinforcing plant defense mechanisms against potential pathogens.

7 Growth Promoter

Chitin emerges as a promising plant growth promoter, influencing nutrient uptake and augmenting plant growth hormone activity. This biostimulatory effect is induced through the interaction with plant growth-promoting rhizobacteria (PGPR), as highlighted by [80]. PGPRs, a group of soil bacteria associated with host plants, benefit plant growth by improving nutrient availability. Their mechanisms include enlarging root surface areas for enhanced nutrient absorption and converting nitrogen into simpler forms for easier plant uptake [61]. Moreover, [80] underscore the involvement of PGPRs in nitrogen-fixing processes, bacterial cell wall breakdown using chitinase, and activating signaling molecules to regulate growth. Introducing chitin into the soil around crops stimulates the growth of antagonistic microbes that inhibit harmful bacteria, thereby reducing crop stress and enhancing nutrient absorption [61]. For instance, chitin addition boosted the growth of *Bacillus subtilis* AF1, suppressing the growth of pathogens like *Aspergillus niger* responsible for crown rot in peanuts and *fusarium* in pigeon peas. Consequently, groundnut and pigeon pea seed germination and biomass significantly improved despite disease pressure [82]. This aligns with findings from the experiment by [44], where chitin-based PGPR formulations enhanced the yield of pepper and tomato. Similarly, chitin-supplemented talc-based PGPR Pseudomonas fluorescence, FP7, elevated the fruit yield of mango, Mangifera indica L. [18]. Elshamy et al. [21] observed a 29% increase in the emergence of pigeon pea seedlings and a 33% rise in dry weight when supplemented with a chitin-enhanced peat formulation of PGPR, Bacillus AF1. Thus, chitin operates as a biostimulant rather than a fertilizer, enhancing growth dynamics [21]. Numerous studies have investigated the impact of chitin on plant growth, providing diverse insights. In one such study, [46] examined the growth of *Arabidopsis thaliana* supplemented with chitin/cellulose nanofibers complex (CCNFC) extracted from spent mushroom substrate (SMS). Concentrations of 0.001, 0.1, 1, and 10 mg/kg were explored, revealing that CCNFC supplementation at 0.01 and 0.1 mg/kg moderately increased the number of leaves and shoot dry weight of A. thaliana. However, plant growth diminished at higher concentrations (1 and 10 mg/kg). Parada et al. [62] similarly observed suppressed growth of cabbage and strawberry at higher doses (10 mg/kg) compared to the control, while a promotion in growth was evident at 0.1 mg/kg. Moreover, applying chitin led to upregulating specific growth-related genes associated with plant growth. For instance, the Nodule inception-like protein 7 (NLP7) gene, crucial for nitrogen utilization in plant roots, exhibited increased expression in tomato roots treated with ChNF [82]. Similarly, the nitrate transporter gene [Solyc08g078950.2 (LeNRT1.1], homologous to the dual-affinity nitrate transporter AtNRT1.1 (AtNPF6.3, AtCHL1, was upregulated in ChNF-treated roots. These findings suggest that chitin application positively influences the expression of growth-related genes, thereby promoting plant growth and enhancing nutrient utilization (Table 1).

8 Chitin-Based Nanoparticles and Soil Health

In the intricate web of sustainable agriculture, soil health stands as a cornerstone, influencing crop productivity and the overall resilience and sustainability of farming ecosystems. With their multifaceted properties, chitin-based nanoparticles have emerged as potent contributors to soil health. Chitosan-coupled copper nanoparticles (ch-CuNPs) offer numerous benefits as a growth enhancer and fungicide, displaying promising characteristics for replacing traditional pesticides and mitigating their detrimental environmental effects. In heavy metal removal, iron/chitin nanoparticles have undergone successful "in vitro" testing for both antimicrobial applications and environmental uses [26]. Composite beads consisting of carboxylated graphene oxide/chitosan/cellulose have effectively removed Cu^{2+} from both water and soil, concurrently reducing its bioaccumulation in wheat plants. Nanomembranes based on novel chitosan formulations show promise in treating industrial wastewater contaminated with chromium [75], while chitosan-based magnetic microgels have proven effective in cleaning cesium-contaminated clay [70]. Nano-fungal chitosan exhibits proficiency in removing Pb^{2+} and Cu^{2+} from water and soil matrices [6], and microfluidically-generated chitosan microspheres adsorb Cu ions from industrial wastewater [98]. Studies involving columns packed with chitosan aimed at contaminated groundwater demonstrate their capability to remove arsenic compounds [67]. Notably, a composite material comprising chitosan and water hyacinth plants (*Eichhornia crassipes*) displays excellent absorption capacities for Cu, Pb, and Cd ions [94].

Table 1 Chitosan's impact on vegetable crops is the subject of examination, exploring various outcomes in plant development and overall performance

Plant	Effects of chitosan	References
Cynara scolymus L.	The use of chitosan enhanced both the germination process and the growth of artichoke plants, leading to a notable reduction in fungal infections	[109]
Hibiscus esculentus L.	Applying chitosan foliar treatment at 100 or 125 ppm during the early stages of growth can result in increased yields	[59]
Phaseolus vulgaris L.	Chitosan enhanced yields in both traditionally timed and delayed sowings The nanocarriers made of alginate/chitosan (ALG/CS) effectively promoted leaf area and elevated levels of chlorophylls and carotenoids	[36, 66]
Zingiber officinale Roscoe	The storage-related rot of ginger rhizomes was inhibited by the application of chitosan and oligochitosan Enhanced activity of defense enzymes in ginger was observed with the use of chitosan and oligochitosan	[52]
Ipomoea batatas L.	Chitosan decelerated cell growth, triggered cell necrosis, and had a substantial impact on the fatty acid composition of Ceratocystis fimbriata	[101]
Cicer arietinum L.	The application of chitosan nanoparticles loaded with thiamine resulted in an enhanced germination percentage and growth in chickpeas. Additionally, foliar application stimulated the protection of chickpea seedlings against wilt disease and increased the production of indole acetic acid (IAA) in the seedlings	[42]
Capsicum frutescence L.	The combined application of chitosan, used as a seed treatment at 1% and as a foliar spray at 0.5%, demonstrated the utmost effectiveness in managing chili anthracnose. This approach not only controlled the disease but also promoted increased yield and positive changes in yield-contributing characteristics	[3]
Cucumis sativus L.	The application of chitosan as a seed treatment provided complete resistance, reaching 100%, against damping off induced by Phytophthora capsici. Additionally, chitosan has the potential to trigger the synthesis of defense-responsive enzymes and stimulate phytohormones in cucumber plants	[110]

(continued)

Table 1 (continued)

Plant	Effects of chitosan	References
Capsicum annuum L. var. grossum (L.) Sendt	Chitosan nanoparticles (CsNPs) demonstrated a notable role in combating biofilm formation by foodborne pathogens The application of chitosan nano-coating (CsNC) extended the freshness of freshly cut bell pepper, prolonging its shelf life Treatments involving chitosan increased germination rates and enhanced seedling growth and emergence, particularly in cold test conditions The combination of CaCl2, tea tree oil (TTO), and low molecular weight chitosan (LMWCS) effectively slowed down microbial growth in freshly cut bell pepper	[29]
Vicia faba L.	The recently developed nanobiocomposites of carboxymethyl chitosan-titania have the potential to mitigate the adverse effects of Bean Yellow Mosaic Virus (BYMV)	[91]
Solanum lycopersicum L.	The combination of chitosan and chelated copper exhibited greater efficacy in activating enzymes linked to pathogenicity compared to the individual actions of chitosan or copper The application of chitosan in conjunction with compost and arbuscular mycorrhizal fungi significantly enhanced the growth of tomatoes	[20]
Allium cepa L.	Utilizing nano chitosan has the potential to enhance the effectiveness of conventional fertilizers and increase the net return per fed The application of chitosan/polyacrylic acid hydrogel nanoparticles (CS/PAA-HNPs) resulted in boosted yield, improved plant growth, and increased nutrient content in onion bulbs	[25]
Pisum sativum L.	Controlling pea powdery mildew involves the modulation of signaling pathways upregulated by Jasmonic acid and Salicylic acid	[64]
Solanum tuberosum L.	The growth and spore germination of Fusarium oxysporum can be inhibited by chitosan, which also triggers resistance	[72]
Ocimum basilicum L.	Applying chitosan lactate foliar treatment has the potential to enhance the accumulation of bioactive substances, boost antioxidant enzyme activity, and improve both photosynthetic rate and overall plant growth	[27]
Solanum melongena L.	The nanocomposites that were synthesized demonstrated enhancements in both nematocidal activity and the systemic immune response of plants	[7]

Additionally, an unconventional method employing chitosan adsorption combined with duckweed (*Lemna gibba* L.) phytoremediation proves efficient in removing boron from freshwater reservoirs [96]. These integrated approaches underscore the potential usefulness of combined methods in wastewater treatment. Remarkably, mushroom stem waste containing chitin and chitosan can be directly employed for the adsorption of pharmaceutical products such as paracetamol and 17 α-ethynyl estradiol in soil and surface water. This approach reduces costs and minimizes environmental impacts [57].

This section delves into the intricate relationship between chitin-based nanoparticles and soil health. It explores their roles in promoting microbial activity, suppressing soil-borne diseases, and fostering a balanced and resilient soil ecosystem. The practical implications of these interactions are illustrated through a relevant example. Soil health is a complex and dynamic concept encompassing various physical, chemical, and biological attributes. A healthy soil ecosystem supports the growth of plants, fosters nutrient cycling, retains water, and provides a habitat for diverse microbial communities [84]. However, conventional agricultural practices, including synthetic fertilizers and pesticides, can often disrupt this delicate balance, leading to degraded soil health over time. Chitin-based nanoparticles offer a unique avenue to restore and enhance soil health due to their interactions with soil microorganisms and their inherent properties. To illustrate the impact of chitin-based nanoparticles on soil health, let's consider a scenario in the cultivation of strawberries.

Strawberries are susceptible to soil-borne diseases, particularly those caused by pathogenic fungi. The use of conventional fungicides can have detrimental effects on soil health, negatively impacting beneficial microorganisms and disrupting the delicate microbial balance. In a research study focused on strawberry cultivation, chitosan-based nanoparticles were employed as a natural alternative to synthetic fungicides [81]. These nanoparticles, known for their antifungal properties, were applied to the soil to suppress pathogenic fungi growth while promoting soil health. The nanoparticles exhibited antifungal properties, creating an environment that deterred the growth of pathogenic fungi responsible for diseases like *Fusarium* wilt and root rot [61]. This natural disease suppression reduced the need for synthetic fungicides, mitigating their potential negative impacts on soil health. Chitin-based nanoparticles stimulated the activity of beneficial soil microorganisms. Microorganisms such as bacteria and fungi are crucial in nutrient cycling and decomposition of organic matter. By enhancing the population and activity of these beneficial microorganisms, the nanoparticles contributed to improved nutrient availability for the strawberry plants and increased the overall resilience of the soil ecosystem. Furthermore, the chitin-based nanoparticles facilitated the breakdown of organic matter, contributing to the formation of humus—a vital component of healthy soil [76]. Humus improves soil structure, water retention, and nutrient-holding capacity. The increased microbial activity, coupled with the decomposition of organic matter facilitated by the nanoparticles, resulted in forming a fertile and well-structured soil environment conducive to plant growth shown in Fig. 3.

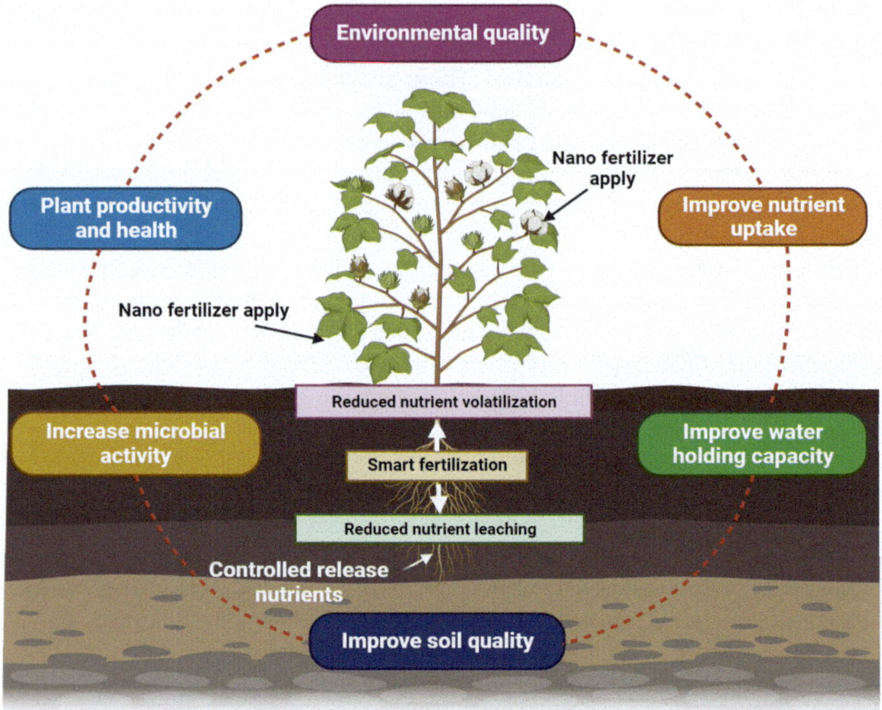

Fig. 3 Fragments derived from chitin and chitosan are pivotal as plant elicitors and growth stimulators

The example of strawberry cultivation underscores the potential of chitin-based nanoparticles to address soil-borne diseases and promote soil health simultaneously. These nanoparticles act as agents of change, fostering a symbiotic relationship between plants and the soil microbiome, ultimately contributing to sustainable and resilient agricultural ecosystems. The positive influence of chitin-based nanoparticles on soil health can be attributed to their interactions with soil microorganisms [37]. Chitin, a major component of the nanoparticles, serves as a carbon source for certain beneficial soil bacteria and fungi. As these microorganisms metabolize chitin, they release by-products that contribute to soil fertility and microbial diversity. Moreover, chitin-based nanoparticles act as a substrate for chitinase enzymes produced by soil microorganisms. Chitinase enzymes play a crucial role in breaking down chitin, releasing nitrogen and other nutrients in forms that plants can readily utilize. This enzymatic activity contributes to nutrient cycling, promoting a sustainable flow of nutrients within the soil ecosystem. The chitin-based nanoparticles have also enhanced mycorrhizal associations in the rhizosphere. Mycorrhizal fungi form symbiotic relationships with plant roots, aiding in nutrient uptake and improving plant resistance to environmental stress, as shown in Table 2 [47]. The presence of chitin-based nanoparticles promotes the colonization of mycorrhizal fungi, further enhancing nutrient acquisition by plants and fortifying their overall health. While the positive impact of chitin-based nanoparticles on soil health is evident, challenges and considerations persist. The dosage and frequency of nanoparticle application must be carefully calibrated to avoid unintended consequences, and the long-term effects of nanoparticles on soil health require continued research [63]. Additionally, the scalability of nanoparticle production, cost considerations, and the need for standardized application protocols must be addressed for practical implementation on a larger scale. Regulatory aspects also warrant attention to ensure chitin-based nanoparticles' safe and responsible use in agriculture.

9 Challenges and Future Prospects

While integrating chitin-based nanoparticles into agricultural practices holds great promise, several challenges must be addressed to ensure their effective and sustainable implementation. Simultaneously, recognizing potential prospects and avenues for further research and development is crucial. This section explores the current challenges and envisions the future prospects of chitin-based nanoparticles in agriculture. Chitin-based nanoparticles have been gaining attention in agriculture for crop nutrition and fertilization [78]. These nanoparticles are derived from chitin, a natural polymer found in the exoskeletons of crustaceans, insects, and certain fungi. The utilization of chitin-based nanoparticles in agriculture poses both challenges and promising future prospects:

9.1 Challenges

9.1.1 Production Cost and Scalability

One of the primary challenges is the cost of producing chitin-based nanoparticles. The extraction and processing methods can be expensive, making it essential to develop cost-effective and scalable production techniques [48].

9.1.2 Biocompatibility and Environmental Impact

Ensuring the biocompatibility of chitin-based nanoparticles is crucial to avoid adverse effects on soil microorganisms, plants, and the surrounding ecosystem. Assessing the long-term environmental impact is also a concern [108].

Table 2 Utilizing chitin- and chitosan-derivatives for plant protection under various stress conditions

Plant species	Stress condition	Traits of defensive molecules and administration procedures	Protective influence	References
Capsicum annuum L.	Biotic stress	1% chitosan applied foliarly	Protection against Phytophthora capsici	[22]
Oryza sativa L.	Biotic stress	Spraying seedlings with 0.3% chitosan oligosaccharide	Protection against Fusarium oxysporum	[54]
Solanum tuberosum L.	Biotic stress	Immersion of tubers in 0.4% chitosan	Protection against Fusarium spp.	[56]
Stone fruit trees	Biotic stress	Foliar application of 0.001% chitosan-Ag nanoparticles	Protection against Pseudomonas syringae	[79]
Sesamum indicum L.	Abiotic stress	Foliar application of 0.00048 and 0.00064% chitosan	Drought stress tolerance	[45]
Zea mays L. cv. White Pearl	Abiotic stress	Addition of 0, 2, and 4 g of chitin to 1 kg of soil	Drought stress tolerance	[50]
Triticum aestivum L., Zea mays L	Abiotic stress	Seed coating with 25, 50, and 75% chitosan	Salt stress tolerance	[68]
Solanum lycopersicum Mill	Abiotic stress	Bacteria Encouraging Growth Aggregated with Chitosan	Salt stress tolerance	[9]
Zea mays L.	Abiotic stress	Soaking seedlings in 0.01% chitosan	Cadmium stress tolerance	[71]
Solanum melongena L.	Abiotic stress	Foliar application of 0.0125, 0.0150, and 0.02% chitosan	Heat stress tolerance	[49]
Capsicum annuum L.	Abiotic stress	Spraying plants with 0.00125, 0.00250, and 0.00375% Chitosan	Heat stress tolerance	[5]
Phaseolus vulgaris cv. Contender	Abiotic stress	Seed priming and foliar application of chitosan nanoparticles loaded with 10% NPK fertilizers	Poor soil tolerance	[1]
Fragaria x ananassa Duch. cv. Elsanta	Abiotic stress	Foliar application of 0.001% chitosan	Poor fertilization tolerance	[92]
Solanum lycopersicum	Abiotic stress	Application of 1 mg per plant of chitosan in the transplant cavity soil	Poor fertilization tolerance	[20]

9.1.3 Stability and Delivery

Maintaining the stability of chitin-based nanoparticles in various soil conditions and ensuring effective delivery to plant roots pose challenges. Developing formulations that enhance stability and enable controlled release is essential.

9.1.4 Regulatory Approval

Obtaining regulatory approval for using chitin-based nanoparticles in agriculture may be a lengthy process. Compliance with safety and efficacy standards is necessary to gain acceptance in the market.

9.2 Future Prospects

9.2.1 Enhanced Nutrient Uptake

Chitin-based nanoparticles have the potential to improve nutrient uptake by plants. Future research can focus on optimizing formulations to enhance the efficiency of nutrient delivery, leading to increased crop yields [23].

9.2.2 Biostimulant Properties

Chitin-based nanoparticles may act as biostimulants, promoting plant growth and stress resistance. Further exploration of these properties can open new avenues for their application in sustainable agriculture [61].

9.2.3 Disease Resistance

Chitin is known to induce resistance against pathogens in plants. Research can delve into developing chitin-based nanoparticles that effectively protect against diseases, reducing the need for chemical pesticides [80].

9.2.4 Customized Formulations

Tailoring chitin-based nanoparticles for specific crops and soil types can optimize their effectiveness. Future prospects involve developing customized formulations that cater to the unique nutritional needs of different plants.

9.2.5 Integration with Smart Agriculture

Incorporating chitin-based nanoparticles into precision agriculture practices can enhance their efficacy [87]. This includes integrating them with sensor technologies and data analytics to provide real-time insights into crop health and nutrient requirements.

9.2.6 Collaborative Research and Innovation

Collaboration between researchers, agricultural experts, and industry stakeholders is crucial for advancing the field. Joint efforts can lead to innovative solutions, address challenges, and accelerate the adoption of chitin-based nanoparticles in agriculture.

10 Conclusion

Chitin-based nanoparticles represent a cutting-edge solution for addressing the challenges of crop nutrition and fertilization in modern agriculture. Their unique properties enable efficient nutrient delivery, controlled release, and improved fertilizer efficiency, contributing to increased crop yields and sustainable farming practices. The role of chitin-based nanoparticles in crop nutrition and fertilization presents both challenges and promising prospects for the future of agriculture. These nanoparticles derived from chitin, a natural polymer, hold immense potential to enhance nutrient uptake, promote plant growth, and reduce reliance on chemical fertilizers and pesticides. However, their widespread adoption faces hurdles such as production costs, ensuring environmental safety, stability in different soil conditions, and regulatory approvals. Overcoming these challenges requires concerted efforts in research, innovation, and collaboration among scientists, agricultural experts, and industry stakeholders. The future prospects for chitin-based nanoparticles in agriculture are encouraging. These prospects include tailored formulations for specific crops, their integration into precision agriculture systems, potential biostimulant and disease-resistant properties, and their role in sustainable farming practices. Continued research, technological advancements, and a deeper understanding of chitin-based nanoparticle interactions with plants and the environment will pave the way for their effective utilization. Chitin-based nanoparticles can revolutionize crop nutrition and fertilization practices by addressing challenges and leveraging their potential benefits, contributing to more sustainable, efficient, and environmentally friendly agricultural systems.

References

1. Abdel-Aziz HMM, Hasaneen MNA, Omer AM (2019) Impact of engineered nanomaterials either alone or loaded with NPK on growth and productivity of French bean plants: seed priming vs foliar application. S Afr J Bot 125:102–108
2. Aklog YF, Egusa M, Kaminaka H, Izawa H, Morimoto M, Saimoto H, Ifuku S (2016) Protein/CaCO3/chitin nanofiber complex prepared from crab shells by simple mechanical treatment and its effect on plant growth. Int J Mol Sci 17(10):1600. https://doi.org/10.3390/ijms17101600
3. Akter J, Jannat R, Hossai MM, Ahmed JU, Rubayet MT (2018) Chitosan for plant growth promotion and disease suppression against anthracnose in chilli. Int J Environ Agric Biotechnol 3:806–817
4. Alamdari S, Mirzaee O, Jahroodi FN, Tafreshi MJ, Ghamsari MS, Shik SS, Ara MHM, Lee KY, Park HH (2022) Green synthesis of multifunctional ZnO/chitosan nanocomposite film using wild Mentha pulegium extract for packaging applications. Surf Interfaces 34:102349
5. Al-Hassan FAM, Majid BH (2019) Effect of arginine, chitosan and agryl mulching on the growth and yield of pepper plant under the conditions of unheated greenhouses. Plant Arch. 19:256–262
6. Alsharari SF, Tayel AA, Moussa SH (2018) Soil emendation with nano-fungal chitosan for heavy metals biosorption. Int J Biol Macromol 118:2265–2268
7. Attia MS, El-Sayyad GS, Abd Elkodous M, Khalil WF, Nofel MM, Abdelaziz AM, Farghali AA, El-Batal AI, El Rouby WMA (2021) Chitosan and EDTA conjugated graphene oxide antinematodes in eggplant: toward improving plant immune response. Int J Biol Macromol 179:333–344
8. Chakravarty J, Edwards TA (2022) Innovation from waste with biomass-derived chitin and chitosan as green and sustainable polymer: a review. Energy Nexus 8:100149
9. Chanratana M, Joe MM, Roy Choudhury A, Anandham R, Krishnamoorthy R, Kim K, Jeon S, Choi J, Choi J, Sa T (2019) Physiological response of tomato plant to chitosan-immobilized aggregated Methylobacterium oryzae CBMB20 inoculation under salinity stress. 3 Biotech 9:397
10. Cheng Y, Wang Y, Han Y, Li D, Zhang Z, Zhu X, Wang H (2019) The stimulatory effects of nanochitin whisker on carbon and nitrogen metabolism and on the enhancement of grain yield and crude protein of winter wheat. Molecules (Basel, Switzerland) 24(9):1752. https://doi.org/10.3390/molecules24091752
11. Choudhary RC, Kumaraswamy RV, Kumari S, Sharma SS, Pal A, Raliya R, Biswas P, Saharan V (2019) Zinc encapsulated chitosan nanoparticle to promote maize crop yield. Int J Biol Macromol 127:126–135
12. Chun SC, Chandrasekaran M (2019) Chitosan and chitosan nanoparticles induced expression of pathogenesis-related proteins genes enhances biotic stress tolerance in tomato. Int J Biol Macromol 125:948–954
13. Dahiya N, Tewari R, Hoondal GS (2006) Biotechnological aspects of chitinolytic enzymes: a review. Appl Microbiol Biotechnol 71(6):773–782. https://doi.org/10.1007/s00253-005-0183-7
14. Das S, Chaudhari AK, Singh VK, Dwivedy AK, Dubey NK (2023) Chitosan based encapsulation of Valeriana officinalis essential oil as edible coating for inhibiting of fungi and aflatoxin B1 contamination, nutritional quality improvement, and shelf-life extension of Citrus sinensis fruits. Int J Biol Macromol 233:123565
15. De Tender C, Vandecasteele B, Verstraeten B, Ommeslag S, De Meyer T, De Visscher J, Debode J (2021) Chitin in strawberry cultivation: foliar growth and defense response promotion, but reduced fruit yield and disease resistance by nutrient imbalances. Mol Plant-Microbe Interact 34(3):227–239. https://doi.org/10.1094/MPMI-08-20-0223-R
16. Divya K, Jisha MS, Vijayan S, Nair SJ (2019) Optimization of chitosan nanoparticle synthesis and its potential application as germination elicitor of Oryza sativa L. Int J Biol Macromol 124:1053–1059

17. Do NHN, Truong QT, Le PK, Ha AC (2022) Recent developments in chitosan hydrogels carrying natural bioactive compounds. Carbohydr Polym 294:119726
18. Donga S, Chanda S (2021) Facile green synthesis of silver nanoparticles using Mangifera indica seed aqueous extract and its antimicrobial, antioxidant and cytotoxic potential (3-in-1 system). Artif Cells Nanomed Biotechnol 49(1):292–302
19. Du M, Yi Y, Yin Y, Cai Z, Cai W, Li J, He G, Zhang J (2023) Bacterial-triggered photodynamic nano-system based on hematoporphyrin-modified chitosan for sustainable plant disease control. Eur Polym J 191:112035
20. El Amerany F, Rhazi M, Wahbi C, Taourirte M, Meddich A (2020) The effect of chitosan, arbuscular mycorrhizal fungi, and compost applied individually or in combination on growth, nutrient uptake, and stem anatomy of tomato. Sci Hortic 261:109015
21. Elshamy MT, Husseiny SM, Farroh KY (2019) Application of nano-chitosan NPK fertilizer on growth and productivity of potato plant. J. Sci. Res. Sci. 36:424–441. https://doi.org/10.21608/JSRS.2019.58522
22. Esyanti RR, Dwivany FM, Mahani S, Nugrahapraja H, Meitha K (2019) Foliar application of chitosan enhances growth and modulates expression of defense genes in chilli pepper (Capsicum annuum L.). Aust J Crop Sci 13:55–60
23. Fan Z, Wang L, Qin Y, Li P (2023) Activity of chitin/chitosan/chitosan oligosaccharide against plant pathogenic nematodes and potential modes of application in agriculture: a review. Carbohydr Polym 306:120592
24. Fellet G, Pilotto L, Marchiol L, Braidot E (2021) Tools for nano-enabled agriculture: fertilizers based on calcium phosphate, silicon, and chitosan nanostructures. Agronomy 11(6):1239. https://doi.org/10.3390/agronomy11061239
25. Geries L, Omnia HSM, Marey RA (2020) Soaking and foliar application with chitosan and nano chitosan to enhancing growth, productivity and quality of onion crop. Plant Arch 20:3584–3591
26. Gomaa EZ (2018) Iron nanoparticles α-chitin nanocomposite for enhanced antimicrobial, dyes degradation and heavy metals removal activities. J Polym Environ 26:3638–3654
27. Hawrylak-Nowak B, Dresler S, Rubinowska K, Matraszek-Gawron R (2021) Eliciting effect of foliar application of chitosan lactate on the phytochemical properties of Ocimum basilicum L. and Melissa officinalis L. Food Chem 342:128358
28. Hossin MA, Al Shaqsi NHK, Al Jouby SSJA, Al Sibani MA (2021) A review of polymeric chitin extraction, characterization, and applications. Arab J Geosci 14:1870
29. Hu X, Saravanakumar K, Sathiyaseelan A, Wang MH (2020) Chitosan nanoparticles as edible surface coating agent to preserve the fresh-cut bell pepper (Capsicum annuum L. var. grossum (L.) Sendt). Int J Biol Macromol 165:948–957
30. Husen A (2022) Engineered nanomaterials for sustainable agricultural production, soil improvement and stress management. Elsevier Inc., Cambridge. https://doi.org/10.1016/C2021-0-00054-7
31. Husen A (2023) Nanomaterials and nanocomposites exposures to plants (response interaction, phytotoxicity and defense mechanisms). Springer Nature Singapore Pte Ltd., Singapore. https://doi.org/10.1007/978-981-99-2419-6
32. Husen A, Iqbal M (2019) Nanomaterials and plant potential. Springer International Publishing, Cham. https://doi.org/10.1007/978-3-030-05569-1
33. Husen A, Jawaid M (2020) Nanomaterials for agriculture and forestry applications. Elsevier Inc., Cambridge. https://doi.org/10.1016/C2018-0-02349-X
34. Husen A, Siddiqi KS (2014) Phytosynthesis of nanoparticles: concept, controversy and application. Nanosc Res Lett 9(229):1–24. https://doi.org/10.1186/1556-276X-9-229
35. Husen A, Siddiqi KS (2014) Carbon and fullerene nanomaterials in plant system. J Nanobiotechnol 12(16):1–10
36. Ibrahim EA, Ramadan WA (2015) Effect of zinc foliar spray alone and combined with humic acid or/and chitosan on growth, nutrient elements content and yield of dry bean (Phaseolus vulgaris L.) plants sown at different dates. Sci Hortic 184:101–105

37. Ikram R, Jan BM, Qadir MA, Sidek A, Stylianakis MM, Kenanakos G (2021) Recent advances in chitin and chitosan/graphene based bio-nanocomposites for energetic applications. Polymers 13:3266
38. Jawad AH, Mubarak NSA, Abdulhameed AS (2020) Tunable Schiff's basecross-linked chitosan composite for the removal of reactive red 120 dye: adsorption and mechanism study. Int J Biol Macromol 142:732–741. https://doi.org/10.1016/j.ijbiomac.2019.10.014
39. Jin T, Liu T, Lam E, Moores A (2021) Chitin and chitosan on the nanoscale. Nanosc Horiz 6:505–542
40. Joseph SM, Krishnamoorthy S, Paranthaman R, Moses JA, Anandharamakrishnan C (2021) A review on source-specific chemistry, functionality, and applications of chitin and chitosan. Carbohydr Polym Technol Appl 2:100036
41. Kabiriyel J, Jeyanthi R, Jayakumar K, Amalraj A, Arjun P, Shanmugarathinam S, Vignesh G, Raja Mohan C (2023) Green synthesis of carboxyl methyl chitosan-based curcumin nanoparticles and its biological activity: influence of size and conductivity. Carbohydr Polym. Technol Appl 5:100260
42. Kaur P, Duhan JS, Thakur R (2018) Comparative pot studies of chitosan and chitosan-metal nanocomposites as nano-agrochemicals against fusarium wilt of chickpea (Cicer arietinum L.). Biocatal Agric Biotechnol 14:466–471
43. Kertmen A, Ehrlich H (2022) Patentology of chitinous biomaterials. Part I: chitin. Carbohydr Polym 282:119102
44. Khati P, Bhatt P, Kumar R, Sharma A (2018) Effect of nanozeolite and plant growth promoting rhizobacteria on maize. 3 Biotech 8:141. https://doi.org/10.1007/s13205-018-1142-1
45. Khordadi Varamin J, Fanoodi F, Sinaki JM, Rezvan S, Damavandi A (2018) Physiological response of sesame (Sesamum indicum L.) to application of chitosan and magnesium-nano fertilizers under irrigation cut-off in a sustainable agriculture system. Iran J Plant Physiol 9:2629–2639
46. Li J, Liu R, Zhang C, Yang J, Lyu L, Shi Z, Man YB, Wu F (2022) Selenium uptake and accumulation in winter wheat as affected by level of phosphate application and arbuscular mycorrhizal fungi. J Hazard Mater 433:128762
47. Li R, He J, Xie H, Wang W, Bose SK, Sun Y, Hu J, Yin H (2019) Effects of chitosan nanoparticles on seed germination and seedling growth of wheat (Triticum aestivum L.). Int J Biol Macromol 126:91–100
48. Liao J, Wang Y, Hou B, Zhang J, Huang H (2023) Nano-chitin reinforced agarose hydrogels: effects of nano-chitin addition and acidic gas-phase coagulation. Carbohydr Polym 313:120902
49. Liaqat A, Ihsan MZ, Rizwan MS, Mehmood A, Ijaz M, Alam M, Abdullah M, Wajid M, Hussain R, Naeem M (2019) Inducing effect of chitosan on the physiological and biochemical indices of eggplant (Solanum melongena L.) genotypes under heat and high irradiance. Appl Ecol Environ Res 17:11273–11287
50. Lin KH, Lin FW, Wu CW, Chang YS (2019) Biostimulation of maize (Zea mays) and irrigation management improved crop growth and water use under controlled environment. Agronomy 9:559
51. Liu Y, Liu M, Yang S, Luo B, Zhou C (2018) Liquid crystalline behaviors of chitin nanocrystals and their reinforcing effect on natural rubber. ACS Sustain Chem & Eng 6(1):325–336. https://doi.org/10.1021/acssuschemeng.7b02586
52. Liu Y, Wisniewski M, Kennedy JF, Jiang Y, Tang J, Liu J (2016) Chitosan and oligochitosan enhance ginger (Zingiber officinale Roscoe) resistance to rhizome rot caused by Fusarium oxysporum in storage. Carbohydr Polym 151:474–479
53. Lv J, Lv X, Ma M, Oh DH, Jiang Z, Fu X (2023) Chitin and chitin-based biomaterials: a review of advances in processing and food applications. Carbohydr Polym 299:120142
54. Ma B, Wan J, Liu C, Hu J, Tan K, Zhao F, Yuan M, Zhang J, Gai Z (2019) Preventive effects of fluoro-substituted benzothiadiazole derivatives and chitosan oligosaccharide against the rice seedling blight induced by Fusarium oxysporum. Plants 8:538

55. Malerba M, Cerana R (2019) Recent applications of chitin-and chitosan-based polymers in plants. Polymers 11(5):839
56. Mejdoub-Trabelsi B, Touihri S, Ammar N, Riahi A, Daami-Remadi M (2020) Effect of chitosan for the control of potato diseases caused by Fusarium species. J Phytopathol 168:18–27
57. Menk JJ, do Nascimento AIS, Leite FG, de Oliveira RA, Jozala AF, de Oliveira Junior JM, Chaud MV, Grotto D (2019) Biosorption of pharmaceutical products by mushroom stem waste. Chemosphere 237:124515
58. Mohan K, Ganesan AR, Ezhilarasi PN, Kondamareddy KK, Rajan DK, Sathishkumar P, Rajarajeswaran J, Conterno L (2022) Green and eco-friendly approaches for the extraction of chitin and chitosan: a review. Carbohydr Polym 287:119349
59. Mondal MMA, Malek MA, Puteh AB, Ismail MR, Ashrafuzzaman M, Naher L (2012) Effect of foliar application of chitosan on growth and yield in okra. Aust J Crop Sci 6:918–921
60. Mujtaba M, Ali Q, Yilmaz BA, Kurubas MS, Ustun H, Erkan M, Kaya M, Cicek M, Oner ET (2023) Understanding the effects of chitosan, chia mucilage, levan based composite coatings on the shelf life of sweet cherry. Food Chem 416:135816
61. Olenska E, Małek W, Wojcik' M, Swiecicka I, Thijs S, Vangronsveld J (2020) Beneficial features of plant growth-promoting rhizobacteria for improving plant growth and health in challenging conditions: a methodical review. Sci Total Environ 743, Article 140682. https://doi.org/10.1016/j.scitotenv.2020.140682
62. Parada RY, Egusa M, Aklog YF, Miura C, Ifuku S, Kaminaka H (2018) Optimization of nanofibrillation degree of chitin for induction of plant disease resistance: elicitor activity and systemic resistance induced by chitin nanofiber in cabbage and strawberry. Int J Biol Macromol 118:2185–2192. https://doi.org/10.1016/j.ijbiomac.2018.07.089
63. Parajuli S, Hasan MJ, Ureña-Benavides EE (2022) Effect of the Interactions between oppositely charged cellulose nanocrystals (CNCs) and chitin nanocrystals (ChNCs) on the enhanced stability of soybean oil-in-water emulsions. Materials 15(19):6673
64. Patel JS, Selvaraj V, Gunupuru LR, Rathor PK, Prithiviraj B (2020) Combined application of Ascophyllum nodosum extract and chitosan synergistically activates host-defense of peas against powdery mildew. BMC Plant Biol 20:1–10
65. Pereira Ad ES, Oliveira HC, Fraceto LF (2019) Polymeric nanoparticles as an alternative for application of gibberellic acid in sustainable agriculture: a field study. Sci Rep 9(1):1–10. https://doi.org/10.1038/s41598-019-43494-y
66. Pereira AES, Silva PM, Oliveira JL, Oliveira HC, Fraceto LF (2017) Chitosan nanoparticles as carrier systems for the plant growth hormone gibberellic acid. Colloids Surf B Biointerfaces 150:141–152
67. Pérez Mora BE, Bellú SE, Mangiameli MF, García SI, González JC (2019) Optimization of continuous arsenic biosorption present in natural contaminated groundwater. J Chem Technol Biotechnol 94:547–555
68. Peykani LS, Sepehr MF (2018) Effect of chitosan on antioxidant enzyme activity, proline, and malondialdehyde content in Triticum aestivum L. and Zea maize L. under salt stress condition. Iran J Plant Physiol 9:2661–2670
69. Pusztahelyi T (2018) Chitin and chitin-related compounds in plant–fungal interactions. Mycology 9(3):189–201. https://doi.org/10.1080/21501203.2018.1473299
70. Qian J, Zhou L, Yang X, Hua D, Wu N (2020) Prussian blue analogue functionalized magnetic microgels with ionized chitosan for the cleaning of cesium-contaminated clay. J Hazard Mater 386:121965
71. Qu DY, Gu WR, Zhang LG, Li CF, Chen XC, Li J, Li LJ, Xie TL, Wei S (2019) Role of chitosan in the regulation of the growth, antioxidant system and photosynthetic characteristics of maize seedlings under cadmium stress. Russ J Plant Physiol 66:140–151
72. Ren J, Tong J, Li P, Huang X, Dong P, Ren M (2021) Chitosan is an effective inhibitor against potato dry rot caused by Fusarium oxysporum. Physiol Mol Plant Pathol 113:101601
73. Rodsamran P, Sothornvit R (2018) Bioactive coconut protein concentrate films incorporated with antioxidant extract of mature coconut water. Food Hydrocoll 79:243–252. https://doi.org/10.1016/j.foodhyd.2017.12.037

74. Sampath L, Ngasotter S, Layana P, Balange AK, Nayak BB, Xavier KM (2022) Effect of chemical treatment duration on physicochemical, rheological, and functional properties of colloidal chitin. Food Hydrocoll Health 2, Article 100091. https://doi.org/10.1016/j.fhfh.2022.100091
75. Sangeetha K, Vinodhini AP, Sudha PN, Faleh AA, Sukumaran A (2019) Novel chitosan based thin sheet nanofiltration membrane for rejection of heavy metal chromium. Int J Biol Macromol 132:939–953
76. Sathiyabama M, Manikandan A (2021) Foliar application of chitosan nanoparticle improves yield, mineral content and boost innate immunity in finger millet plants. Carbohydr Polym 258, Article 117691. https://doi.org/10.1016/j.carbpol.2021.117691
77. Shahrajabian MH, Petropoulos SA, Sun W (2023) Survey of the influences of microbial biostimulants on horticultural crops: case studies and successful paradigms. Horticulturae 9:193
78. Shahrajabian MH, Sun W, Cheng Q (2022) Foliar application of nutrients on medicinal and aromatic plants, the sustainable approaches for higher and better production. Beni-Suef Univ J Basic Appl Sci 11:26
79. Shahryari F, Rabiei Z, Sadighian S (2020) Antibacterial activity of synthesized silver nanoparticles by sumac aqueous extract and silver-chitosan nanocomposite against pseudomonas syringae pv. syringae. J Plant Pathol
80. Shamshina JL, Rogers RD (2019) Applications of chitin in agriculture. In: Sustainable agriculture reviews, vol 36. Springer, Cham, pp 125–146. https://doi.org/10.1007/978-3-030-16581-9_4
81. Shamshina JL, Kelly A, Oldham T, Rogers RD (2020) Agricultural uses of chitin polymers. Environ Chem Lett 18(1):53–60. https://doi.org/10.1007/s10311-019-00934-5
82. Shang H, Ma C, Li C, Zhao J, Elmer W, White JC (2021) Copper oxide nanoparticle-embedded hydrogels enhance nutrient supply and growth of lettuce (Lactuca sativa) infected with Fusarium oxysporum f. sp. lactucae. Environ Sci Technol 55:13432–13442. https://doi.org/10.1021/acs.est.1c00777
83. Sharp RG (2013) A review of the applications of chitin and its derivatives in agriculture to modify plant-microbial interactions and improve crop yields. Agronomy 3(4):757–793. https://doi.org/10.3390/agronomy3040757
84. Sheikhalipour M, Mohammadi SA, Esmaielpour B, Spanos A, Mahmoudi R, Mahdavinia GR, Milani MH, Kahnamoei A, Nouraein M, Antoniou C (2023) Seedling nanopriming with selenium-chitosan nanoparticles mitigates the adverse effects of salt stress by inducing multiple defence pathways in bitter melon plants. Int J Biol Macromol 242:124923
85. Siddiqi KS, Husen A (2022) Plant response to silver nanoparticles: a critical review. Crit Rev Biotechnol 42(7):773–990. https://doi.org/10.1080/07388551.2021.1975091
86. Sikder A, Pearce AK, Parkinson S, Napier RM, O'Reilly RK (2021) Recent trends in advanced polymer materials in agriculture related applications. ACS Appl Polym Mater 3:1203–1217. https://doi.org/10.1021/acsapm.0c00982
87. Singh A, Dhiman N, Kar AK, Singh D, Purohit MP, Ghosh D (2020) Advances in controlled release pesticide formulations: prospects to safer integrated pest management and sustainable agriculture. J Hazard Mat 385:121525. https://doi.org/10.1016/j.jhazmat.2019.121525
88. Singh SP, Endley N (2020) Fabrication of nano-silica from agricultural residue and their application. In: Husen A, Jawaid M (eds) Nanomaterials for agriculture and forestry applications. Elsevier, Amsterdam, pp 107–134
89. Sinha T, Bhagwatwar P, Krishnamoorthy C, Chidambaram R (2019) Polymer based micro- and nanoencapsulation of agrochemicals. In: Polymers for agri-food applications. Springer, Cham, pp 5–28. https://doi.org/10.1007/978-3-030-19416-1_2
90. Sivakami MS, Gomathi T, Venkatesan J, Jeong HS, Kim SK, Sudha PN (2013) Preparation and characterization of nano chitosan for treatment wastewaters. Int J Biol Macromol 57:204–212. https://doi.org/10.1016/j.ijbiomac.2013.03.005
91. Sofy AR, Hmed AA, Alnaggar AEAM, Dawoud RA, Elshaarawy RFM, Sofy MR (2020) Mitigating effects of Bean yellow mosaic virus infection in faba bean using new carboxymethyl chitosan-titania nanobiocomposites. Int J Biol Macromol 163:1261–1275

92. Soppelsa S, Kelderer M, Casera C, Bassi M, Robatscher P, Matteazzi A, Andreotti C (2019) Foliar applications of biostimulants promote growth, yield and fruit quality of strawberry plants grown under nutrient limitation. Agronomy 9:483
93. Suwanchaikasen P, Nie S, Idnurm A, Selby-Pham J, Walker R, Boughton BA (2023) Effects of chitin and chitosan on root growth, biochemical defense response and exudate proteome of Cannabis sativa. Plant Environ Interact 4:115–133
94. Taha AA, Hameed NJ, Rashid FH (2019) Preparation and characterization of (hyacinth plant/chitosan) composite as a heavy metal removal. Baghdad Sci J 16:865–870
95. Tang Y, Zhan T, Fan, G, Huang J, Zhou Y, Yuan X, Hu C (2022) Selenium combined with chitin reduced phosphorus leaching in soil with pomelo by driving soil phosphorus cycle via microbial community. J Environ Chem Eng 10(1), Article 107060. https://doi.org/10.1016/j.jece.2021.107060
96. Türker OC, Baran TA (2019) Combination method based on chitosan adsorption and duckweed (Lemna gibba L.) phytoremediation for boron (B) removal from drinking water. Int J Phytoremed 20:175–183
97. Vel'asquez CL, Pirela MR (2016) Biochemical aspects of the chitin fungicidal activity in agricultural uses. Chitosan in the Preservation of Agricultural Commodities, pp 279–298. https://doi.org/10.1016/B978-0-12-802735-6.00010-0
98. Wang B, Bai Z, Jiang H, Prinsen P, Luque R, Zhao S, Xuan J (2019) Selective heavy metal removal and water purification by microfluidically generated-chitosan microspheres: characteristics, modeling and application. J Hazard Mater 364:192–205
99. Wang SL, Chang WT (1997) Purification and characterization of two bifunctional chitinases/lysozymes extracellularly produced by pseudomonas aeruginosa K-187 in a shrimp and crab shell powder medium. Appl Environ Microbiol 63(2):380–386. https://doi.org/10.1128/aem.63.2.380-386.1997
100. Wang Y, Yang L, Zhou X, Wang Y, Liang Y, Luo B, Dai Y, Wei Z, Li S, He R (2023) Molecular mechanism of plant elicitor daphnetin-carboxymethyl chitosan nanoparticles against Ralstonia solanacearum by activating plant system resistance. Int J Biol Macromol 241:124580
101. Xing K, Li TJ, Liu YF, Zhang J, Zhang Y, Shen XQ, Li XY, Miao XM, Feng ZZ, Pen X (2018) Antifungal and eliciting properties of chitosan against Ceratocystis fimbriata in sweet potato. Food Chem 268:188–195
102. Xue W, Han Y, Tan J, Wang Y, Wang G, Wang H (2017) Effects of nanochitin on the enhancement of the grain yield and quality of winter wheat. J Agric Food Chem 66(26):6637–6645. https://doi.org/10.1021/acs.jafc.7b00641
103. Yihun FA (2022) Nanochitin preparation and its application in polymer nanocomposites: a review. Emergent Mater 5:1–30. https://doi.org/10.1007/s42247-022-00427-2
104. Yin W, Yan R, Zhou X, Li X, Sang S, McClements DJ, Chen L, Long J, Jiao A, Wang J (2023) Preparation of robust, water-resistant, antibacterial, and antioxidant chitosan-based films by incorporation of cinnamaldehyde-tannin acid-zinc acetate nanoparticles. Food Chem 419:136004
105. Zeljkovic S, Paradikovic N, Vinkovic T, Tkalec M, Maksimovic I, Haramija J (2013) Nutrient status, growth and proline concentration of French marigold (Tagetes patula L.) as affected by biostimulant treatment. J Food Agric Environ 11:232
106. Zhan T, Hu C, Kong Q, Shi G, Tang Y, Zhou Y, Zhao X (2021) Chitin combined with selenium reduced nitrogen loss in soil and improved nitrogen uptake efficiency in Guanxi pomelo orchard. Sci Total Environ 799, Article 149414. https://doi.org/10.1016/j.scitotenv.2021.149414
107. Zhang Y, Zhao M, Cheng Q, Wang C, Li H, Han X, Fan Z, Su G, Pan D, Li Z (2021) Research progress of adsorption and removal of heavy metals by chitosan and its derivatives: a review. Chemosphere 279, Article 130927. https://doi.org/10.1016/j.chemosphere.2021.130927
108. Zhao L, Guan X, Baowei Y, Di N, Liu X, Ma Q, Yang S, Yilihamu A, Yang ST (2019) Carboxylated graphene oxide-chitosan spheres immobilize Cu2+ in soil and reduce its bioaccumulation in wheat plants. Environ Int 133:105208

109. Ziani K, Ursúa B, Maté JI (2010) Application of bioactive coatings based on chitosan for artichoke seed protection. Crop Prot 29:853–859
110. Zohara F, Surovy MZ, Khatun A, Prince MFRK, Akanda MAM, Rahman M, Islam MT (2019) Chitosan biostimulant controls infection of cucumber by Phytophthora capsici through suppression of asexual reproduction of the pathogen. Acta Agrobot 72

Chitin-Based Nanoparticles for Crop Plant Growth Promotion

Om Prakash, Babli Kumari, Priti Pal, Maharudra Pratap Singh, Anand Prakash, Brijesh Pandey, Vishnu Agarwal, and Akhilesh Kumar Singh

Abstract The utilization of nanotechnology in agriculture has emerged as a promising avenue for enhancing plant growth and productivity. Chitin, a naturally occurring polysaccharide abundant in the exoskeletons of crustaceans and insects, has gained attention as a sustainable and eco-friendly source for nanoparticle synthesis. This chapter explores the role of chitin-based nanoparticles (CNPs) in promoting plant growth and improving crop yields. CNPs exhibit unique physicochemical properties, including high surface area, biocompatibility, and the ability to encapsulate various bioactive compounds. These characteristics make CNPs ideal candidates for novel agricultural applications. The interaction between CNPs and plants involves complex mechanisms, such as improved nutrient uptake, enhanced stress tolerance, and modulation of plant–microbe interactions. Studies have demonstrated that CNPs can serve as efficient carriers for essential nutrients and growth-promoting substances, delivering them to plant roots in a controlled manner. Furthermore, CNPs have been reported to stimulate plant defence mechanisms, leading to increased resistance against pathogens and environmental stressors. The biodegradability of chitin ensures minimal environmental impact, aligning with the growing demand for sustainable agricultural practices. This chapter reviews recent research on the application of chitin-based nanoparticles in agriculture, emphasizing their potential to revolutionize plant growth promotion strategies. The integration of CNPs into existing agricultural practices holds promise for sustainable and eco-friendly crop management, addressing global challenges such as food security and environmental sustainability. Future research directions and challenges in optimizing the use of CNPs for plant growth promotion are also discussed.

O. Prakash · B. Kumari · M. P. Singh · A. Prakash · B. Pandey · A. K. Singh (✉)
Department of Biotechnology, School of Life Sciences, Mahatma Gandhi Central University, Motihari 845401, Bihar, India
e-mail: akhileshsingh@mgcub.ac.in

P. Pal
Environmental Engineering, Shri Ramswaroop Memorial College of Engineering and Management, Tewariganj, Faizabad, Road, Lucknow 226028, India

V. Agarwal
Department of Biotechnology, Motilal Nehru National Institute of Technology Allahabad, Prayagraj 211004, India

R. K. Bachheti et al. (eds.), *Chitin-Based Nanoparticles for the Agriculture Sectors*, Smart Nanomaterials Technology, https://doi.org/10.1007/978-981-96-0920-8_8

Keywords Agriculture · Biocide · Elicitor · Plant growth regulator · Sustainable development · And chitin fertilizer

1 Introduction

Chitin, characterized by its β-(1–4)-poly-N-acetyl-D-glucosamine structure, is widely prevalent in the natural world and ranks as the second most abundant polysaccharide, surpassed only by cellulose. In its natural occurrence, chitin forms organized macrofibrils and serves as the primary structural element in the exoskeletons of crustaceans such as crabs and shrimps, as well as in the cell walls of fungi. Due to stringent regulations overseeing the utilization of plant growth regulators, fertilizers, and pesticides, there is an increasing market demand for organic alternatives [4]. Among the potential alternatives, chitin stands out as a promising option [65]. Chitin, a biopolymer consisting of N-acetylglucosamine (GlcNAc), is naturally occurring and plays a crucial role in the survival and reproduction of many pests, pathogenic fungi, and oomycetes, posing a significant challenge to agricultural productivity [85]. It is an essential component of the exoskeletons of insects, crustaceans, and fungi, making it the second most abundant biopolymer after cellulose on Earth. Chitin's mechanical properties, chemical stability, biodegradability, and non-toxic nature make it highly sought after [57]. It is found in the exoskeletons of arthropods like crabs, lobsters, and prawns, forming a semicrystalline polymer with a microfibrillar structure [30]. However, despite its abundance and importance, there is still a lack of understanding regarding its properties in soilless plant cultivation compared to soil ecosystems [14, 15]. Chitin has been identified as a general elicitor in plants, capable of inducing biological responses and actions based on size and composition [61]. It has shown potential as a sustainable fungicide, and ongoing research is exploring its role in enhancing the gene-related defense response in plants [19]. While it signaling mechanisms have been studied in A. thaliana, there is a need for further research on commercially important plants [19]. Chitin and its derivatives have been found to aid in crop growth under various stresses, and their potential as bio-stimulatory agents is a subject of intense research [1, 20, 69]. Apart from agricultural applications, chitin is also used in food, paper, pharmaceuticals, textiles, cosmetics, and as a cleaner of hazardous contaminants [2, 59]. It has been employed in various plant production systems, either alone or in combination with other elements and polymers, for tasks like seed covering, foliage spraying, soil incorporation, and post-harvest fruit protection [26, 53, 74]. Chitin's ability to influence plant nutrition, stimulate plant defense, and enhance beneficial microorganism activity has positioned it as an eco-friendly biopolymer in agriculture [69]. It interacts with pattern recognition receptors on plant cell surfaces, triggering pathogen-associated molecular pattern (PAMP-triggered immunity (PTI against potential invaders [1, 21]. Beyond its agricultural applications, chitin and its derivatives have shown promise in food and biomedical fields. For instance, chitosan, a derivative of chitin, has been found to lower cholesterol levels by preventing the absorption of fats, aiding in weight loss and body fat reduction. Additionally, chitin and its derivatives exhibit biodegradation, biocompatibility, anticancer, antibacterial, non-toxic, immune-stimulating, haemostatic, wound healing, tissue engineering, and drug delivery properties [69]. In recent years, researchers have harnessed the unique properties of chitin to synthesize nanoparticles with applications in diverse fields, including medicine, environmental remediation, and agriculture. The utilization of CNPs in plant growth promotion represents a paradigm shift in sustainable agriculture, as it combines the benefits of nanotechnology with the inherent ecological advantages of chitin.

2 Chitin and Chitin-Based Nanoparticles

The second most prevalent naturally occurring polysaccharide, chitin, is present in fungus cell walls, the exoskeletons of insects and crustaceans, and even the beaks of octopuses [16]. It consists of 1,4-β-glycosidic linkages connecting N-acetyl-D-glucosamine units, with a little quantity of deacetylated units (2-amino-2-deoxy-D-glucose) [60]. The crystalline structure of chitin contains intramolecular hydrogen bonds, which explain why it is insoluble in water and many other organic solvents. Chitin undergoes deacetylation to become chitosan, an acid-soluble heteropolymer mostly made of D-glucosamine units. The physicochemical and functional characteristics of chitin-based materials and their uses are greatly influenced by the degree of acetylation. Since the 1930s, chitin has been the subject of increased academic research, which has led to its use in many different disciplines. They are utilized in food packaging for film and coating applications and in medical for scaffolds and bandages. Various uses for chitin-based nanomaterials, including nanoparticles, hydrogels, and nanofibers, have been suggested. The preparation techniques for chitin-based nanoparticles have been disclosed; we will first review these techniques and then look at potential uses. Chitin nanomaterials, including chitin nanocrystals (ChNCs) or chitin nanowhiskers (ChNWs) characterized by dimensions ranging from 100 to 800 nm in length and 6 to 60 nm in width, as well as chitin nanofibers (ChNFs) with lengths in the micrometer range [71], can be isolated. The specific structure and size of these chitin nanomaterials are contingent upon factors such as the chitin source and extraction method employed. The breakdown of natural chitin into ChNCs or ChNFs results in the creation of nanomaterials exhibiting a notably high aspect ratio. This elevated aspect ratio of chitin nanomaterials imparts distinct advantages, enhancing properties such as transparency, reactivity, and mechanical strength in materials [40]. The preparation methods of ChNCs/ChNWs, along with their respective yield, size, and crystallinity index, play a crucial role in determining the characteristics of these nanomaterials. For instance, the incorporation of ChNCs with a high aspect ratio into natural rubber (NR) latex to form NR/ChNCs composites has been observed to significantly improve the mechanical properties and thermal stability of the resultant rubber composites [62]. Therefore, understanding the relationship between preparation methods, dimensions, and properties of chitin nanomaterials is essential for harnessing their potential in various applications.

3 Extraction of Chitin

Over the years, numerous techniques have been put up and employed to produce pure chitin, yet no universal technique has been accepted. Chemical or enzymatic treatments Table 1 could be used for demineralization as well as deproteinization.

3.1 *Chemical Extraction*

3.1.1 Chemical Deproteinization

Deproteinization is a crucial step in chitin processing as it involves breaking the chemical bonds that link chitin to proteins. This process is essential for biomedical applications, as the protein component in shellfish can cause allergies in some individuals [73]. Initially, chemical deproteinization was the primary method employed, using a variety of reagents such as NaOH, Na_2CO_3, $NaHCO_3$, KOH, K_2CO_3, Ca $(OH)_2$, Na_2SO_3, $NaHSO_3$, $CaHSO_3$, Na_3PO_4, and Na_2S. However, the reaction conditions varied widely across different studies, with NaOH concentrations ranging from 0.125 to 5.0 M, treatment durations spanning from minutes to days, and temperatures reaching up to 160 °C. A common side effect of the hydrolysis process with NaOH is a reduction in the molecular weight of the biopolymer, leading to deacetylation of chitin, deproteinization, and hydrolysis. Commercial-scale production of chitin and its derivatives involves two main chemical processes: demineralization through acidic treatment at high temperatures and deproteinization through alkaline treatment. Subsequently, a decolorization stage is employed to remove pigments and lipids from the product [43]. Recent studies have explored the use of stirred bioreactors at room temperature, demonstrating improvements in product quality and reduced processing times [33]. These innovations hold promise for enhancing the efficiency and effectiveness of the deproteinization process in chitin production.

3.1.2 Chemical Demineralization

Demineralization is a crucial process in chitin extraction, primarily targeting the removal of minerals, especially calcium carbonate, from the raw material. Acidic solutions such as HCl, HNO_3, H_2SO_4, CH_3COOH, and HCOOH are commonly employed in demineralization processes [49]. Among these acids, dilute hydrochloric acid is often preferred due to its effectiveness. The demineralization process involves the breakdown of calcium carbonate into water-soluble calcium salts and the release of carbon dioxide, making it a relatively straightforward process. Several factors, including the degree of mineralization in the shell, extraction duration, temperature, particle size, acid concentration, and solvent ratio, influence the efficiency of demineralization treatments. However, these parameters often require empirical

Table 1 Comparison between chemical and biological methods for chitin extraction [27]

Extraction methods	Treatment	Advantages	Disadvantages	Quality of chitin
Traditional chemical extraction	**Acidic treatment**: demineralisation with decalcifying agents (HCl, HNO_3, H_2SO_4, CH_3COOH, and HCOOH) **Alkali treatment**: deproteinization wuth NaOH	Used at the industrial scale for chitin preparation Short processing time	Environmentally unfriendly, high cost because of the effluent treatment generated after acid and alkaline reagents. Removed proteins and minerals cannot be used as human and animal food supplements after being damaged by the acidic and alkali treatment	Despite the complete removal of organic salts, there can be deacetylation and depolymerisation reactions
Biological extraction	**Lactic acid treatment**: demineralisation with lactic acidproducing bacteria **Proteases treatments:** deproteinsation with proteases producing bacteria	Environmentally safe, low cost because of the absence of effluent—Solubilized proteins and minerals may be used as human and animal nutrients	Limited to laboratory scale studies only	Homogeneous and high-quality final products

adjustment to achieve optimal results. The stoichiometry of the reaction between acid and calcium carbonate is an essential consideration, as it requires two molecules of HCl to convert one molecule of calcium carbonate into calcium chloride. Therefore, the acid intake must be sufficient to ensure a complete reaction [57]. However, due to the heterogeneous nature of the material, achieving complete demineralization can be challenging, often requiring larger volumes or more concentrated acid solutions. These factors highlight the complexity of demineralization processes and the need for careful optimization to achieve the desired outcomes in chitin extraction.

3.2 Biological Extraction

Chemical extraction of chitin has several drawbacks that affect its overall quality and the efficiency of the purification process:

- Degradation of Physicochemical Qualities: Chemical extraction can degrade the physicochemical properties of chitin, leading to decreases in molecular weight (MW) and deacetylation (DA). These changes can negatively impact the intrinsic qualities of the purified chitin.
- Impact on Wastewater: Chemical extraction processes can result in wastewater effluent containing harmful chemicals, posing environmental challenges.
- Increased Cost: The use of chemicals in extraction can raise the overall cost of chitin purification procedures.

In response to these challenges, there is a growing interest in developing green extraction methods based on the principles of "green chemistry." These methods use microbes and enzymes for chitin extraction, offering potential advantages over traditional chemical methods. Comparative studies, such as those conducted by [34], have shown promising results for biological methods of chitin extraction. In their study, the biological method using microbes preserved the chitin structure better than the chemical method. Additionally, research by [6] demonstrated that chitin produced from deproteinizing prawn shells using proteolytic bacteria had a higher molecular weight compared to chemically produced chitin from shellfish. Biological extraction methods offer several benefits, including improved repeatability with less effort, reduced solvent usage, lower energy input, and easier manipulation. However, these biological approaches are currently limited to laboratory-based research. Recent reviews have highlighted enzymatic deproteinization and fermentation as the two most popular biological techniques for chitin extraction [35, 52], indicating the growing interest and potential of biological methods in chitin extraction processes.

3.2.1 Enzymatic Deproteinization

Proteases are required for the extraction of chitin. The primary sources of proteolytic enzymes include microorganisms, plants, and animals. Numerous proteases, including trypsin, alcalase, pepsin, papain, pancreatin, devolves, and pepsin, reduce deacetylation and depolymerization during chitin extraction by removing proteins from crustacean shells. This treatment affects the accessibility for the reactants and can be applied either before or after the solid material goes through the demineralization stage. In the deproteinization stage, both crudely extracted and purified proteases are utilized. On the other hand, crude proteases which are less expensive and more effective since coexisting proteases are present are more efficient than commercially refined enzymes. The two main sources of crude proteases are bacteria and fish viscera, with bacterial proteases being the most prevalent. The functional groups of enzymes found in animal tissues—which may be extracted for commercial use in both active and stable forms—are also found in marine creatures. About 50% of the seafood harvest in a few of the major fish-producing nations is made up of by-products [44]. Most of these commodities are wasted and misused. Therefore, it may be interesting to apply these unrefined enzymes to the chitin extraction process to lower expenses and protect the environment.

It should be mentioned that enzymatic procedures are less effective than chemical ones, usually leaving 5%–10% of residual protein connected to the extracted chitin. To boost its purity and maintain the structure of the final isolated chitin, it could then undergo a second NaOH treatment (under kinder circumstances and for a shorter period of time).

3.2.2 Fermentation

Deproteinization by fermentation, which can be accomplished by endogenous microorganisms (referred to as auto-fermentation) or by adding specific strains of microorganisms, can reduce the expense of utilizing enzymes. The latter can be accomplished through co-fermentation, sequential fermentation, single-stage fermentation, or two-stage fermentation. According to [47, 48] summary numerous microbe species were suggested for the fermentation of crab shells [47, 48]. There are two main types of fermentation methods: non-lactic acid fermentation and lactic acid fermentation.

Lactic Acid Fermentation

A specific strain of *Lactobacillus* sp. that produces both lactic acid and proteases can be used as an inoculum to perform the fermentation of crustacean shells. Lactic acid is produced by converting glucose, which lowers the pH of silage and inhibits the growth of bacteria that cause spoiling. After the calcium carbonate and lactic acid react, a precipitate of calcium lactate is produced. The lighter shells are retrieved and then washed with water. Either pure crustacean shells or the entire waste product from shrimp, including the heads and viscera, can be used for this operation. Therefore, proteases generated by other strains, gut bacteria found in the treated shrimps' intestinal systems, or proteases found in the biowaste itself could all have contributed to the deproteinization and concomitant liquefaction of the proteins. Numerous variables affect the effectiveness of lactic acid fermentation, including temperature, the length of the fermentation process, the type and quantity of inoculums, the carbon supply and its concentration, the starting pH and pH evolution during fermentation, and temperature [38].

Non-lactic-Acid Fermentation

For the fermentation of crustacean shells in non-lactic acid fermentation, both bacteria and fungi were employed, such as *Bacillus* spp., *Pseudomonas* spp and *Aspergillus* spp. [31]. Six proteolytic Bacillus strains—*Bacillus pumilus A1, B. mojavensis A21, B. licheniformis RP1, B. cereus SV1, B. amyloliquefaciens An6 and B. subtilis A26*—were assessed by [67, 68] on the fermentation of shrimp waste. The *Bacillus* The findings showed that all strains could deproteinize shrimp faces. *B. cereus* SV1 was used to achieve the highest possible level of deproteinization. These scientists also examined the impact of a higher glucose concentration on fermentation, and they found that glucose enhanced demineralization while having no discernible impact on the degree of deproteinization [31].

4 Role of Chitin Nanoparticles in Plant Growth Promotion

Chitin nanoparticles have gained attention for their potential role in promoting plant growth due to their unique properties and ability to interact with plants at a cellular level. Here are some of the ways chitin nanoparticles are believed to contribute to plant growth promotion.

4.1 Enhanced Nutrient Uptake

Chitin nanoparticles can improve the availability and uptake of essential nutrients by plants. Their small size and high surface area allow them to interact with soil nutrients, making them more accessible to plant roots. Chitin nanoparticles have been investigated for their potential to enhance plant nutrient uptake. Their small size and high surface area make them effective at interacting with soil nutrients and facilitating their uptake by plant roots. This enhanced nutrient uptake can improve plant growth and productivity [29]. Chitin nanoparticles may also act as carriers for essential nutrients, protecting them from leaching or immobilization in the soil and ensuring their availability to plants. Additionally, chitin nanoparticles have been shown to promote the growth of beneficial microorganisms in the soil, which can further enhance plant nutrient availability through processes such as nitrogen fixation and nutrient mineralization. Overall, using chitin nanoparticles in agriculture can improve nutrient uptake efficiency in plants, leading to healthier and more productive crops [8].

4.2 Stimulation of Plant Immunity

Chitin nanoparticles can act as elicitors of plant immune responses. When applied to plants, they can trigger the production of defense-related compounds and proteins, enhancing the plant's resistance to pathogens and environmental stresses. Chitin nanoparticles have been found to stimulate plant immunity by acting as elicitors of plant defense responses [8]. When applied to plants, chitin nanoparticles are recognized by specific receptors on the plant cell surface, triggering a series of biochemical and molecular events that lead to the activation of defense mechanisms. One of the key responses triggered by chitin nanoparticles is the production of defense-related compounds such as phytoalexins, pathogenesis-related (PR) proteins, and reactive oxygen species (ROS). These compounds play crucial roles in plant defense against pathogens by inhibiting their growth and spread. Chitin nanoparticles can also induce the expression of genes involved in plant defense pathways, activating systemic acquired resistance (SAR) and induced systemic resistance (ISR). SAR and ISR are mechanisms through which plants enhance their resistance to a broad spectrum of pathogens following exposure to certain elicitors, including chitin nanoparticles. Furthermore, chitin nanoparticles have been shown to enhance the activity of plant growth regulators such as jasmonic acid (JA) and salicylic acid (SA), which are key signaling molecules involved in plant defense responses. This enhanced signaling can lead to a more robust and effective immune response in plants. Overall, the stimulation of plant immunity by chitin nanoparticles can improve plant health and reduce susceptibility to diseases, ultimately leading to enhanced crop productivity and quality [47, 48].

4.3 Stress Tolerance

Chitin nanoparticles have been shown to improve plant tolerance to various environmental stresses, such as drought, salinity, and heavy metals. They can help plants maintain cellular homeostasis and reduce oxidative stress under adverse conditions. Chitin nanoparticles show promise in enhancing plant stress tolerance, offering potential benefits for agriculture. They have been studied for their ability to help plants cope with environmental stresses like drought, salinity, and heavy metal toxicity. These nanoparticles can improve water retention in soil, regulate ion balance in plant cells, and reduce the toxicity of heavy metals. Additionally, chitin nanoparticles may stimulate the production of antioxidants and other protective compounds in plants, aiding in the mitigation of oxidative stress. By enhancing stress tolerance, chitin nanoparticles could improve crop resilience and productivity in challenging growing conditions. Ongoing research aims to further understand the mechanisms behind these stress-alleviating effects and explore their practical applications in agriculture [10].

4.4 Biocontrol Agents:

Chitin nanoparticles can carry biocontrol agents, such as beneficial microbes or bioactive compounds. By encapsulating these agents, chitin nanoparticles can protect them from degradation and deliver them to plants more effectively, promoting plant growth and health. Chitin nanoparticles have also shown potential as carriers for biocontrol agents, beneficial microorganisms, or bioactive compounds that can help protect plants from pathogens. By encapsulating these agents, chitin nanoparticles can protect them from degradation and deliver them more effectively to plants. This targeted delivery system can enhance the bioavailability and efficacy of biocontrol agents, improving their ability to suppress harmful pathogens and promote plant health. Additionally, chitin nanoparticles themselves have been reported to stimulate the growth of beneficial microbes in the soil, further contributing to their biocontrol properties. Overall, using chitin nanoparticles as carriers for biocontrol agents represents a promising strategy for sustainable pest management in agriculture [45].

4.5 Enhanced Plant Growth Regulators

Chitin nanoparticles can potentially enhance the effectiveness of plant growth regulators, such as hormones or signaling molecules. Chitin nanoparticles can influence plant growth and development by facilitating the uptake and delivery of these regulators to plant cells. Chitin nanoparticles have been investigated for their potential to enhance the effectiveness of plant growth regulators (PGRs). These nanoparticles can serve as carriers for PGRs, protecting them from degradation and facilitating their targeted delivery to plant tissues [8]. By improving the bioavailability and stability of PGRs, chitin nanoparticles can enhance their efficacy in promoting plant growth and development. Furthermore, chitin nanoparticles themselves have been shown to stimulate the production of endogenous plant growth regulators, such as auxins, cytokinins, and gibberellins, which play key roles in regulating various aspects of plant growth, including cell division, elongation, and differentiation. By modulating the levels of these endogenous hormones, chitin nanoparticles can influence plant growth processes, leading to improved plant growth and development. Overall, using chitin nanoparticles in conjunction with plant growth regulators holds promise for enhancing plant growth and productivity in agriculture. Further research is needed to fully understand the mechanisms underlying the interaction between chitin nanoparticles and PGRs and to optimize their application in agricultural systems [29].

5 Application of Chitin Nanoparticle

Chitin nanoparticles have found diverse applications across various fields due to their unique properties.

5.1 Agriculture Applications

The agricultural use of chitin is limited due to its insolubility in water and traditional solvents, which hinders its application in the agricultural sector. This challenge has led to the wide adoption of water-soluble chitin derivatives like nanochitin, chitosan, and chitin oligosaccharides in agricultural practices. Studies have shown these derivatives effectively promote plant growth, acting as fertilizers, antimicrobials for controlling plant diseases, antitranspirants, and seed coatings. Chitin has been identified for its potential to enhance plant growth hormone activity and nutrient uptake, largely attributed to the biostimulation induced by plant growth-promoting rhizobacterium (PGPR) [63]. PGPRs, a group of soil bacteria associated with host plants, have been linked to enhanced plant growth by increasing nutrient availability. They achieve this by improving nitrogen bioavailability, either by increasing root surface area for nutrient absorption or by converting nitrogen into a more easily absorbable form [9]. Additionally, PGPRs are involved in nitrogen-fixing processes, breakdown of bacterial cell walls using chitinase, and activation of growth-controlling signalling molecules. Studies by [9], have demonstrated that the introduction of chitin into the soil surrounding crops can stimulate the growth of beneficial microbes that suppress harmful bacteria, leading to reduced crop stress and improved nutrient uptake [9, 67, 68]. For example, the addition of chitin has been shown to enhance the growth of Bacillus subtilis AF 1, which in turn inhibited the growth of harmful fungi such as *Aspergillus niger,* responsible for fusarium in pigeon peas and crown rot in peanuts. This resulted in increased biomass, seed germination, and overall improved crop yield [41]. Further studies have supported these findings, showing that applying chitin-based PGPR formulations can enhance the yield of crops such as pepper, tomato, and mango [37, 78]. Additionally, research by [41] revealed that chitin-supplemented formulations of PGPR led to significant improvements in the growth of pigeon pea, enhancing both dry weight and seedling emergence. These findings suggest that chitin acts as a biostimulant rather than a fertilizer, promoting plant growth through various mechanisms. Recent studies have also investigated the impact of chitin on plant growth. [3] conducted experiments with Arabidopsis thaliana supplemented with chitin/cellulose nanofibers complex (CCNFC) derived from wasted mushroom substrate (SMS). They found that low concentrations of

CCNFC enhanced leaf number and shoot dry weight, while higher concentrations had adverse effects on plant growth. Similarly, [54] observed that strawberry and cabbage growth was promoted at lower doses but inhibited at higher doses of chitin supplementation. Furthermore, chitin has been shown to influence the expression of specific genes associated with plant growth [82] found that tomato roots treated with chitin nanofibers (ChNF) exhibited increased expression of the Nodule inception-like protein 7 (NLP7) gene, which is involved in nitrogen uptake in plant roots. Similarly, observed elevated expression of the dual-affinity nitrate transporter gene in roots treated with ChNF, indicating enhanced nutrient uptake and growth promotion. In summary, chitin and its derivatives have shown great potential in promoting plant growth and enhancing nutrient uptake in agriculture. Their ability to stimulate beneficial microbial activity, influence gene expression related to growth, and improve nutrient availability makes them valuable tools for sustainable crop production. Further research and optimization of chitin-based formulations and applications are necessary to harness their potential for agricultural use fully [18].

5.2 *Fertilizers*

Fertilizers play a crucial role in providing essential nutrients to plants for their growth and immunity [64], Chitin, recognized as an organic nitrogen-type fertilizer (C), is a natural component used in fertilizer formulations due to its low C/N ratio and high nitrogen concentration [23]. This biopolymer constitutes a significant portion (20–58%) of the dry weight of shellfish, such as shrimp, crab, and krill, making up approximately 75% of their discarded weight [79]. The shells of these shellfish, composed mainly of calcium carbonate and chitin, ranging in size from 0.05 to 0.15 in., can be directly added to the soil as a soil conditioner or fertilizer [56]. Due to their rich micronutrient, phosphorus, and nitrogen content, these shells can effectively serve as fertilizers for crops and plants [12]. When these shells are added to the soil, the chitin within them is gradually broken down by soil organisms, releasing nitrogen as a supplement to the soil. The calcium compound, $CaCO_3$, acts similarly to agricultural lime, enhancing the soil's nitrogen concentration over an extended period [56]. This fertilizing effect is attributed to the promotion of specific microorganisms' growth or the biodegradation of chitin into ammonia-derived compounds, which act as fertilizers due to their amino groups [13]. Studies by [15], 14. have demonstrated that adding chitin to the soil stimulates the growth of beneficial soil microorganisms, such as fungi and chitinase-producing bacteria, which contribute to the breakdown of chitin. This microbial activity increases the availability of calcium and nitrogen in the soil, benefiting plant growth. Additionally, research by [84] suggests that adding chitin to the soil can reduce the need for phosphate and nitrogen fertilizers by 50 and 25%, respectively, while also enhancing soil microbial activity and nutrient availability. Chitin also contains about 6.9% nitrogen, contributing to its fertilizer properties [56]. For example, shrimp meal fertilizer contains 1% potassium, 2.6%

phosphorus, and 8.5% nitrogen, with an N-P-K ratio of 8.5–2.6–1 [25]. This combination of nutrients makes chitin and its derivatives valuable in enhancing soil fertility and promoting plant growth. Moreover, nanochitin has emerged as a promising fertilizer for agriculture [80] demonstrated that treating winter wheat with nanochitin resulted in significant increases in growth rates and nutrient levels, particularly for large spike wheat (LSW) and multi-spike wheat (MSW). These findings indicate the potential of nanochitin in improving soil fertility and enhancing plant development. In conclusion, chitin and nanochitin show great promise as effective and sustainable fertilizers for agriculture, offering a natural and environmentally friendly alternative to traditional chemical fertilizers. Further research and development in this field can lead to the widespread adoption of chitin-based fertilizers, contributing to sustainable agricultural practices and improved crop yields [80].

5.3 Antimicrobial and Eliciting Activity

While chitin's structure lacks free amino groups, it can still play a significant role in plant defense mechanisms by indirectly stimulating the plant's immune response [63, 77]. When a plant is invaded by a pathogen, chitinase, an enzyme that breaks down chitin into oligomeric fragments, is activated, triggering the production of chitinolytic enzymes [58]. These enzymes break down chitin into oligomers, which act as microbe-associated molecular patterns (MAMPs), initiating defense responses in plants [50, 58]. The activation of defense mechanisms includes producing and activating various compounds such as phytoalexins, phenolics, terpenes, and reactive oxygen species (ROS) [67, 68]. Additionally, physical barriers like lignin and callose deposition, as well as tylose production, are established to prevent pathogen invasion [67, 68]. Chitin supplementation has been shown to enhance the expression of defense-related genes in plants. For example, in strawberry plants treated with chitin, the expression of genes FaPR1 and FaWRKY1 was upregulated, providing protection against fungal infections [14]. Similarly, cabbage and strawberry plants treated with chitin nanofibrils showed upregulation of defense-related genes, such as mitogen-activated protein kinase (MPK3) and pathogenesis-related gene (PR2) [54]. Moreover, chitin and its derivatives can act as a shield against viral infections by blocking virus attachment to plant surfaces and transmitting systemic resistance to other plant organs. They can also serve as nematicides against root-knot nematodes by releasing ammonia during soil degradation, inhibiting nematode infectivity [70]. Chitin treatments have been shown to reduce nematode populations in plants, leading to increased shoot and root weights [70]. This reduction in nematode populations is attributed to the increased activity of chitinase-producing microorganisms in the soil, which parasitize nematode eggs and egg sacs, limiting their reproductive capacity [22]. However, it's important to note that higher concentrations of chitin can have phytotoxic effects. Therefore, while chitin and its derivatives show promise as effective agents in plant defense and immunity, their application requires careful consideration of dosage and environmental impact [22].

5.4 Seed Treatment

Plant seeds are often treated with agents for two primary purposes: protection against microbial and insect degradation and promotion of germination and growth. These treatments can take two forms: contact agents, which remain on the seed's surface, and systematic agents, which penetrate the seed shell and affect internal seed processes [63]. Chitin has shown promise as a seed treatment agent that can enhance grass seed germination, increase plant carbohydrate production, and improve frost resistance [63]. For instance, pigeon pea seedlings treated with a chitin-supplemented peat formulation of *B. subtilis* AF 1 exhibited a 29% increase in dry weight and a 33% increase in emergence [41]. Similarly, [76] found that adding a substrate containing 1 and 2% chitin, along with 200, 500, and 1000 ppm gibberellic acid (GA3), increased the germination rate of chervil (*Anthriscus cerefolium L.*) seeds. However, reported contrasting results, noting that while lower levels of chitin improved basil seed germination and reduced fungal activity, higher chitin concentrations led to decreased germination rates [32].

6 Biomedical Applications

While unmodified chitin has limited intrinsic benefits in medicinal applications, chitin derivatives such as chitosan and chitin nanoparticles show promise for a variety of biomedical uses [5, 17]. For instance, chitosan derived from squid pens exhibits favourable properties for manufacturing contact lenses, including oxygen permeability, mechanical stability, immunological compatibility, optical clarity, and wettability for ocular applications. Its ability to form films, along with its antibacterial and wound-healing properties, makes it particularly promising for the development of ocular bandage lenses, which could be beneficial for severe eye injuries. In addition to contact lenses, various forms of chitosan, such as hydrogels, nanofibers, and nanoparticles, are being explored for targeted drug delivery in cancer treatments. Chitosan polymers possess properties that can help chemotherapy drugs target different types of cancer cells more effectively. Furthermore, chitin nanoparticles have been investigated for several biomedical applications, including wound dressing, tissue engineering, anticancer treatments, antibacterial treatments, and drug delivery [51].

6.1 Wound Dressing

Several studies have highlighted the use of nanochitin in wound dressings to enhance skin wound regeneration [28, 36, 46]. Hydrogels made from chitin nanofibers (ChNFs) have been shown to promote the differentiation of bone marrow stromal cells (BMSCs) into angiogenic cells and fibroblasts, aiding in the healing of skin wounds. In rabbit wound healing trials, nonwoven chitin fabric, which retains the α-chitin structure and inherent biocompatibility of chitin, demonstrated superior wound healing outcomes compared to standard gauze [28]. Moreover, research by [46] demonstrated that ChNFs sheets have the capability to distribute and adsorb silver nanoparticles (Ag NPs) efficiently. When combined, ChNFs and Ag NPs exhibited potent antibacterial activity against *Pseudomonas aeruginosa, Escherichia coli,* and the H1N1 influenza A virus. This indicates their potential for use as anti-infective wound dressings, despite ChNFs' relatively modest antimicrobial activity [46]. Additionally, [36] investigated the impact of Ag NPs/ChNFs on fibroblast cytotoxicity in vitro and the rate of wound healing in vivo. They found that cleaning wounds covered with Ag NPs/ChNFs using saline could reduce the delay in wound regeneration observed in diabetic mice, suggesting that Ag NPs/ChNFs remain effective as anti-infective dressings for healing-impaired wounds [36].

6.2 Drug Delivery

Another application of nanochitin is in drug delivery systems. Nanofiber drug systems offer advantages such as a higher overall release rate, shorter diffusion channels, and a larger specific surface area compared to bulk materials [75]. Kinoda et al. [36], studied the effects of amorphous chitin nanoparticles (AC–NPs) in combination with curcumin (CUR), docetaxel (DOC), and 5-fluorouracil (5-FU). Their research revealed that three different types of AC-NPs exhibited entrapment and drug loading efficiencies of 98.1, 77.2, and 47.12%, respectively. These drug-loaded AC-NPs demonstrated a potential to induce a stronger response in gastric adenocarcinoma (AGS) cells, potentially leading to cell death. Given the hydrophobic nature of CUR, CUR-AC-NPs exhibited higher drug encapsulation and loading efficiency compared to DOC-AC-NPs and 5-FU-AC-NPs. Therefore, CUR-AC-NPs are considered the most effective carrier among the AC-NPs. This research suggests that in future drug delivery methods, CUR-AC-NPs could be a preferred or screened option for delivering drugs effectively [36].

6.3 Anticancer Treatment

In recent years, the use of nanochitin in anticancer therapy has increased significantly. Apoptosis in cancer cells may also result from it. It is not hazardous to non-cancerous cells, but it is clearly detrimental to many types of cancer cells [76]. Sharif et al. [66] evaluated the toxicity of chitin nanoparticles (ChNP), silver nanoparticles (AgNP), copper nanoparticles (CuNP), chitin/silver nanocomposites (ChNP/AgNP), and chitin/copper nanocomposites (ChNP/CuNP) using human breast cancer (MCF-7) cells as a model. ChNP-containing metal nanocomposites were more cytotoxic to

MCF-7 cells than AgNP and CuNP alone, and it was evident that ChNP/AgNP caused apoptosis. The considerable efficiency of the composite material against cancer cells was further demonstrated by the addition of ChNP, which reduced the toxicity of AgNP and CuNP on non-cancerous HEK-293 T cells. The development of new anticancer medications may be based on metal nanocomposites that have integrated chitin nanostructure [66].

6.4 Antibacterial Treatments

There is no evidence of nanochitin's antibacterial qualities on its own; instead, chitosan plays a major role in terms of antibacterial capabilities. This is because the negatively charged residues of proteins, lipids, and carbohydrates present on the surfaces of bacteria can interact with the different polycationic amines found in chitosan. Bacterial growth is stopped by this interaction [55]. For example, by breaking down the cell walls and membranes of *E. coli* and *Staphylococcus aureus*, [11] demonstrated that chitosan is a natural bactericide. Generally speaking, DA, MW, pH, and bacterial strain all impact chitosan's degree of antibacterial activity [76]. Nevertheless, DA and pH have less impact on chitosan's MW. Since the MW can impact both cell penetration and electrostatic interactions, it should be considered when DA and pH are under control. Following the identification of the target microbe, the DA and pH conditions are combined. Depending on the strain, the MW of outstanding antimicrobial activity varies, a higher MW is preferable for Gram-positive bacteria, while a lower MW is preferable for Gram-negative bacteria [39]. Through the use of an in vivo β-galactosidase expression assay, the anti-fungal effects of chitosan were demonstrated in a yeast model, resulting in a disruption of protein synthesis [42]. In conclusion, chitin and its derivatives hold significant potential across various industries, from agriculture to food, pharmaceuticals, and biomedicine, owing to their diverse properties and ecological benefits Fig. 1. Further research into their applications and mechanisms of action is essential for harnessing their full potential in sustainable development.

7 Future Prospects

In order to combat plant illnesses, our research suggests that Chitin might be mixed with other substances to form a new type of bactericide. Theoretical and practical advances in this area have been promising in recent years, but further study is needed to fully comprehend how these chemicals work [7]. Concerning the mode of action, further investigation into the antibacterial properties of chitosan should centre on clarifying the molecular details of the underlying mechanisms [83]. The participation and collaboration of government regulatory agencies, businesses, and research institutes is crucial for the widespread implementation of the antimicrobial mechanism

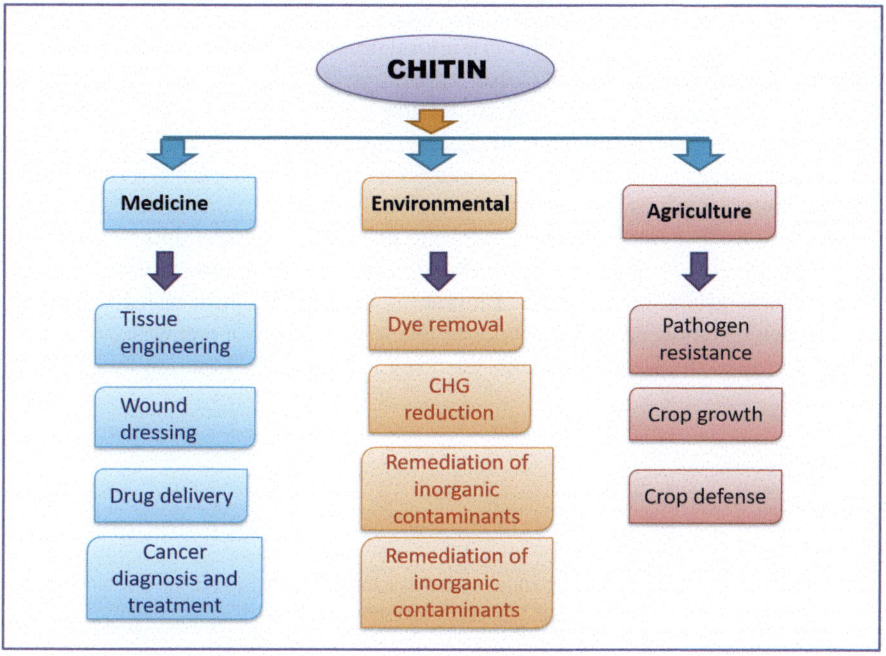

Fig. 1 Various application of chitin in plant growth promotion

[72]. Consequently, further study should centre on understanding their molecular structure to hasten their development and aid in pursuing sustainable agriculture. To what extent does this hold the key to unlocking the mysteries of chitin's biochemical roles [24]. In order to further their growth and the cause of sustainable agriculture, future research should concentrate on understanding their molecular structure. This could help shed light on the biochemical roles of chitin, which are now unknown [81]

8 Conclusions

The utilization of chitin-based nanoparticles (CNPs) represents a promising frontier in the realm of sustainable agriculture, offering innovative solutions for plant growth promotion. The unique properties of chitin, coupled with the advancements in nanotechnology, have enabled the synthesis of CNPs with distinct advantages for enhancing crop yields and mitigating environmental stressors. A comprehensive review of the literature shows that CNPs play a multifaceted role in agriculture, influencing nutrient uptake, stress tolerance, and overall plant health. The interaction between CNPs and plants involves intricate mechanisms beyond traditional fertilization approaches. The controlled release of nutrients facilitated by CNPs, their ability to modulate plant defence mechanisms, and their eco-friendly nature underscore their potential as catalysts for sustainable agricultural practices. Moreover, the biodegradability of chitin aligns with the global imperative for environmentally conscious farming. As we navigate the challenges of feeding an ever-growing population amidst environmental constraints, understanding and harnessing the full potential of CNPs is crucial. Future research endeavors should focus on optimizing CNP formulations, exploring diverse applications across crop varieties, and addressing potential ecological implications. The integration of CNPs into mainstream agricultural practices has the potential to revolutionize plant growth promotion, contributing to a more resilient and sustainable global food system. Embracing the role of chitin-based nanoparticles in agriculture not only aligns with the principles of green technology but also signifies a transformative step towards ensuring food security in a rapidly changing world.

References

1. Abbas SM (2013) The influence of biostimulants on the growth and on the biochemical composition of Vicia fabaCV. Giza 3 beans. Rom Biotechnol Lett 18:8061–8068
2. Abhinaya M, Parthiban R, Kumar PS, Vo DVN (2021) A review on cleaner strategies for extraction of chitosan and its application in toxic pollutant removal. Environ Res 196:110996 [CrossRef]
3. Ahmad A, Mubarak NM, Naseem K, Tabassum H, Rizwan M, Najda A, Kashif M, Bin-Jumah M, Hussain A, Shaheen A, Abdel-Daim MM (2020) Recent advancement and development of chitin and chitosan-based nanocomposite for drug delivery: critical approach to clinical research. Arab J Chem 13(12):8935–8964
4. Aktar MdW, Sengupta D, Chowdhury A (2009) Impact of pesticides use in agriculture: their benefits and hazards. Interdiscip Toxicol 2:1
5. Bays HE, Evans JL, Maki KC, Evans M, Maquet V, Cooper R, Anderson JW (2013) Chitin-glucan fiber effects on oxidized low-density lipoprotein: a randomized controlled trial. Eur J Clin Nutr 67(1):2–7. https://doi.org/10.1038/ejcn.2012.121
6. Bustos RO, Healy MG (1994) Microbial deproteinization of waste prawn shell. In: Proceedings of the second international symposium on environmental biotechnology, vol 94. Biotechnology, Brighton, UK, pp 15–25 [Google Scholar]
7. Chae SY, Jang MK, Nah JW (2005) Influence of molecular weight on oral absorption of water soluble chitosans. J Control Release 102(2):383–394
8. Chakraborty M, Hasanuzzaman M, Rahman M, Khan MAR, Bhowmik P, Mahmud NU, Islam T (2020) Mechanism of plant growth promotion and disease suppression by chitosan biopolymer. Agriculture 10(12):624
9. Chowdary VT, Baba SR, Divya J, Varma PK (2022) Role of chitosan in eco-friendly management of plant diseases for sustainable agriculture. New Future Dev Microb Biotechnol Bioeng 1:219–244
10. Chun SC, Chandrasekaran M (2019) Chitosan and chitosan nanoparticles induced expression of pathogenesis-related proteins genes enhances biotic stress tolerance in tomato. Int J Biol Macromol 125:948–954
11. Chung YC, Chen CY (2008) Antibacterial characteristics and activity of acid- soluble chitosan. Bioresour Technol 99(8):2806–2814. https://doi.org/10.1016/j.biortech.2007.06.044
12. Costa RE (1977) The fertilizer value of shrimp and crab processing wastes. Master's thesis, Oregon State University

13. Dahiya N, Tewari R, Hoondal GS (2006) Biotechnological aspects of chitinolytic enzymes: a review. Appl Microbiol Biotechnol 71(6):773–782. https://doi.org/10.1007/s00253-005-0183-7
14. De Tender C, Mesuere B, Van der Jeugt F, Haegeman A, Ruttink T, Vandecasteele B, Kuramae EE (2019) Peat substrate amended with chitin modulates the N-cycle, siderophore and chitinase responses in the lettuce rhizobiome. Sci Rep 9(1):9890. https://doi.org/10.1038/s41598-019-46106-x
15. Debode J, De Tender C, Soltaninejad S, Van Malderghem C, Haegeman A, Van der Linden I, Maes M (2016) Chitin mixed in potting soil alters lettuce growth, the survival of zoonotic bacteria on the leaves and associated rhizosphere microbiology. Front Microbiol 7:565. https://doi.org/10.3389/fmicb.2016.00565
16. Díaz-Santana-Iturrios M, Zepeda-Benitez V, Pacheco-Ovando R, Cornejo CF, Cerda JM, Salinas-Zavala CA, Granados-Amores J (2022) Northeastern Pacific Octopus beak shape for species-level detection. Malacologia 64(2):169–184
17. Dong L, Wichers HJ, Govers C (2019) Beneficial health effects of chitin and chitosan. Chitin Chitosan: Prop Appl 31:145–167
18. Egusa M, Matsukawa S, Miura C, Nakatani S, Yamada J, Endo T, Kaminaka H (2020) Improving nitrogen uptake efficiency by chitin nanofiber promotes growth in tomato. Int J Biol Macromol 151:1322–1331. https://doi.org/10.1016/j.ijbiomac.2019.10.178
19. Egusa M, Watanabe S, Li H, Zewude DA, Ifuku S, Kaminaka H (2023) Production of copper nanoparticle-immobilized chitin nanofibers and their role in plant disease control. J Pestic Sci 48(3):86–92
20. Ekin Z (2019) Integrated use of humic acid and plant growth promoting rhizobacteria to ensure higher potato productivity in sustainable agriculture. Sustainability 11(12):3417
21. Gemin LG, Mogor ÁF, Oliveira De, Amatussi J, Mogor G (2019) Microalgae associated to humic acid as a novel biostimulant improving onion growth and yield. Sci Hortic 256:108560 [CrossRef]
22. Godoy G, Rodriguez-Kabana R, Shelby RA, Morgan-Jones G (1983) Chitinamendments for control of Meloidogyne arenaria in infested soil. II. Effects onmicrobial population. Nematropica 63–74
23. Gooday GW (1990) The ecology of chitin decomposition. Adv Microb Ecol 11:387–430. https://doi.org/10.1007/978-1-4684-7612-5_10
24. Gornik K, Grzesik M, Duda BR (2008) The effect of chitosan on rooting of gravevine cuttings and on subsequent plant growth under drought and temperature stress. J Fruit Ornam Plant Res 16:333–343
25. Gravel V, Dorais M, Menard C (2012) Organic fertilization and its effect on development of sweet pepper transplants. HortScience: Publ Am Soc Horticul Sci 47(2):198–204. https://doi.org/10.21273/HORTSCI.47.2.198
26. Guerrero YR, Gonzalez LM, Rodriguez AF, Vazquez MN (2017) Effect of seed treatment with chitosan on the growth of rice (Oryza Sativa L.) seedlings cv. Inca LP-5 in saline medium. Cultiv Trop 36:143–150
27. Hamed I, Özogul F, Regenstein JM (2016) Industrial applications of crustacean by-products (chitin, chitosan, and chitooligosaccharides): a review. Trends Food Sci Technol 48:40–50. https://doi.org/10.1016/j.tifs.2015.11.007
28. Huang Y, Zhong Z, Duan B, Zhang L, Yang Z, Wang Y, Ye Q (2014) Novel fibers fabricated directly from chitin solution and their application as wound dressing. J Mater Chem B 2(22):3427–3432. https://doi.org/10.1039/c4tb00098f
29. Ingle PU, Shende SS, Shingote PR, Mishra SS, Sarda V, Wasule DL, Gade A (2022) Chitosan nanoparticles (ChNPs): a versatile growth promoter in modern agricultural production. Heliyon Int J Food Microbiol 164(1):108–112. https://doi.org/10.1016/j.ijfoodmicro.2013.03.025
30. Jayakumar R, Nwe N, Tokura S, Tamura H (2007) Sulfated chitin and chitosan as novel biomaterials. Int J Biol Macromol 40(3):175–181 [PubMed] [Google Scholar]
31. Jin T, Liu T, Lam E, Moores A (2021) Chitin and chitosan on the nanoscale. Nanosc Horiz 6(7):505–542

32. Kanawi MA, AL Haydar M, Radhi WN (2021) Effect of chitin and chitosan in improvement of plant growth and anti-fungal activity. Egypt J Botany 61(2):513–519. https://doi.org/10.21608/ejbo.2021.41084.1549
33. Kaur S, Dhillon GS (2015) Recent trends in biological extraction of chitin from marine shell wastes: a review. Crit Rev Biotechnol 35:44–61 [PubMed] [Google Scholar]
34. Khanafari A, Marandi REZA, Sanati SH (2009) Recovery of chitin and chitosan from shrimp waste by chemical and microbial methods
35. Kim Y, Park RD (2015) Progress in bioextraction processes of chitin from crustacean biowastes. J Korean Soc Appl Biol Chem 58:545–554
36. Kinoda J, Ishihara M, Hattori H, Nakamura S, Fukuda K, Yokoe H (2016) Cytotoxicity of silver nanoparticle and chitin-nanofiber sheet composites caused by oxidative stress. Nanomaterials 6(10):189. https://doi.org/10.3390/nano6100189
37. Kokalis-Burelle N, Vavrina CS, Rosskopf EN, Shelby RA (2002) Field evaluation of plant growth-promoting rhizobacteria amended transplant mixes and soil solarization for tomato and pepper production in Florida. Plant Soil 238(2):257–266. https://doi.org/10.1023/A:1014464716261
38. Kumar V, Sangeetha K, Ajitha P, Aisverya S, Sashikala S, Sudha PN (2018) Chitin and chitosan: the defense booster in agricultural field. In: Handbook of biopolymers. Jenny Stanford Publishing, pp 93–134
39. Lee S, Hao LT, Park J, Oh DX, Hwang DS (2023) Nanochitin and nanochitosan: chitin nanostructure engineering with multiscale properties for biomedical and environmental applications. Adv Mater 35(4), Article 2203325. https://doi.org/10.1002/adma.202203325
40. Malerba M, Cerana R (2020) Chitin-and chitosan-based derivatives in plant protection against biotic and abiotic stresses and in recovery of contaminated soil and water. Polysaccharides 1(1):21–30
41. Manjula K, Podile AR (2005) Increase in seedling emergence and dry weight of pigeon pea in the field with chitin-supplemented formulations of Bacillus subtilis AF 1. World J Microbiol Biotechnol 21(6):1057–1062. https://doi.org/10.1007/s11274-004-8148-z
42. Marquez IG, Akuaku J, Cruz I, Cheetham J, Golshani A, Smith ML (2013) Disruption of protein synthesis as antifungal mode of action by chitosan
43. Marquis-Duval FO (2008) Isolation et valorisation des constituants de la carapace de la crevette nordique. PhD dissertation, Laval University, Quebec, Canada [Google Scholar]
44. Mehta MR, Biradar SP, Mahajan HP, Bankhele RR, Hivrale AU (2022) Chitosan and chitosan-based nanoparticles in horticulture: past, present and future prospects. Role of chitosan and chitosan-based nanomaterials in plant sciences 1:453–74
45. Mittal D, Kumar A, Balasubramaniam B, Thakur R, Siwal SS, Saini RV, Saini AK (2022) Synthesis of biogenic silver nanoparticles using plant growth-promoting bacteria: potential use as biocontrol agent against phytopathogens. Biomater Polym Horiz 1(1):22–31
46. Nguyen VQ, Ishihara M, Kinoda J, Hattori H, Nakamura S, Ono T, Miyahira Y, Matsui T (2014) Development of antimicrobial biomaterials produced from chitin-nanofiber sheet/silver nanoparticle composites. J Nanobiotechnol 12(1):1–9. https://doi.org/10.1186/s12951-014-0049-1
47. Nisar N, Tsuzuki T, Lowe A, Rossiter JT, Javaid A, Powell G, Iqbal M (2021) Chitin nanofibers trigger membrane bound defence signaling and induce elicitor activity in plants. Int J Biol Macromol 178:253–262
48. Nisar N, Tsuzuki T, Lowe A, Rossiter JT, Javaid A, Powell G, Waseem R, Al-Mijalli SH, Iqbal M (2021) Chitin nanofibers trigger membrane bound defense signaling and induce elicitor activity in plants. Int J Biol Macromol 1(178):253–262
49. No HK, Hur EY (1998) Control of foam formation by antifoam during demineralization of crustacean shell in preparation of chitin. J Agric Food Chem 46:3844–3846
50. Okada M, Matsumura M, Ito Y, Shibuya N (2002) High-affinity binding proteins for N-acetylchitooligosaccharide elicitor in the plasma membranes from wheat, barley and carrot cells: conserved presence and correlation with the responsiveness to the elicitor. Plant Cell Physiol 43(5):505–512. https://doi.org/10.1093/pcp/pcf060

51. Oyatogun GM, Esan TA, Akpan EI, Adeosun SO, Popoola API, Imasogie BI, Akinluwade KJ (2020) Chitin, chitosan, marine to market. Handbook of chitin and chitosan. Elsevier, pp 335–376. https://doi.org/10.1016/B978-0-12-817970-3.00011-0
52. Palma-Guerrero J, Larriba E, Güerri-Agulló B, Jansson HB, Salinas J, Lopez-Llorca LV (2010) Chitosan increases conidiation in fungal pathogens of invertebrates. Appl Microbiol Biotechnol 87:2237–2245
53. Pandey P, Kumar Verma M, De N (2018) Chitosan in agricultural context-a review. Bull Environ Pharmacol Life Sci 7:87–96
54. Parada R Y, Egusa M, Aklog Y F, Miura C, Ifuku S, Kaminaka H (2018) Optimization of nanofibrillation degree of chitin for induction of plant disease resistance: Elicitor activity and systemic resistance induced by chitin nanofiber in cabbage and strawberry. International Journal of Biological Macromolecules, 118, 2185–2192. https://doi.org/10.1016/j.ijbiomac.2018.07.089
55. Park BK, Kim MM (2010) Applications of chitin and its derivatives in biological medicine. Int J Mol Sci 11(12):5152–5164. https://www.mdpi.com/1422-0067/11/12/5152
56. Peniston QP, Johnson EL (1980) U.S. patent no. 4,199,496. U.S: Patent and Trademark Office, Washington, DC
57. Percot A, Viton C, Domard A (2003) Optimization of chitin extraction from shrimp shells. Biomacromol 4(1):12–18
58. Pusztahelyi T (2018) Chitin and chitin-related compounds in plant–fungal interactions. Mycology 9(3):189–201. https://doi.org/10.1080/21501203.2018.1473299
59. Saheed IO, Da Oh W, Suah FBM (2021) Chitosan modifications for adsorption of pollutants—a review. J Hazard Mater 408:124889 [CrossRef]
60. Salaberria AM, Labidi J, Fernandes SC (2015) Different routes to turn chitin into stunning nano-objects. Eur Polymer J 1(68):503–515
61. Sashiwa H (2014) Chemical aspects of chitin and chitosan derivatives. In: Kim S-K (ed) Chitin and chitosan derivatives: advances in drug discovery and developments, pp 93–111
62. Shahrajabian MH, Chaski C, Polyzos N, Tzortzakis N, Petropoulos SA (2021) Sustainable agriculture systems in vegetable production using chitin and chitosan as plant biostimulants. Biomolecules 11(6):819
63. Shamshina JL, Rogers RD (2019) Applications of chitin in agriculture. In: Sustainable agriculture reviews, vol 36. Springer, Cham, pp 125–146. https://doi.org/10.1007/978-3-030-16581-9_4
64. Shamshina JL, Kelly A, Oldham T, Rogers RD (2020) Agricultural uses of chitin polymers. Environ Chem Lett 18(1):53–60. https://doi.org/10.1007/s10311-019-00934-5
65. Shamshina JL, Rogers RD (2021) 3D printing of cellulose and chitin from ionic liquids for drug delivery: a mini-review. In: Application of ionic liquids in drug delivery, pp 71–90
66. Sharif R, Mujtaba M, Ur Rahman M, Shalmani A, Ahmad H, Anwar T, Tianchan D, Wang X (2018) The multifunctional role of chitosan in horticultural crops; a review. Molecules 23(4):872
67. Sharp RG (2013) A review of the applications of chitin and its derivatives in agriculture to modify plant-microbial interactions and improve crop yields. Agronomy 3(4):757–793
68. Sharp RG (2013) A review of the applications of chitin and its derivatives in agriculture to modify plant-microbial interactions and improve crop yields. Agronomy 3(4):757–793. https://doi.org/10.3390/agronomy3040757
69. Shukla PS, Mantin EG, Adil M, Bajpai S, Critchley AT, Prithiviraj B (2019) Ascophyllum nodosum-based biostimulants: Sustainable applications in agriculture for the stimulation of plant growth, stress tolerance, and disease management. Front Plant Sci
70. Spiegel Y, Chet I, Cohn E (1987) Use of chitin for controlling plant plant-parasitic nematodes: II. Mode of action. Plant Soil 98:337–345. https://doi.org/10.1007/BF02378355
71. Sun W, Shahrajabian MH, Petropoulos SA, Shahrajabian N (2023) Developing sustainable agriculture systems in medicinal and aromatic plant production by using chitosan and chitin-based biostimulants. Plants 12(13):2469

72. Tanaka Y, Tanioka S, Tanaka M, Tanigawa T, Kitamura Y (1997) Effect of chitin and chitosan particles on BALB/c mice by oral and parenteral administration. Biomaterials 18(8):591–595
73. Tolaimate A (2020) Exploration des gisementschitineux de la faune marine marocaine. Procédéd'extraction de chitinesfortementacétylées. Préparation de chitosanes à caractéristiquescontrôlées. PhD dissertation, Cadi Ayyad University, Marrakech, Maroc [Google Scholar]
74. Toscano S, Romano D, Massa D, Bulgari R, Franzoni G, Ferrante A (2018) Biostimulant applications in low input horticultural cultivation systems. Italus Hortus [CrossRef]
75. Usman A, Zia KM, Zuber M, Tabasum S, Rehman S, Zia F (2016) Chitin and chitosan-based polyurethanes: a review of recent advances and prospective biomedical applications. Int J Biol Macromol 1(86):630–645
76. Vedula SS, Yadav D (2021) Chitosan-based membranes preparation and applications: challenges and opportunities. J Indian Chem Soc 98(2):100017
77. Velasquez CL, Pirela MR (2016) Biochemical aspects of the chitin fungicidal activity in agricultural uses. Chitosan in the Preservation of Agricultural Commodities, pp 279–298. https://doi.org/10.1016/B978-0-12-802735-6.00010-0
78. Vivekananthan R, Ravi M, Ramanathan A, Samiyappan R (2004) Lytic enzymes induced by pseudomonas fluorescens and other biocontrol organisms mediate defence against the anthracnose pathogen in mango. World J Microbiol Biotechnol 20(3):235–244. https://doi.org/10.1023/B:WIBI.0000023826.30426.f5
79. Wang SL, Chang WT (1997) Purification and characterization of two bifunctional chitinases/lysozymes extracellularly produced by pseudomonas aeruginosa K-187 in a shrimp and crab shell powder medium. Appl Environ Microbiol 63(2):380–386. https://doi.org/10.1128/aem.63.2.380-386.1997
80. Xue W, Han Y, Tan J, Wang Y, Wang G, Wang H (2017) Effects of nanochitin on the enhancement of the grain yield and quality of winter wheat. J Agric Food Chem 66(26):6637–6645. https://doi.org/10.1021/acs.jafc.7b00641
81. Yanat M, Schroën K (2021) Preparation methods and applications of chitosan nanoparticles; with an outlook toward reinforcement of biodegradable packaging. React Funct Polym 161:104849 [CrossRef]
82. Yu LH, Wu J, Tang H, Yuan Y, Wang SM, Wang YP, Xiang CB (2016) Overexpression of Arabidopsis NLP7 improves plant growth under both nitrogen-limiting and-sufficient conditions by enhancing nitrogen and carbon assimilation. Sci Rep 6(1):27795. https://doi.org/10.1038/srep27795
83. Zeng LT, Qin CQ, Wang W, Chi WL, Li W (2008) Absorption and distribution of chitosan in mice after oral administration. Carbohydr Polym 71(3):435–440
84. Zhan J, Qin Y, Gao K, Fan Z, Wang L, Xing R, Li P (2021) Efficacy of a chitin-based water-soluble derivative in inducing Purpureocilliumlilacinumagainst nematode disease (Meloidogyne incognita). Int J Mol Sci 22(13):6870. https://doi.org/10.3390/ijms22136870
85. Zhu KY, Merzendorfer H, Zhang W, Zhang J, Muthukrishnan S (2016) Biosynthesis, turnover, and functions of chitin in insects. Annu Rev Entomol 61:177–196

Application of Chitin/Chitosan Based Agro Nano Chemicals in Crop Protection

Ashwani Kumar Sanghi, Rajnish Kumar, Kundan Kumar Chaubey, Anish Kumar Pal, Shalini Sharma, Deen Dayal, Ashok Kumar, and Deepak Kumar Verma

Abstract The agriculture sector is an integral part of the global economy after the industrial sector, alongside the vast majority of people in the world depend on agriculture for their livelihood along with food. Challenges such as climate changes, deficiency of clean water and cultivable land, emerging plant pest species, and global warming have been increasing and affecting agriculture crop production. In the outlook of protecting agriculture or crops from the environment through chemical, mechanical, and scientific encroachment, scientists and researchers began to search for solutions for the protection of crops. Chitin and Chitosan have distinct characteristics such as biocompatibility, nontoxicity, biodegradability, and low allergenicity, making them useful for agriculture or crop protection in the industrial and agriculture sectors. Chitin and Chitosan possess several other unique biological features, including anti-cancer, antifungal, immunity-boosting, antioxidant, and secondary

A. K. Sanghi · R. Kumar
School of Allied and Health Sciences, MVN University, Palwal 121105, Harryana, India

K. K. Chaubey (✉)
Department of Biotechnology, School of Basic and Applied Sciences, Sanskriti University, Mathura 281401, India
e-mail: kundan2006chaubey@gmail.com

A. K. Pal
Division of Research and Innovation, School of Applied and Life Sciences, Uttaranchal University, Dehradun 248007, India

S. Sharma
Department of Veterinary Physiology and Biochemistry, Lala Lajpat Rai University of Veterinary and Animal Sciences, Hisar 125004, Haryana, India

D. Dayal
Department of Biotechnology, Institute of Applied Sciences and Humanities, GLA University, Mathura, Uttar Pradesh, India

A. Kumar
Division of Animal Health, ICAR—Central Institute for Research on Goats, Makhdoom, Mathura 281122, India

D. K. Verma
Centre for Health Research and Innovation (CHRI), New Delhi, India

R. K. Bachheti et al. (eds.), *Chitin-Based Nanoparticles for the Agriculture Sectors*, Smart Nanomaterials Technology, https://doi.org/10.1007/978-981-96-0920-8_9

metabolic activities. Chitin and Chitosan are both known to have biological properties that are exclusive to them. Therefore, chitin and its deacetylated derivative, which is Chitosan, as well as their derivatives, such as agro nano chemicals, can now be investigated in their nano form as a biodegradable material that can replace compounds that are poisonous and non-biodegradable for crop protection. For crop protection, this chapter summarizes that using chitin and Chitosan, along with their agro nano chemicals, is safe, more accessible, and more environmentally friendly.

Keywords Chitin · Chitosan · Agro nano chemicals · Crop protection

1 Introduction

The agricultural field provides wood and raw materials for industries and food for the survival of life. Agriculture also plays the most crucial role in supporting the developing country's economy. There are many challenges, such as contamination of agricultural land and water resources with various dangerous environmental pollutants such as pesticides and chemical fertilizers, random climate change, and enriching food demands with a constant increase in global population [78]. In 2019, the FAO (Food and Agricultural Organization) predicted that global demand would increase by 25–70% by 2050 for agricultural production to tackle the food crisis in the rising human population. During the last few decades, advancements in nanotechnology research and the use of NPs have significantly impacted agriculture development and precision farming. The use of NPs and their products improved the agriculture output while minimizing agrochemical input (pesticides, herbicides, and fertilizers) in the agriculture field [79]. NPs can be synthesized by physical, chemical, and biological methods [99]. Nanotechnology is a modern approach to developing new products, so, in recent times, there has been a huge demand for nanotechnology research. Nanotechnology has improved the everyday lives of each person by imparting its prominent footprint in society [101]. Nanotechnology has a wide range of applications in nearly all industries, such as chemical, pharmaceuticals, cosmetics, diagnosis, therapeutics, electronics, and agricultural production industries [4, 10, 9, 29–33, 52, 53, 58, 63, 83, 98].

2 Nanoparticles (NPs)

The integration of NPs and nanotechnology into various consumer products has demonstrated remarkable outcomes compared to their larger equivalents. Consequently, the industrial and agriculture sectors have shown clear interest in NPs and nanotechnology because of their distinct and increasing features [37]. NP materials with unique chemical and physical properties, like nanoclays, nanotubes, and nanowires, can show increased sensitivity and faster reaction times [51, 111]. Most farmers use fertilizers to increase their harvests. However, these chemicals deplete soil fertility by disrupting the soil's mineral equilibrium [97]. The utilization of pesticides and fertilizers in agricultural production significantly contributes to the high production costs, necessitating the implementation of control measures. The utilization of NPs or Nanochemicals in agriculture can enhance production yields, lower production costs, and minimize nutrient losses, hence lowering the amount of required plant protection chemicals [103]. NPs or Nanochemicals possess distinct attributes and serve specific purposes. Nanochemicals are commonly employed as nano herbicides, nano pesticides, and nano fertilizers, as well as nanosensors for detecting climate variations and soil and agricultural field nutrient needs. Nanochemicals are used as catalysts in soil remediation and groundwater remediation, and these nanochemicals also exhibit targeted action and supply sufficient active components in agriculture [13, 47, 86]. By use of agro nanochemicals together with iron, silicon, zinc, titanium, and selenium agricultural yield and the ability of plants to resist environmental stress has been increased [11, 41, 82].

3 Chitin and Chitosan

The More abundant naturally occurring mucopolysaccharide chitin and Chitosan are essential components of many exoskeletons, living things, and seafood-related products [14, 45, 88]. The naturally occurring mucopolysaccharide chitin comprises acetyl-D-glucosamine monomer unit and has the molecular formula $C_8H_{13}O_5N$. Chitin is also known as a "coat of mail Hatchett obtained chitin." from mollusks, which comprises prawns, crabs, lobsters, and crayfish, among other mollusks, by exposing their shells to mineral acids in 1799. Chitin was initially discovered in 1811 by Henri Braconnot, who named it "fungine" since it was found in the cell walls of mushrooms. Although Hatchett detected a little bubbling, the materials under observation were supple, moldable, and had a yellowish hue reminiscent of cartilage. Also, they kept their form even after a short while [27]. This is the first record of invertebrates having cemented chitin. Through the discovery of it in the cell walls of the mushroom, chemist Henri Braconnot used the name "fungine" to describe chitin when he initially identified it in 1811. Odier created a hornlike material in 1823 by combining cockchafer elytra with potassium hydroxide. Chitin is the name that this substance eventually took [46, 50, 72].

Chitin is the second most abundant amino-polysaccharide biopolymer, following cellulose. It comprises repeated units of N-acetyl-D-glucosamine monomers linked together in a β-(1 → 4) arrangement [112]. Chitin is transparent, pliable, and highly durable in its natural form. Chitin and calcium carbonate constitute approximately 40% of the waste generated by crustaceans, which include shells of invertebrates like crabs and shrimp [5, 17, 89]. Chitin is sometimes mistaken for cellulose, but unlike cellulose, which has the carbonyl group of chitin, you will find acetamide groups (CH3CONH2) (Fig. 1) [18]. The substance comprises β (1,4)-linked N-acetyl-2-amino-2-deoxy-D-glucose (GlcNAc) with traces of 2-amino-2-deoxy-D-glucose. It possesses white color, nitrogen content, lack of elasticity, and strong physical qualities. Chitin is regarded as a key marine renewable resource due to its widespread availability as waste from the global maritime sector. The estimated chitin abundance is expected to exceed 10 billion tonnes per year, according to studies conducted by [14, 107, 113]. Nature contains three distinct crystalline allomorphic forms of chitin: α-chitin, β-chitin, and γ-chitin. The exoskeletons of crustaceans, such as crabs and shrimp, include a sophisticated hierarchical composition consisting of crystalline, anti-parallel α-chitin nanofibers, minerals, and other proteins [87, 110]. Diatom spines, squid pens, and pogonophoran tube-dwelling worms consist of parallel, monocyclic crystals of the β-chitin chain [71]. The γ-chitin chains in fungi, yeasts, and insect cocoons can be either parallel or antiparallel. The properties of γ-chitin are comparable to those of α-chitin and β-chitin, as previously reported by [26, 36, 40]. No other forms of Chitosan can be transformed from the α-form, only the β-form may be converted.

Chitin and Chitosan, agro nano chemical compounds are derived from agriculture and possess abundant possible agricultural and industrial applications such as in the food industry, papermaking, bioengineering, and textile industries [3] (Fig. 2). Chitin and Chitosan, agricultural nano compounds has generated interest and significant attention among the research scientists and industry professionals due to their broad range of applications, their non-toxic properties and their favorable biocompatibility properties [2, 35, 84]. At the same time, synthetic polymers like nylon, polyethylene, polyester, Teflon, and epoxy have yet to generate interest and significant attention among research scientists and industry professionals due to their limited range of applications compared to chitin and Chitosan.

Felix Hoppe-Seyler introduced the term "chitosan" in 1894. Chitosan, a positively charged hetero polymer, is derived from the deacetylation of chitin. Chitin is predominantly detected in arthropod, crustacean, insect, and mollusk exoskeletons [44]. Chitin is extracted and isolated through deproteinization, demineralization, and decolorization [22, 94]. The chitosan production requires the deacetylation reaction of chitin. These agro-nano-chemical-biopolymers can be processed and improved

Fig. 1 Structure of chitin

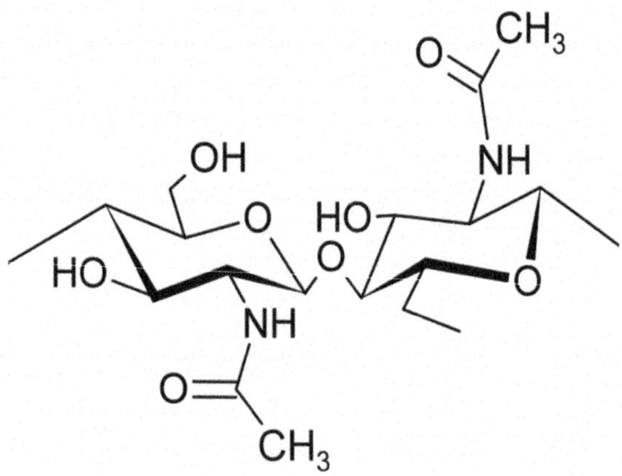

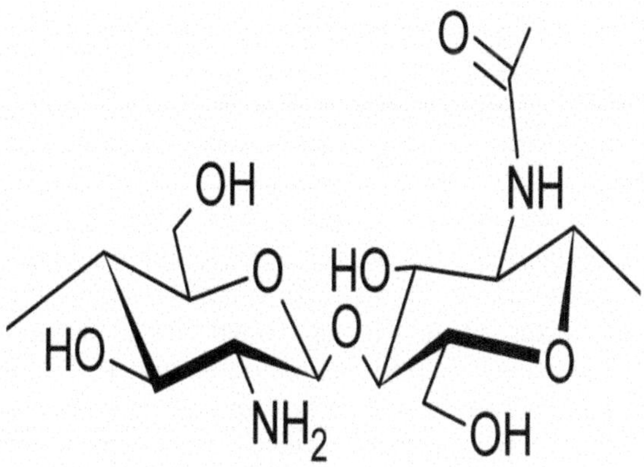

Fig. 2 Structure of Chitosan

by biological extraction techniques or by microbial utilization techniques. Fungi have been used to extract chitin and Chitosan [26]. Chitosan and its derivatives are more appropriate for their usage in various applications, such as beads, films, gels, membranes, and sponges [44, 105]. The discovery of chitin predates the finding of Chitosan by 40 years.

Charles Rouget synthesized Chitosan from chitin by heating the chitin in an alkaline solution in 1859, and up to 1950 the chemical structure of Chitosan still needed to be identified [85]. The industrial commercialization and utilization of chitin and Chitosan started in numerous countries such as Australia, China, Canada, France, Germany, Norway, Korea, Poland, the United States, and the United Kingdom in the 1970s. The unique characteristics, e biocompatibility, biodegradability, crucial chemical composition, and macromolecular structure of chitin and Chitosan played a key role in industrial commercialization and industrial utilization [16, 23, 26, 69]. The industrial annual consumption rate of Chitosan is approximately 700–800 tonnes [65].

Agriculture research scientists have received attention from Chitin and Chitosan due to their easy availability, marketability, and versatility characteristics. The many biological, chemical, and physical characteristics of chitin and Chitosan contribute to their immense value on a global scale. The agricultural and industrial sectors also find several uses for it. The use of chitin, Chitosan, and agro nano compounds, which are derivatives of these materials, shows potential for crop protection. The nitrogen-rich composition and favorable C/N ratio of chitin and Chitosan make them promising biofertilizers that could increase agricultural production. When added to soil, chitin increases the activity of microbes and helps reduce the spread of plant diseases.

Cell plasma membrane receptors, in particular Pathogen-Associated Molecular Pattern (PAMP) receptors, are responsible for chitin identification in the agricultural sector. In agricultural settings, PAMP receptors help protect crops from bacterial, fungal, and viral infections by activating the immune system and triggering defensive reactions [80]. Both academic and agricultural researchers can benefit from chitin due to its insolubility in water and its exploitable activities. Because of its nontoxicity, compatibility with living things, and capacity to break down naturally occurring chemicals, chitin is an ideal crop protection agent for use in agricultural systems. Chitin derived from yeast cell walls by postharvest therapy enhances the resistance of tomato fruits from gray mold rot disease caused by Botrytis cinerea. Chitin treatment promotes disease resistance in tomato fruit by increasing reactive oxygen species (ROS) and protects it from callose deposition and defense-related enzyme activity like catalase, chitinase, peroxidase, and superoxide dismutase (Figs. 3 and 4).

Furthermore, Chitin treatments also increase the gene expression [100]. The chitin oligosaccharide products derived from shrimp shells through sonication, autoclaving, and mechanical grinding methods stimulate the expression of genes associated with vegetative growth and carbon and nitrogen metabolism in Arabidopsis thaliana (*A. thaliana*). Chitin oligosaccharide treatment in plants promotes an increase in fresh weight, radical length, and total carbon and nitrogen content compared to untreated plants [104].

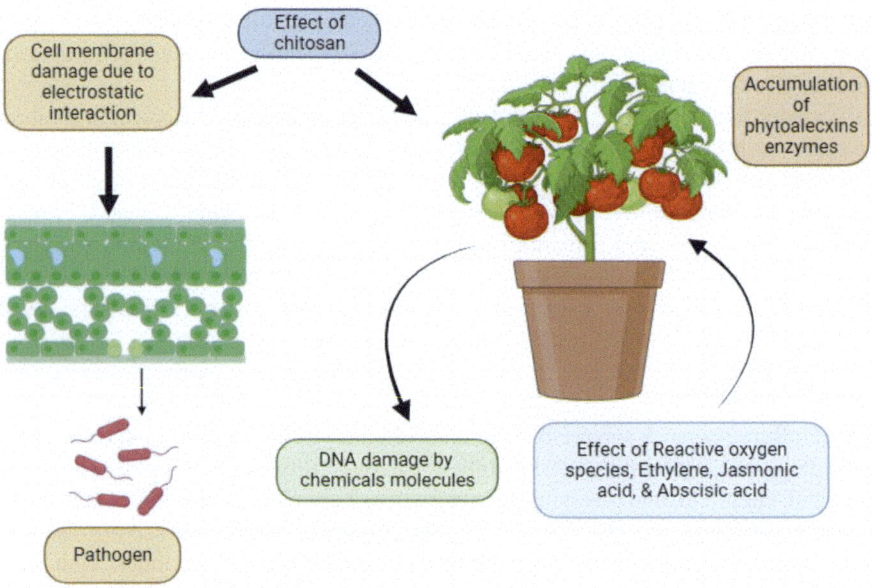

Fig. 3 Role of chitosan act as a natural compound to manage disease of plant

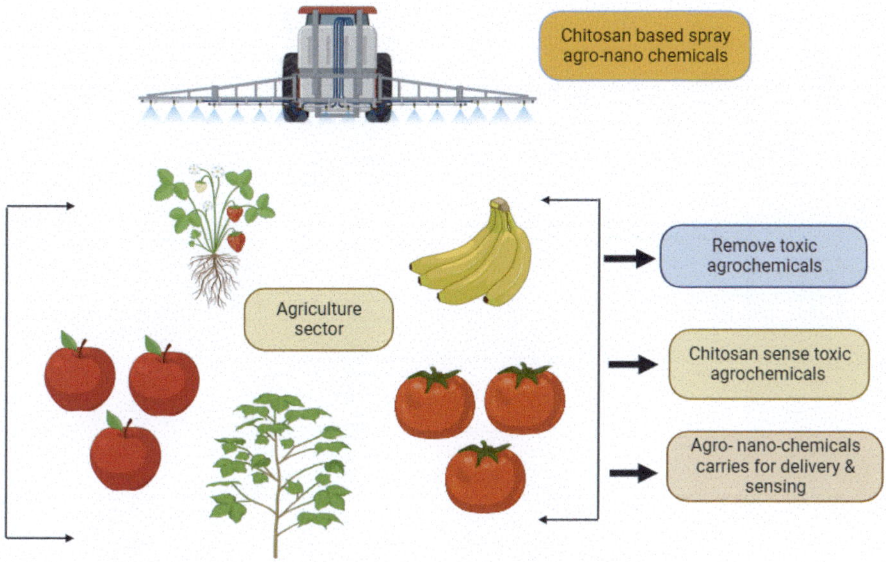

Fig. 4 Chitosan based agronanochemicals spray in agriculture and protect crop

4 Chitosan and Agricultural Nano-Chemicals Derived from Chitosan

The utilization of chitin and its derivatives, namely agro nano-chemicals derived from shrimp shells through mechanical grinding, autoclaving, and sonication, stimulated the activation of genes associated with vegetative development as well as carbon and nitrogen metabolism in *A. thaliana.* Agro nano-chemicals derived from crab shells, such as chitin and Chitosan, are currently utilized in grinder pretreatment and high-pressure water jets. According to Egusa et al. [21], these compounds enhance disease resistance in *A. thaliana* and rice crops. When applied to strawberry and cabbage crops, chitin and its compounds effectively prevent fungal infections. An investigation carried out in 2018 by Parada et al. [75] found that a soil-chitin nano-chemical compound mixture significantly reduced the frequency of Alternaria brassicicola spots and the size of Colletotrichum fructicola lesions in strawberry and cabbage crops when compared to crops that were not treated. Compared to untreated plants, tomato plants treated with chitin or its derivatives exhibit significantly better resistance to Fusarium wilt disease and have more robust growth stimulation. Research has shown that chitin and its derivatives can effectively prevent diseases and protect crops when applied to soil [21] (Fig. 5).

The growth of Fusarium pseudograminearum and Fusarium graminearum can be strongly inhibited by chitin and its derivatives, such as nanochitin whisker (NCW). According to [54, 55], these fungi are well-known to produce crown rot, a severe disease that impacts wheat crops. Tobacco root rot can be efficiently suppressed by

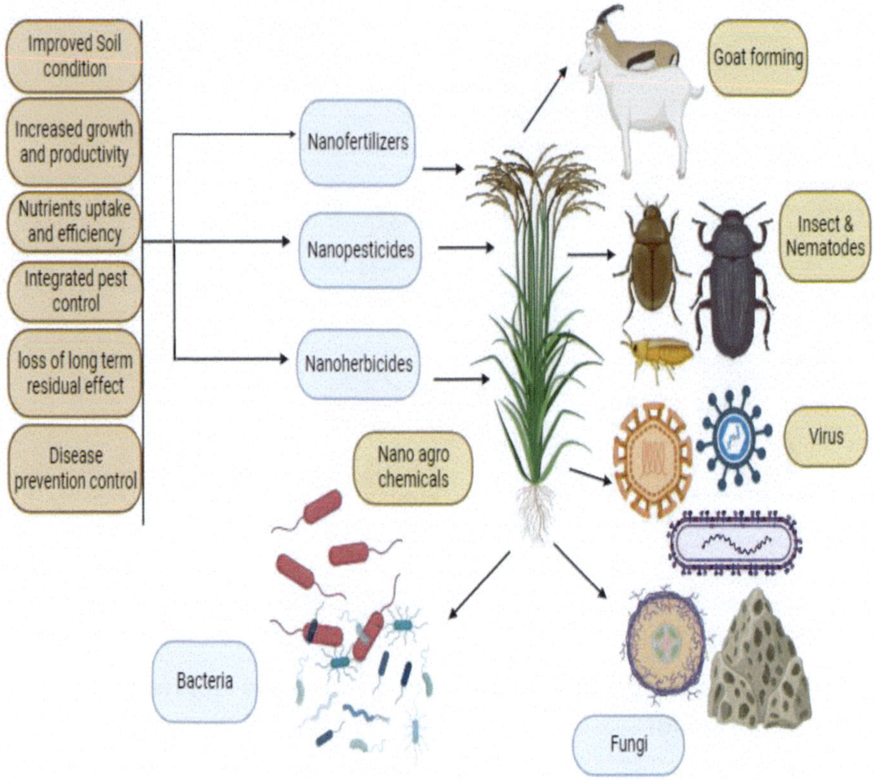

Fig. 5 Chitosan nanochemicals applied on plant and effect on plant growth & physiology

combining chitin and its derivatives with fungicides such metalaxyl, mancozeb, and thiophanate methyl. This combination has a surprising synergistic effect. Tobacco farms could benefit significantly from using these compounds instead of artificial fungicides, according to Zhou et al. [115]. Nanochitin whisker (NCW) is one of several chitin derivatives that have shown antimicrobial efficacy against Salmonella typhi, Escherichia coli, Pseudomonas aeruginosa, and maybe Fusarium oxysporum. Nanochitin whisker (NCW) is one of several chitin derivatives that have been shown to effectively prevent the growth of various bacterial and fungal diseases [25]. Research has shown that Chitosan and its agronanochemical derivatives can make agricultural commodities more resistant to diseases while decreasing the additional agrochemicals needed. However, greater amounts of Chitosan and similar substances may be necessary, as hinted at by [48]. The low cost and lack of danger associated with agro-nano chemicals give them great promise for practical use. When it comes to protecting crops, Chitosan—a bio-agro nano-chemical—stands out thanks to its nontoxicity, biodegradability, and compatibility with plants.

Because of its adaptability and the possibility of improving its biophysical and chemical characteristics, Chitosan—a very important agro nano-chemical—has numerous potential uses. Researchers in agricultural science have long acknowledged chitosan and agro nano-chemicals derived from it for the wide variety of antibacterial and regulatory capabilities they contain. Chitosan and chitosan-based agro nano-chemicals help in the protection of tomato crops by inhibiting the growth of *F. oxysporum* in Zingiber officinale; Chitosan, and chitosan-based agro nano-chemicals provide resistance against *F. dandiya,* and also stimulating the production of pathogenesis-related proteins, [7, 15]. Chitosan and chitosan-based agro nano-chemicals help protect finger millet crops from *Pyricularia grisea* and enhance their resistance against blast disease. Chitosan and chitosan-based agro nano-chemicals protect chickpea crops from *Fusarium oxysporum* prevent *Fusarium* wilt, a seed-borne and soil-borne disease [39, 91]. Chitosan-based agro nano-chemicals provide protection in rubber tree crops as well as in rice against leaf fall disease caused by *Corynesporacassiicola* [54, 55, 70].

Chitosan and harpin-chitosan agro nano-chemicals provide protection in tomato plants against the infection caused by *R. solani* by enhancing their defense response. Agro nano-chemicals containing thiamine-chitosan protect the chickpea seedlings against *F. oxysporum.* Agro nano-chemicals containing salicylic acid-chitosan protect the lettuce seedlings. Chitosan based agro nano-chemicals containing tripolyphosphate promote the growth of Capsicum annuum plants through micropropagation [49, 62, 68, 90]. Chitosan-based agro nano-chemicals containing Cymbopogon martinii essential oil have antifungal and anti-mycotoxin properties. These chemicals protect the grain crops against *F. graminearum* [38]. Table 1 displays the utilization of chitin and chitosan derived agro nano compounds for crop protection (Table 2).

5 Chitin/Chitosan Based Agro Nano Chemicals in Crop Protection

Chitin and Chitosan, deacetylated polymer of N-acetyl glucosamine possesses amine and hydroxyl groups, which provide it numerous unique features and make it safe and non-toxic. Chitin/Chitosan form complexes and gels by its interaction with numerous polyanions. By electrostatic interaction, the positive amine group of Chitin/Chitosan provides its stability with anionic material such as genes, proteins, and drugs. These properties of Chitin/Chitosan based Agro nano chemicals make their use in crop protection [12]. Chitin and Chitosan are powerful bio-stimulants that farmers use to boost crop yields and lessen the influence of pests and illnesses. They are used as bio-stimulants for managing plant diseases and protecting crops and provide an attractive substitute for chemical pesticides. As biodegradable substitutes for harmful and non-biodegradable chemicals, chitin and Chitosan obtained from natural sources could be utilized for crop protection [67].

You can combat multiple abiotic stressors with agronano chemicals derived from chitin and Chitosan. Finding strategies to make plants more resistant to abiotic stress is crucial for maintaining the production of healthy food in a safe environment,

Table 1 Recent application of chitin/chitosan based agro nano chemicals in crop protection

Plant crop species	Chitin nano-chemicals characteristics	Chitin/chitosan nano-chemicals effect	Key references
Solanum lycopersicum	From *Saccharomyces cerevisiae* cell wall	Resistance to *Botrytis cinerea*	[100]
Arabidopsis Thaliana	Chitin oligomers/tetramers	Resistance to *Alternaria brassicicola* and *Pseudomonas syringae*	[19]
Oryza sativa japonica	Chitin oligomers	Resistance to *Alternaria brassicicola*and *Pseudomonas syringae*	[19]
Brassica oleracea	Chitin nano-chemicals	Resistance to *Alternaria brassicicola* and *Colletotrichum fructicola*	[75]
Fragaria x ananassa	Chitin nanofibers resistance	Resistance to *Alternaria brassicicola* and *Pseudomonas syringae*	[75]
Solanum lycopersicum	Protein/CaCO3/chitin nano-chemicals or protein/ chitin nano-chemicals	Resistance to *Fusarium oxysporum*	[20]
Triticun aestivum	Nanochitin whisker	Resistance against *Fusarium pseudograminearum* and *Fusariumgraminearum*	[54, 55]
Nicotiana tabacum L.	Nanochitin whisker	Resistance against *Fusarium spp.*	[115]
Gossypium spp.	Zn or Cu chitosan nano-chemicals	Resistance against *Rhizoctonia solani*	[1]
Eleusine coracana	Chitosan nano-chemicals	Resistance against *Pyricola grisea*	[91]
Cicer arietinum L	CuO, ZnO or Ag Chitosan nano-chemicals	Resistance against Fusarium oxysporum	[39]
Hevea brasiliensis	Ag chitosan nano-chemicals	Resistance against Corynespora cassiicola	[70]
Piper nigrum	Chitosan nano-chemicals	Resistance against Phytophthora capsica	[57]
Oryza sativa L	Ag chitosan nano-chemicals	Enhanced disease resistance	[54, 55]
Solanum lycopersicum	Harpin chitosan nano-chemicals	Resistance against Rhizoctonia solani	[68]
Cicer arietinum L., Zea mays L.	Salicylic acid chitosan nano-chemicals	Enhanced disease resistance	[49, 62]
Zea mays L.	*Cymbopogon martini essential oil* Chitosan nano-chemicals	*Resistance to Fusarium graminearum*	[38]

(continued)

Table 1 (continued)

Plant crop species	Chitin nano-chemicals characteristics	Chitin/chitosan nano-chemicals effect	Key references
Hevea brasiliensis	*Coumarin*	*Resistance to Alternaria solanii and Fusarium spp*	[109]
Triticum aestivum L.	*Vanillin and salicylic acid*	*Resistance to Puccinia recondite and Cochliobolus sativus*	[77]
Solanum lycopersicum	*Bentonite*	*Resistance to Fusarium solani and Pseudomonas syringe*	[64]
Saccarum spp.	*S-nitrosoglutathione*	*Enhanced tolerating drought*	[95]

especially in light of the current conditions of global climate change and the ever-increasing global population. Seawater, heat, heavy metals, UV light, and salt are some of the abiotic variables causing farmers to lose ground globally. Environmentally friendly agro nanochemicals derived from chitin and Chitosan and new agronomic approaches and crop genotype selection require more investigation [28] (Figs. 6 and 7).

Heavy metals and other contaminants from human activities are becoming a major environmental issue. These non-biodegradable heavy metals are toxic, mutagenic, and carcinogenic, even in low quantities. Therefore, developing efficient strategies for their removal from the environment is crucial for farm protection. It is necessary to eliminate contaminants for crop protection and reuse precious resources such as cultivable soil and fresh water. Chitin and chitosan-based agro nano chemicals can easily integrate with certain ions, molecules, and materials to create complex crop production and protection structures. Chitin and chitosan-based agro nano chemicals are biocompatible, biodegradable, resistant to mechanical and antimicrobial attack, and do not produce potentially harmful secondary end products, making them useful in crop protection [60, 81] (Fig. 8).

Agriculture production is decreasing because of biotic stress (pathogen attack, herbivores, wounding) and abiotic stress (water scarcity or surplus, extreme heat or cold, high or low salinity, UV radiation, heavy metals, and other harmful pollutants) harm plant growth and development [28]. It becomes necessary to protect the crop and enhance crop production because of the increase in food demand for the growing population. Various approaches have been tested for crop protection, such as natural chemicals such as chitin/chitosan-based agro nanochemicals and synthetic chemicals such as fertilizers and pesticides. (Fig. 3) In today's scenario, there is large-scale use of synthetic chemicals in crop production and crop protection, and the excessive use of these chemicals in agriculture production has adverse effects on crops and living organisms. So, the search and use of eco-friendly natural chemicals such as Chitin/Chitosan based Agro nano chemicals in crop protection has been encouraged. Chitin/Chitosan based Agro nano chemicals seem to be promising tools in solving this problem in crop protection [59] (Fig. 9).

Table 2 Recent application of chitin/chitosan based agro nano chemicals in crop protection

Chitin/chitosan agro nanochemicals	Plant pathogen	Plant species	Effect of plant species	Key references
Based on chitosan agronanochemicals	*Fusarium oxysporum*	*Tomato, eggplant and pepper*	Highly inhibit on mycelial growth	[6]
Chitosan based agronanochemicals polyacrylamide acid	*Aphis gossypii*	*Cotton*	The number of female plant reduce in the laboratory conditions	[34]
Chitosan based agronanochemicals polyacrylamide acid	*Callosobruchus chinensis*	*Castor leaves*	The number of female plant reduce in the laboratory conditions	[76]
Chitosan based agronanochemicals polyacrylamide acid	*Callosobruchus maculatus*	*Soybean*	The number of female plant reduce in the laboratory conditions and store in 84%	[102]
Zn based chitosan	*Fusarium graminearum fungicides*	Wheat, barley & other types of grain	The chitosan effect on the plant of concentration of the treatment and environment	[108]
Chitosan based copper and salicylic acids	*Fusarium verticillioids*	Maize	Treatment after 8 h disease has been reduced 50%	[73]
Agronanochemicals chitosan based act as fungicides	*Pyricularia grisea*	Rice, finger millet	Pathogen attack a 2-fold increase in chitinase induce resistance	[114]
Agronanochemicals oleoyl	*Verticillium dahliae*	Cotton	Nanochemicals internalized the fungal cells increased to deformation of spore and hyphae	[74]
Nano protocatechuic acids	*Pyricularia oryzae*	Rice plant	Increase the zone of inhibition compared to Chitosan based nanochemicals	[56]

(continued)

Table 2 (continued)

Chitin/chitosan agro nanochemicals	Plant pathogen	Plant species	Effect of plant species	Key references
Copper based agronanochemicals	*Curvularia lunata*	Maize, *Zea mays*	Increase defense response against a significants response of super oxide dismutase	[42, 43]
Fungicide Pyraclostrobin agro nanochemicals	*Colletotrichum Gloeosporioides* and *Alternaria* spp.	Pomegranate	Alternaria and Colletotrichum alone fruit rots of pomegranate are caused by multiple species	[61]
Polyacrylic acid NPs insecticides	*Callosobruchus chinensis*	Red kidney bean, green gram	Reduce the nutritious content of grains that have been kept and result in significant quantitative and qualitative losses	[96]
Nano chitosan agro nanochemicals used in fungicides	*Alternaria solani, Fusarium oxysporum,* and *Pyricularia grisea,*	Tomato	Inhibit the mycelial growth	[66]
Fungicide, Hexaconazole NPs	*Rhizoctonia solani*	Wheat and rice	The primary targets of Rhizoctonia solani are the leaf sheath, leaf blades, and in extreme circumstances, the entire plant	[93]
Fungicide, Hexaconazole NPs	*Ganoderma boninense*	Palm oil	On oil palms, it results in basal stem rot disease	[42]
Fungicide, Tebuconazole porphyrinic-pectin NPs	*Xanthomonas campestris,*	Drumstick leaves	When it gets to the intended location in the plant, it causes damage to the cells, which causes the tissues to get macerated	[24]
Agro nanochemicals Nematicide Avermectin	*Caenorhabditis elegans*	Chinese dodder, Chinese rubber tree	Mortality rate increase of nematodes	[92]

(continued)

Table 2 (continued)

Chitin/chitosan agro nanochemicals	Plant pathogen	Plant species	Effect of plant species	Key references
Herbicide, Paraquat	*Eichhornia crassipes*	Water hyacinth	Upsetting nutrient cycles and food chains, which modifies the structure and function of ecosystems	[8]
Fungicide, Tebuconazole	*Pseudomonas syringae*	Almond tree	Effect on plant species caused bacterial canker	[106]

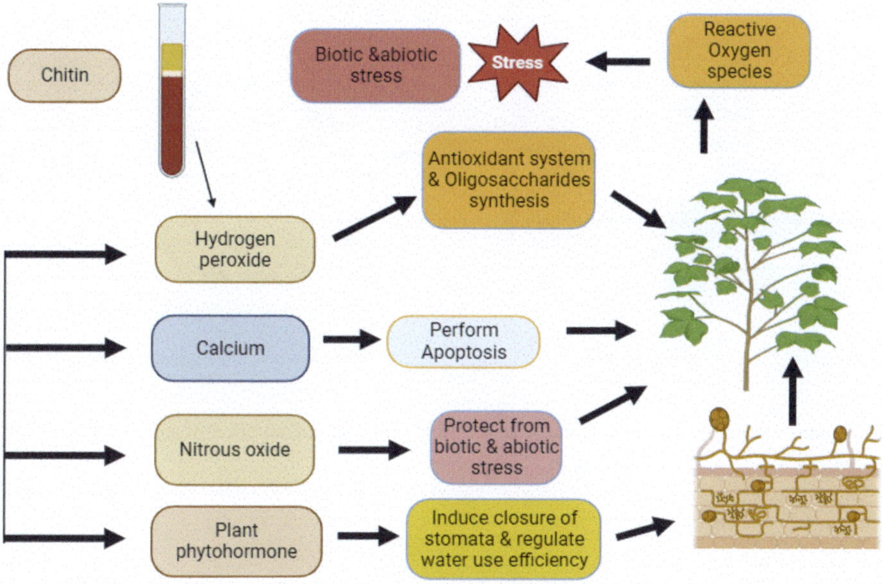

Fig. 6 Chitin regulate on plant growth and generate ROS plant resistant to biotic and abiotic stress

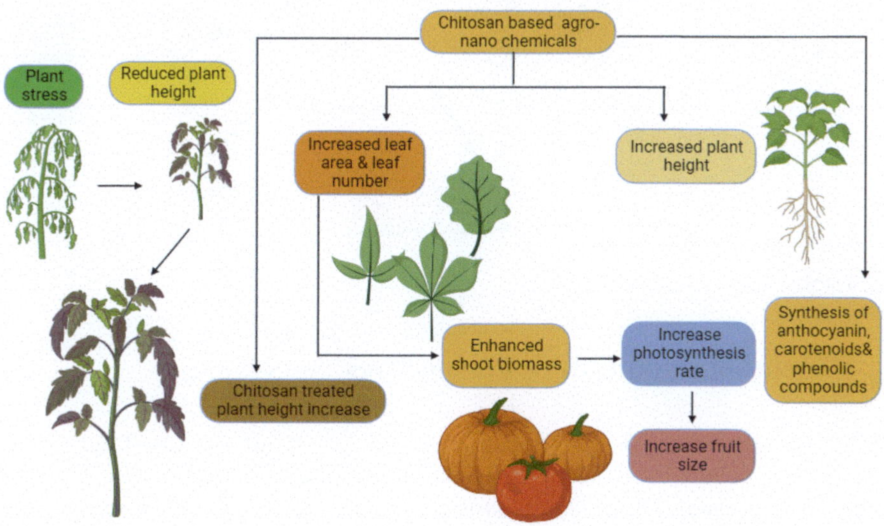

Fig. 7 Chitosan based agro nano chemicals regulate plant growth & fruit size under stressful conditions

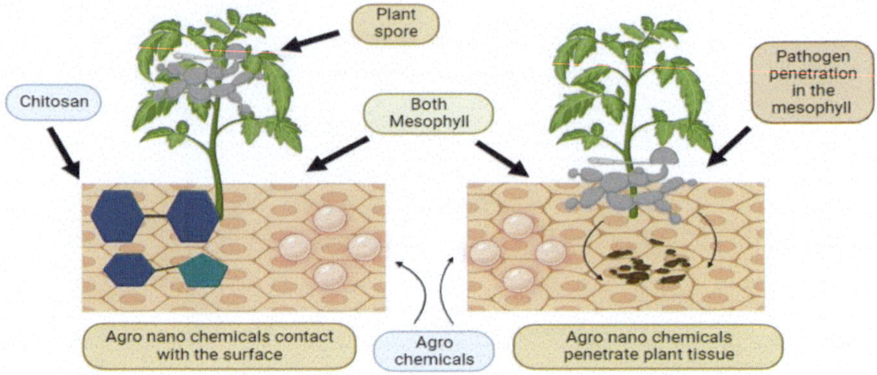

Fig. 8 Agro nanochemicals sprey on the plant leaf surface and penetration of soil surface

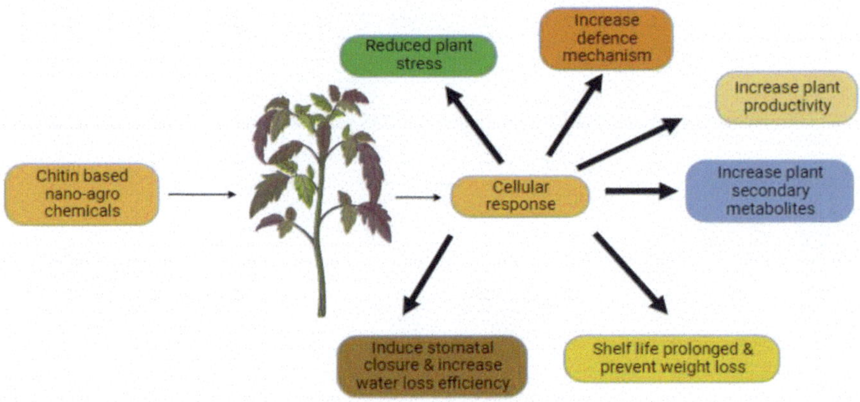

Fig. 9 Chitin act as biostimulant in plant

6 Conclusions and Future Perspectives

The distinct antibacterial characteristics of chitin and chitosan-based agro nano chemicals stimulate researchers to consider their application in crop protection and biomedical research. Chitin and Chitosan and their agro nano chemical derivatives have proved their efficiency in both the industrial and agricultural sectors. Chitin and Chitosan have proved their efficiency to enhance crop production and to provide protection against infectious diseases. Many agriculture research scientists and scholars have shown interest in utilizing chitin and Chitosan and their agro nano chemical derivatives for crop protection and to enhance crop production in the agricultural sectors. This interest is expected to result in rapid improvement and the prolonged application of these chemicals in crop production and crop protection.

Chitin and Chitosan and their agronano chemical derivatives can serve as a future substitute for harmful and non-biodegradable substances. Chitin and Chitosan and their agronano chemical derivatives exhibit biocatalytic and biodegradable properties. The available research suggests that chitin and chitosan-based agro nano chemicals have the immense potential to be environmentally friendly, safe, and versatile in nature. Chitin and Chitosan and their agronano chemical derivatives have a wide range of applications in the protection of crops. Employing chitin and chitosan-based agro nano compounds will specifically benefit agriculture research scientists and scholars involved in enhancing crop production and crop protection. Chitin and chitosan-based agro nano chemicals have an application as a cost-effective substitute for harmful and hazardous chemicals in the agricultural sectors, specifically for the protection of crops. Therefore, there is no doubt that using chitin and chitosan-based agro nano chemicals for crop protection and development of crop production will be a great future prospect in the agriculture sector and industry.

References

1. Abd-Elsalam KA, Vasilkov AY, Said-Galiev EE, Rubina MS, Khokhlov AR, Naumkin AR, Shtykova EV, Alghuthaymi MA (2018) Bimetallic blends and chitosan nanocomposites: novel antifungal agents against cotton seedling damping-off. Eur J Plant Pathol 151:57–72
2. Abo Elsoud MM, El Kady EM (2019) Current trends in fungal biosynthesis of chitin and chitosan. Bull Natl Res Centre 43(1):59. https://doi.org/10.1186/s42269-019-0105-y
3. Akila RM (2014) Fermentative production of fungal chitosan, a versatile biopolymer (perspectives and its applications). Adv Appl Sci Res 5(4):157–170
4. Angelopoulou P, Giaouris E, Gardikis K (2022) Applications and prospects of nanotechnology in food and cosmetics preservation. Nanomaterials 12:1196
5. Arbia W, Arbia L, Adour L, Amrane A (2013) Chitin extraction from crustacean shells using biological methods–a review. Food Technol Biotechnol 51(1):12–25
6. Arie T (2019) Fusarium diseases of cultivated plants, control, diagnosis, and molecular and genetic studies. J Pestic Sci 44(4):275–281
7. Athawale V, Paralikar P, Avinash P, Ingle AP, Rai M (2018) Biogenically engineered NPs inhibit Fusarium oxysporum causing soft-rot of ginger. IET Nanobiotechnol 12:1084–1089
8. Ayanda OI, Ajayi T, Asuwaju FP (2020) Eichhornia crassipes (Mart.) Solms: uses, challenges, threats, and prospects. Sci World J 1:3452172
9. Bachheti RK, Bachheti A, Husen A (2024) Metal and metal-oxide based nanomaterials (Synthesis, agricultural, biomedical and environmental interventions). Springer Nature Singapore Pte Ltd., Singapore
10. Bachheti A, Bachheti RK, Husen A (2024) Carbon-based smart nanomaterials (Synthesis, agricultural, biomedical and environmental interventions). Springer Nature Singapore Pte Ltd., Singapore. https://doi.org/10.1007/978-981-97-0240-4, ISBN: 978-981-97-0239-8
11. Bansal K, Hooda V, Verma N, Kharewal T, Tehri N, Dhull V, Gahlaut A (2022) Stress alleviation and crop improvement using silicon NPs in agriculture. A review. SILICON 14:10173–10186
12. Bharani RSA, Namasivayam SKR, Shankar SS (2014) Biocompatible chitosan NPs incorporated pesticidal protein Beauvericin (Csnp-Bv) preparation for the improved pesticidal activity against major groundnut defoliator *Spodoptera Litura* (Fab.) (Lepidoptera; Noctuidae). Int J ChemTech Res 6(12):5007–5012

13. Bhatia M (2020) Encapsulation of active molecules and their delivery system a review on application of encapsulation in agricultural processes. Elsevier, Amsterdam, The Netherlands, pp 131–140
14. Casadidio C, Peregrina DV, Gigliobianco MR, Deng S, Censi R (2019) Chitin and chitosans: characteristics, eco-friendly processes, and applications in cosmetic science. Mar Drugs 17(6):369. https://doi.org/10.3390/md17060369
15. Chun SC, Chandrasekaran M (2019) Chitosan and chitosan NPs induced expression of pathogenesis-related proteins genes enhances biotic stress tolerance in tomato. Int J Biol Macromol 125:948–954
16. Crini G (2019) Historical review on chitin and chitosan biopolymers. Environ Chem Lett 17(4):1623–1643. https://doi.org/10.1007/s10311-019-00901-0
17. Dayakar B, Xavier KM, Ngasotter S, Layana P, Balange AK, Priyadarshini B, Nayak BB (2022) Characterization of spray-dried carotenoprotein powder from Pacific white shrimp (Litopenaeusvannamei) shells and head waste extracted using papain, Antioxidant, spectroscopic and microstructural properties. LWT 159, Article 113188. https://doi.org/10.1016/j.lwt.2022.113188
18. Dutta J, Tripathi S, Dutta PK (2012) Progress in antimicrobial activities of chitin, chitosan and its oligosaccharides: a systematic study needs for food applications. Food Sci Technol Int 18(1):3–34. https://doi.org/10.1177/1082013211399195
19. Egusa M, Matsui H, Urakami T, Okuda S, Ifuku S, Nakagami H, Kaminaka H (2015) Chitin nanofibers elucidates the elicitor activity of polymeric chitin in plants. Front Plant Sci 6:1098–1105
20. Egusa M, Parada RY, Aklog YF, Ifuku S, Kaminaka H (2019) Nanofibrillation enhances the protective effect of crab shells against Fusarium wilt disease in tomato. Int J Biol Macromol 128:22–27
21. Egusa M, Matsukawa S, Miura C, Nakatani S, Yamada J, Endo T, Ifuku S, Kaminaka H (2020) Improving nitrogen uptake efficiency by chitin nanofiber promotes growth in tomato. Int J Biol Macromol 151:1322–1331
22. Ehrlich H, Shaala LA, Youssef DT, Żółtowska-Aksamitowska S, Tsurkan M (2018) Discovery of chitin in skeletons of non-verongiid Red Sea demosponges. PLoS ONE 13(5):e0195803. https://doi.org/10.1371/journal.pone.0195803
23. Ferraro V, Cruz IB, Jorge RF, Malcata FX, Pintado ME (2010) Valorisation of natural extracts from marine source focused on marine by-products: a review. Food Res Int 43(9):2221–2233. https://doi.org/10.1016/j.foodres.2010.07.034
24. Fontana R, Caproni A, Buzzi R, Sicurella M, Buratto M, Salvatori F, Marconi P (2021) Effects of Moringa oleifera leaf extracts on Xanthomonas campestris pv. campestris. Microorganisms 9(11):2244
25. Gomaa ZE (2018) Iron NPs a-chitin nanocomposite for enhanced antimicrobial, dyes degradation and heavy metals removal activities. J Polym Environ 26:3638–3654
26. Hahn T, Roth A, Ji R, Schmitt E, Zibek S (2020) Chitosan production with larval exoskeletons derived from the insect protein production. J Biotechnol 310:62–67. https://doi.org/10.1016/j.jbiotec.2019.12.015
27. Hatchett C (1799) Experiments and observations on shell and bone. In: Bowyer W, Nichols J (eds) Transactions of the royal society of London, vol 89. Royal Society, London, pp 315–334
28. He M, He CQ, Ding NZ (2018) Abiotic stresses: General defenses of land plants and chances for engineering multistress tolerance. Front Plant Sci 9:1771
29. Husen A (2020) Introduction and techniques in nanomaterials formulation. In: Husen A, Jawaid M (eds) Nanomaterials for agriculture and forestry applications. Elsevier Inc., Cambridge, pp. 1–14. https://doi.org/10.1016/B978-0-12-817852-2.00001-9
30. Husen A (2022) Engineered nanomaterials for sustainable agricultural production, soil improvement and stress management. Elsevier Inc., Cambridge. https://doi.org/10.1016/C2021-0-00054-7
31. Husen A (2023) Secondary metabolites based green synthesis of nanomaterials and their applications. Springer Nature Singapore Pte Ltd., Singapore. https://doi.org/10.1007/978-981-99-0927-8

32. Husen A (2023) Nanomaterials from agricultural and horticultural products. Springer Nature Singapore Pte Ltd., Singapore. https://link.springer.com/book/9789819934348
33. Husen A (2023) Nanomaterials and nanocomposites exposures to plants (Response, interaction, phytotoxicity and defense mechanisms). Springer Nature Singapore Pte Ltd., Singapore. https://doi.org/10.1007/978-981-99-2419-6
34. Im Y, Park SE, Lee SY, Kim JC, Kim JS (2022) Early-stage defense mechanism of the cotton aphid Aphis gossypii against infection with the insect-killing fungus Beauveria bassiana JEF-544. Front Immunol 13:907088
35. Islam Bhuiya MAR, Islam MN (2017) Chitin and chitosan: structure, properties and applications in biomedical engineering. J Polym Environ 25(3):854–866. https://doi.org/10.1007/s10924-016-0865-5
36. Jang MK, Kong BG, Jeong YI, Lee CH, Nah JW (2004) Physicochemical characterization of α-chitin, β-chitin, and γ-chitin separated from natural resources. J Polym Sci Part A-Polym Chem 42(14):3423–3432. https://doi.org/10.1002/(ISSN)1099-0518
37. Jeevanandam J, Barhoum A, Chan YS, Dufresne A, Danquah MK (2018) Review on NPs and nanostructured materials: history, sources, toxicity and regulations. Beilstein J Nanotechnol 9:1050–1074. https://doi.org/10.3762/bjnano.9.98
38. Kalagatur NK, Ghosh OSN, Sundararaj N, Mudili V (2018) Antifungal activity of chitosan NPs encapsulated with Cymbopogon martinii essential oil on plant pathogenic fungi Fusarium graminearum. Front Pharmacol 9:610. https://doi.org/10.3389/fphar.2018.00610
39. Kaur P, Duhan JS, Thakur R (2018) Comparative pot studies of chitosan and chitosan-metal nanocomposites as nano-agrochemicals against fusarium wilt of chickpea (Cicer arietinum L.). Biocatal Agric Biotechnol 14:466–471
40. Kaya M, Lelesius E, Nagrockaite R, Sargin I, Arslan G (2015) Differentiations of chitin content and surface morphologies of chitins extracted from male and female grasshopper species. PLoS ONE 10(1):e0115531. https://doi.org/10.1371/journal.pone.0115531
41. Keerthana P, Vijayakumar S, Vidhya E, Punitha VN, Nilavukkarasi M, Praseetha PK (2021) Biogenesis of ZnO NPs for revolutionizing agriculture: a step towards anti-infection and growth promotion in plants. Ind Crops Prod 170:113762
42. Khoo YW, Baadu R, Shaiddin A, Tan HT, Khaw YS, Li SF, Chong KP (2023) First report of Curvularia lunata causing leaf spot on Oryza sativa in Sabah, Malaysian Borneo. Plant Dis 107(7):2233
43. Khoo YW, Chong KP (2023) Ganoderma boninense: general characteristics of pathogenicity and methods of control. Front Plant Sci 14:1156869
44. Klinger C, Żółtowska-Aksamitowska S, Wysokowski M, Tsurkan MV, Galli R (2019) Express method for isolation of ready-to-use 3D chitin scaffolds from Aplysinaarcheri (Aplysineidae: Verongiida) demosponge. Mar Drugs 17(3):131. https://doi.org/10.3390/md17020131
45. El Knidri H, Belaabed R, Addaou A, Laajeb A, Lahsini A (2018) Extraction, chemical modification and characterization of chitin and chitosan. Int J Biol Macromol 120:1181–1189. https://doi.org/10.1016/j.ijbiomac.2018.08.139
46. Knorr D (1984) Use of chitinous polymer in food–a challenge for food research and development. Food Technol 38(1):85–97
47. Kour D, Rana KL, Yadav AN, Yadav N, Kumar M, Kumar V, Saxena AK (2020) Microbial biofertilizers: bioresources and eco-friendly technologies for agricultural and environmental sustainability. ISBAB. 23:101487
48. Kumaraswamy RV, Kumari S, Choudhary RC, Pal A, Raliya R, Biswas P, Saharan V (2018) Engineered chitosan based nanomaterials: bioactivities, mechanisms and perspectives in plant protection and growth. Int J Biol Macromol 113:494–506
49. Kumaraswamy RV, Kumari S, Choudhary RC, Sharma SS, Pal A, Raliya R, Biswas P, Saharan V (2019) Salicylic acid functionalized chitosan nanoparticle: a sustainable biostimulant for plant. Int J Biol Macromol 123:59–69
50. Kumari S, Rath PK (2014) Extraction and characterization of chitin and chitosan from (Labeorohit) fish scales. Procedia Mater Sci 6:482–489. https://doi.org/10.1016/j.mspro.2014.07.062

51. Kumbhakar P, Ray SS, Stepanov AL (2014) Optical properties of NPs and nanocomposites. J Nanomater 181365. https://doi.org/10.1155/2014/181365
52. Leso V, Fontana L, Iavicoli I (2019) Biomedical nanotechnology: occupational views. Nano Today 24:10–14. https://doi.org/10.1016/j.nantod.2018.11.002
53. Li WQ, Qing T, Li CC, Li F, Ge F, Fei JJ (2020). Integration of subcellular partitioning and chemical forms to understand silver NPs toxicity to lettuce (Lactuca sativa L.) under different exposure pathways. Chemosphere 258:127349. https://doi.org/10.1016/j.chemosphere.2020.127349
54. Liang R, Li X, Yuan W, Jin S, Hou S, Wang M, Wang H (2018) Antifungal activity of nanochitin whisker against crown rot diseases of wheat. J Agric Food Chem 66:9907–9913
55. Liang W, Yu A, Wang G, Zheng F, Hu P, Jia J, Xu H (2018) A novel water-based chitosan-La pesticide nanocarrier enhancing defense responses in rice (*Oryza sativa* L) growth. Carbohydr Polym 199:437–444
56. Longya A, Talumphai S, Jantasuriyarat C (2020) Morphological characterization and genetic diversity of rice blast fungus, Pyricularia oryzae, from Thailand using ISSR and SRAP markers. J Fungi 6(1):38
57. Luan LQ, Xo DH (2018) In vitro and in vivo fungicidal effects of γ-irradiation synthesized silver NPs against Phytophthora capsici causing the foot rot disease on pepper plant. J Plant Pathol 100:241–248
58. Madhusudhan A, Purohit SD, Prasad R, Husen A (2024) Functional smart nanomaterials and their theranostics approaches. Springer Nature Singapore Pte Ltd., Singapore
59. Malerba M, Cerana R (2016) Chitosan effects on plant systems. Int J Mol Sci 17:996–1010
60. Malerba M, Cerana R (2020) Chitin- and chitosan-based derivatives in plant protection against biotic and abiotic stresses and in recovery of contaminated soil and water. Polysaccharides 1(1):21–30. https://doi.org/10.3390/polysaccharides1010003
61. Manjunatha N, Sharma J, Pokhare SS, Agarrwal R, Patil P G, Sirsat JD, Marathe RA (2022) Characterization of Alternaria and Colletotrichum species associated with pomegranate (Punica granatum L.) in Maharashtra state of India. J Fungi 8(10):1040
62. Martin-Saldaña S, Chevalier MT, Iglesias MJ, Colman SL, Casalongué CA, Álvarez VA, Chevalier AA (2018) Salicylic acid loaded chitosan microparticles applied to lettuce seedlings: recycling shrimp fishing industry waste. Carbohydr Polym 200:321–331
63. Mazayen ZM, Ghoneim AM, Elbatanony RS, Basalious EB, Bendas ER (2022) Pharmaceutical nanotechnology: from the bench to the market. Future J Pharm Sci 8:12
64. Merino D, Mansilla AY, Casalongué CA, Alvarez VA (2018) Preparation, characterization and in vitro testing of nanoclay antimicrobial activities and elicitor capacity. J Agric Food Chem 66:3101–3109
65. Morin-Crini N, Lichtfouse E, Torri G, Crini G (2019) Applications of chitosan in food, pharmaceuticals, medicine, cosmetics, agriculture, textiles, pulp and paper biotechnology and environmental chemistry. Environ Chem Lett 17(4):1667–1692. https://doi.org/10.1007/s10311-019-00904-x
66. Mugao LG, Muturi PW, Gichimu BM, Njoroge EK (2020) In vitro control of Phytophthora infestans and Alternaria solani using crude extracts and essential oils from selected plants. Int J Agronomy 1–10
67. Muley AB, Shingote PR, Patil AP, Dalvi SG, Suprasanna P (2019) Gamma radiation degradation of chitosan for application in growth promotion and induction of stress tolerance in potato (*Solanum tuberosum* L.). Carbohydr Polym 210:289–301
68. Nadendla SR, Rani TS, Vaikuntapu PR, Maddu RR, Podile AR (2018) HarpinPss encapsulation in chitosan NPs for improved bioavailability and disease resistance in tomato. Carbohydr Polym 199:11–19
69. New N, Furuike T, Tamura H (2011) Chitosan from aquatic and terrestrial organisms and microorganisms. Production, properties and applications. In: Johnson BM, Berkel ZE (eds) Biodegradable materials. Nova Science Publishers Inc., New York, pp 29–50.
70. Nhien LTA, Luong ND, Tien LTT, Luan LQ (2018) Radiation synthesis of silver NPs/chitosan for controlling leaf fall disease on rubber trees causing by Corynesporacassiicola. J Nanomater 1:7121549. https://doi.org/10.1155/2018/7121549

71. Nimah YL, Harmami H, Ulfin I, Suprapto S, Welny Saleh C (2019) Water-soluble chitosan preparation from marine sources. Malays J Fundam Appl Sci 15(2):159–163. https://doi.org/10.11113/mjfas.v15n2.971
72. Odier A (1823) Mémoir sur la composition chimique des parties cornées des insectes. Mémoirs de la Societéd'Histoire Naturelle 1:29–42
73. Omotayo OP, Babalola OO (2023) Fusarium verticillioides of maize plant: potentials of propitious phytomicrobiome as biocontrol agents. Front Fungal Biol 4:1095765
74. Palanga KK, Liu R, Ge Q, Gong J, Li J, Lu Q, Gong W (2021) Current advances in pathogen-plant interaction between Verticillium dahliae and cotton provide new insight in the disease management. J Cotton Res 4(1):1–13
75. Parada RY, Egusa M, Aklog YF, Miura C, Ifuku S, Kaminaka H (2018) Optimization of nanofibrillation degree of chitin for induction of plant disease resistance: elicitor activity and systemic resistance induced by chitin nanofiber in cabbage and strawberry. Int J Biol Macromol 118:2185–2192
76. Pathan FL, Deshmukh RR, Annapure US (2021) Potential of cold plasma to control Callosobruchus chinensis (Chrysomelidae: Bruchinae) in chickpea cultivars during four year storage. Sci Rep 11(1):13425
77. Popova EV, Domnina NS, Kovalenko NM, Sokornova SV, Tyuterev SL (2018) Influence of chitosan hybrid derivatives on induced wheat resistance to pathogens with different nutrition strategies. Appl Biochem Microbiol 54:535–539
78. Pouratashi M, Iravani H (2012) Farmers' knowledge of integrated pest management and learning style preferences: implications for information delivery. Int J Pest Manag 58:347–353. https://doi.org/10.1080/09670874.2012.724468
79. Pramanik P, Krishnan P, Maity A, Mridha N, Mukherjee A, Rai V (2020) Application of nanotechnology in agriculture. Environ Nanotechnol 4:317348
80. Pusztahelyi T (2018) Chitin and chitin-related compounds in plant–fungal interactions. Mycology 9:189–201
81. Quesada HB, Baptista ATA, Cusioli LF, Seibert D, Bezerra CO, Bergamasco R (2019) Surface water pollution by pharmaceuticals and an alternative of removal by low-cost adsorbents: a review. Chemosphere 222:766–780
82. Rai P, Singh VP, Sharma S, Tripathi DK, Sharma S (2022) Iron oxide NPs impart cross tolerance to arsenate stress in rice roots through involvement of nitric oxide. Environ Pollut 307:119320
83. Rath SK, Dwivedi V, Husen A, Akhtar N (2024) Smart nanomaterials for drug delivery and neurological diseases management. Springer Nature Singapore Pte Ltd., Singapore. https://doi.org/10.1007/978-981-97-0308-1, ISBN: 978-981-97-0307-4
84. Roohallah SR, Mozhgan GV, John K (2023) A review: the application of chitosan as a carrier for fertilizer. Int J Biol Macromol 252:126483
85. Ruiz GAM, Corrales HFZ (2017) Chitosan, chitosan derivatives and their biomedical applications. In: Shalaby E (eds) Biological activities and application of marine polysaccharides. In Tech, London, UK, pp 87–106
86. Sahu PK, Brahmaprakash GP (2016) Microbial inoculants in sustainable agricultural productivity formulations of biofertilizers approaches and advances. Funct Appl 2:179–198
87. Sajomsang W, Gonil P (2010) Preparation and characterization of ʊ-chitin from cicada sloughs. Mater Sci Eng 30(3):357–363. https://doi.org/10.1016/j.msec.2009.11.014
88. Sampantamit T, Ho L, Lachat C, Sutummawong N, Sorgeloos P (2020) Aquaculture production and its environmental sustainability in Thailand: challenges and potential solutions. Sustainability 12(5):2010. https://doi.org/10.3390/su12052010
89. Sampath L, Ngasotter S, Layana P, Balange AK, Nayak BB, Xavier KM (2022) Effect of chemical treatment duration on physicochemical, rheological, and functional properties of colloidal chitin. Food Hydrocolloids Health 2: Article 100091. https://doi.org/10.1016/j.fhfh.2022.100091
90. Sathiyabama M, Indhumathi M, Muthukumar S (2019) Chitosan NPs loaded with thiamine stimulate growth and enhances protection against wilt disease in Chickpea. Carbohydr Polym 212:169–177

91. Sathiyabama M, Manikandan A (2018) Application of copper-chitosan NPs stimulate growth and induce resistance in finger millet (Eleusine coracana Gaertn.) plants against blast disease. J Agric Food Chem 66:1784–1790
92. Sayed S M, Siems K, Schmitz-Linneweber C, Luyten W, Saul N (2021) Enhanced healthspan in Caenorhabditis elegans treated with extracts from the traditional Chinese medicine plants Cuscuta chinensis Lam. and Eucommia ulmoides Oliv. Front Pharmacol 12, 604435
93. Senapati M, Tiwari A, Sharma N, Chandra P, Bashyal BM, Ellur RK, Krishnan SG (2022) Rhizoctonia solani Kühn pathophysiology: status and prospects of sheath blight disease management in rice. Front Plant Sci 13:881116
94. Shaala LA, Asfour HZ, Youssef DT, Żółtowska-Aksamitowska S, Wysokowski M (2019) New source of 3D chitin scaffolds: The Red Sea demosponge pseudoceratinaarabica (Pseudoceratinidae, Verongiida). Mar Drugs 17(2):92. https://doi.org/10.3390/md17020092
95. Silveira NM, Seabra AB, Marcos FCC, Pelegrino MT, Machado EC, Ribeiro RV (2019) Encapsulation of S-nitrosoglutathione into chitosan NPs improves drought tolerance of sugarcane plants. Nitric Oxide 84:38–44
96. Singh D, Boopathi T (2022) Callosobruchus chinensis (Coleoptera: Chrysomelidae): biology, life table parameters, host preferences, and evaluation of green gram germplasm for resistance. J Stored Prod Res 95:101912
97. Solanki P, Bhargava A, Chhipa H, Jain N, Panwar J (2015) Nano-fertilizers and their smart delivery system. In: Rai M (ed) Nanotechnologies in food and agriculture. Springer, Cham, pp 81–101. https://doi.org/10.1007/978-3-319-14024-74
98. Stefan L, Monchaud D (2019) Applications of guanine quartets in nanotechnology and chemical biology. Nat Rev Chem 3:650–668. https://doi.org/10.1038/s41570-019-0132-0
99. Sudha PN, Sangeetha K, Vijayalakshmi K, Barhoum A (2018) Nanomaterials history, classification, unique properties, production and market. In: Emerging applications of NPs and architecture nanostructures. Elsevier, Amsterdam, The Netherlands, pp 341–384
100. Sun C, Fu D, Jin L, Chen M, Zheng X, Yu T (2018) Chitin isolated from yeast cell wall induces the resistance of tomato fruit to Botrytis cinerea. Carbohydr Polym 199:341–352
101. Surendranath A, Mohanan PV (2020) Impact of NPs in balancing the ecosystem. Biointerface Res Appl Chem 11:10461–10481
102. Tiroesele B, Thomas K, Seketeme S (2014) Control of cowpea weevil, Callosobruchus maculatus (F.) (Coleoptera: Bruchidae), using natural plant products. Insects 6(1):77–84
103. Usman M, Farooq M, Wakeel A, Nawaz A, Cheema SA, UR Rehman H (2020) Nanotechnology in agriculture: current status, challenges and future opportunities. Sci Total Environ 721:137778. https://doi.org/10.1016/j.scitotenv.2020.137778
104. Winkler AJ, Dominguez-Nuñez JA, Aranaz I, Poza-Carrión C, Ramonell K, Somerville S, Berrocal-Lobo M (2017) Short-chain chitin oligomers: promoters of plant growth. Mar Drugs 15:40
105. Wysokowski M, Bazhenov VV, Tsurkan MV, Galli R, Stelling AL (2013) Isolation and identification of chitin in three-dimensional skeleton of aplysinafistularis marine sponge. Int J Biol Macromol 62:94–100. https://doi.org/10.1016/j.ijbiomac.2013.08.039
106. Xin XF, Kvitko B, He SY (2018) Pseudomonas syringae: what it takes to be a pathogen. Nat Rev Microbiol 16(5):316–328
107. Yan N, Chen X (2015) Sustainability: don't waste seafood waste. Nature 524(7564):155–157. https://doi.org/10.1038/524155a
108. Yang F, Jacobsen S, Jørgensen HJ, Collinge DB, Svensson B, Finnie C (2013) Fusarium graminearum and its interactions with cereal heads: studies in the proteomics era. Front Plant Sci 4:37
109. Yang G, Jin O, Xu C, Fan S, Wang C, Xie P (2018) Synthesis, characterization and antifungal activity of coumarin-functionalized chitosan derivatives. Int J Biol Macromol 106:179–184
110. Yang X, Liu J, Pei Y, Zheng X, Tang K (2020) Recent progress in preparation and application of nano-chitin materials. Energy & Environ Mater 3(4):492–515. https://doi.org/10.1002/eem2.12079

111. Yaqoob AA, Parveen T, Umar K, Mohamad MN (2020) Role of nanomaterials in the treatment of wastewater: a review. Water 12:495. https://doi.org/10.3390/w12020495
112. Yihun FA (2022) Nanochitin preparation and its application in polymer nanocomposites: a review. Emergent Mater 5:1–30. https://doi.org/10.1007/s42247-022-00427-2
113. Zargar V, Asghari M, Dashti A (2015) A review on chitin and chitosan polymers: structure, chemistry, solubility, derivatives, and applications. Chem Bio Eng Rev 2(3):204–226. https://doi.org/10.1002/cben.201400025
114. Zhang H, Zheng X, Zhang Z (2016) The Magnaporthe grisea species complex and plant pathogenesis. Mol Plant Pathol 17(6):796
115. Zhou Y, Jiang S, Jiao Y, Wang H (2017) Synergistic effects of nanochitin on inhibition of tobacco root rot disease. Int J Biol Macromol 99:205–221

Challenges and Limitations of Chitin-Based Nanoparticles in Agriculture

Meseret Zebeaman Birhanu, Rahel Gebeyehu Shiferaw, Rakesh Kumar Bachheti, and Ibrahim Nasser Ahmed

Abstract Synthetic fertilizers and pesticides are causing environmental pollution through eutrophication and persistence. Due to this, the synthetic chemicals used in the agriculture sector need to be replaced with natural sources. Several researches indicated chitin, the second most abundant natural polymer next to cellulose, and its derivative, nanochitn, have shown promising uses in the agricultural sector. The main potential areas are growth promoters, fertilizers, insecticidal, antifungal, seed treatment, and antitranspirant or leaf sealant. However, chitin still needs to be fully utilized because of various challenges and limitations. These include chitin isolation and nano chitin preparation difficulties, insect pests' nature and resistance, and chitin's phytotoxicity. In addition, Chitin's low water solubility and adsorption of heavy metals limited its use in agriculture; for that, its nano form, nano chitin, was reported to solve these drawbacks. Therefore, the application to the limitation of chitin nanoparticles is reviewed in this chapter.

Keywords Chitin · Nanochitin · Agricultural application · Limitation

1 Introduction

Although much work is needed, Chitin nanoparticles use in agriculture is at a promising stage for future agriculture advancement. This is because chitin is natural, the future substitute for problematic synthetic agricultural chemicals. For example,

M. Z. Birhanu · I. N. Ahmed (✉)
Nanotechnology Center of Excellence, Addis Ababa Science and Technology University, P.O. Box 16417, Addis Ababa, Ethiopia
e-mail: ibrahim.nasser@aastu.edu.et

M. Z. Birhanu · R. K. Bachheti · I. N. Ahmed
Department of Industrial Chemistry, College of Natural and Applied Sciences, Addis Ababa Science and Technology University, P.O. Box 16417, Addis Ababa, Ethiopia

R. G. Shiferaw
Communicable and Non-Communicable Diseases Research Directorate, Armauer Hansen Research Institute, P. O. Box 1005, Addis Ababa, Ethiopia

R. K. Bachheti et al. (eds.), *Chitin-Based Nanoparticles for the Agriculture Sectors*, Smart Nanomaterials Technology, https://doi.org/10.1007/978-981-96-0920-8_10

25 and 50% of crop yield loss in developed and developing countries is due to plant pathogens (fungi and pests) [22]. Naturally derived insecticidal and antifungal products protect plant growth and increase crop yield. Surprisingly, antagonist bacteria that produce chitinase are used to kill plant fungi. This enzyme, chitinase, hydrolyzes fungi cell walls. For example, *Streptomyces rimosus* exhibited in vitro antifungal properties against *Fusarium solani* and *Alternaria alternate* [3]. Similarly, the enzyme chitinase is also used to kill plant pests or insects with the same mechanism of action. For example, a gram-positive bacteria called *Bacillus thuringiensis* is one of the known insecticidals used as a bio-control agent [22]. The problem with using antagonistic bacteria as a biocontrol insecticide and antifungal is that the antagonist bacteria is pathogenic to human beings. Therefore, an alternative like chitin-based nanoparticle is in high need. Chitin is abundant, and its nanoparticles are much more biocompatible than chitin.

1.1 Overview of Chitin-Based Nanoparticles in Agriculture

Chitin powder is the starting point for chitin based nanoparticle preparation. Chitin powder preparation from the exoskeleton of marine animals can be done either following biological or chemical/conventional methods. The chitin nanocrystal or chitin nanoparticle is made from chitin powder. They are also called chitin nanofibers or chitin nano whiskers. The other main derivative of chitin is chitosan, which is made by deacetylation of chitin. Therefore, chitin-based nanoparticle encompasses all chitin and chitosan-based nanoparticles. In a broader sense, combining chitin or chitosan nanoparticles with other bioactive compounds or elements can exert their physiological effect on the target dangerous species. For instance, from the many uses of nanoparticles, cationic nano chitin shows promising insecticidal activity against insect pests. It is described later in this chapter.

Although chitosan is derived from chitin, the publication made from 2002 up to 2022 in a Web of Science database indicates that chitosan-based nanoparticle increment, compared to chitin, is quite unbelievable. Of the total of 34 910 chitin nanoparticle types of publications, almost 32,554 are all about chitosan nanoparticles, whereas the remaining 2,356 are only about chitin nanoparticles [24]. Therefore, for convenience, since more work is found on the agricultural use of chitosan-based nanoparticles than chitin-based nanoparticles, in this chapter, most chitin-based nanoparticle research work is reviewed than chitosan nanoparticles.

Generally, applications of chitin-based nanoparticles in agriculture are classified into three major areas. The first use is as drug delivery. A study done by [12] indicates that cationic nano chitin used as delivery of commercial pesticide against white aphides insects reduces the amount of commercial pesticide use by half. The other application is as a stabilizing agent. For example, a study done by [11] indicates that the biopesticide Abamectin became more stable when nano chitin was used as a stabilizing agent. The third application is miscellaneous. For example, antifungal [13], heavy metal removal [6]. A study done by [13] showed cationic nano chitin crystal, which has a rod-like structure, is effective against plant fungi that cause Crown rot. Crown rot is a disease of wheat plants.

1.2 Importance of Understanding Challenges and Limitations

The reason why more chitosan-based nanoparticle publication is found than chitin nanoparticles worldwide is that chitosan is more soluble in acidic, basic, and neutral solutions than chitin [8]. The second reason may be an acetyl group attached to the nitrogen atom in chitin. The third reason may be that Chitin has high nitrogen content and low C/N ratio. Therefore, chitin can be directly used as a fertilizer to enhance crop growth [16]. Due to the above probable reasons, less publication of chitin nanoparticles regarding its agricultural use has been found.

Similarly, even though chitin is the second most abundant biomass than cellulose, the research done and real-world application of chitin and chitin nanoparticles is much less than cellulose and cellulose nanocrystals [29]. Nevertheless, the challenges and the limitations of chitin nanoparticles in agriculture are explained in Sects. 4 and 5 of this chapter. Understanding the challenge and limitations of original chitin powder extraction and its non-solubility up to applying chitin nanomaterial is crucial for advancing the agricultural use of chitin powder and chitin nanocrystals. However, the structural difference between chitin and chitosan's effect on the stability of their respective nanomaterial has yet to be studied, and no research data has been found on why more publication is done on chitosan than chitin.

2 Properties of Chitin-Based Nanoparticles

2.1 Brief Overview of Chitin and Its Derivatives

Chitin is the second most abundant natural polymer next to cellulose. While cellulose is mainly found in plants, chitin is found chiefly in sea animals as exoskeletons of crustacean shells. Annually, 10^{11} tonnes of chitin are produced. The production time takes a total of 17–72 h. Although chitin and cellulose are chemically similar, chitin is not derived from cellulose or vice versa [5]. Currently, the primary commercial sources of chitin are the shells of crustaceans such as crabs, shrimps, and lobsters. Chemically chitin is an amino polysaccharide consisting of repeating $\beta(1 \rightarrow 4)$-linked N-acetyl-D-glucosamine units [25]. Chitin has three crystalline allomorphs: α, β, and γ [5]. The structural difference between cellulose and chitin is depicted in Fig. 1.

Chitin adsorbs various heavy metals, pollutants, pesticides, and insecticides from soil or water; chitin and chitosan properties make chitin and chitosan properties

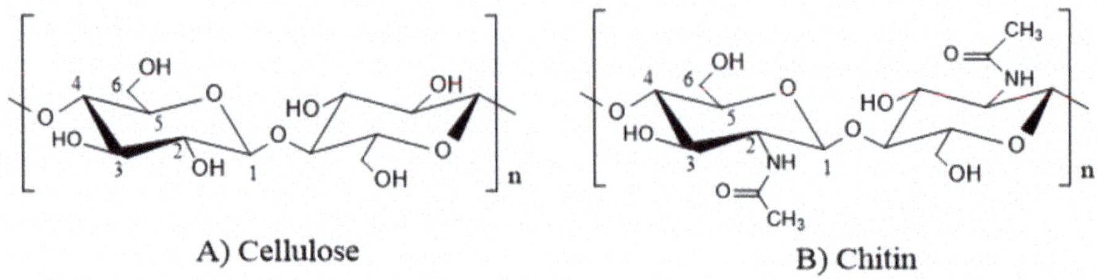

Fig. 1 Chemical structure of **a** cellulose and **b** chitin

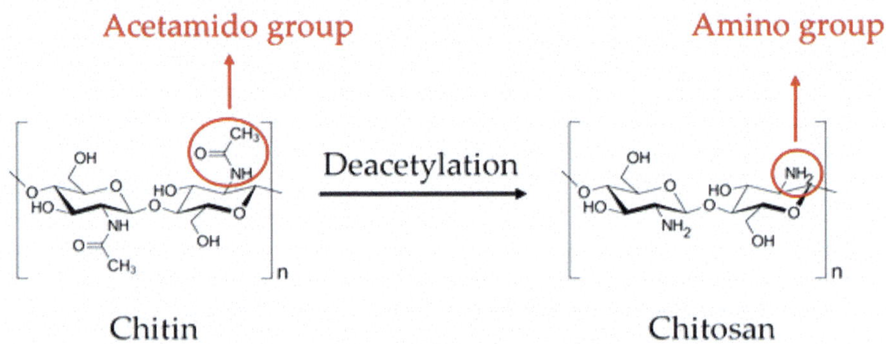

Fig. 2 Chemical conversion of chitin to chitosan [24]

make it inapplicable for agricultural purposes [15]. The other drawback of chitin that hinders agricultural application is its insolubility in water and other conventional solvents. More devastatingly, the specific solvents used to dissolve chitin are also not used in the agricultural sector. However, the nanomaterial of chitin and its deacylation derivative, called chitosan, is soluble in water and other conventional solvents, and hence they can be applied in agriculture instead of the normal chitin. The chemical conversion of chitin to chitosan and their chemical structure difference is shown in Fig. 2. The known nanomaterials of chitin used in agriculture are nano chitin and chitosan or chitin oligosaccharides nanomaterial. Based on research findings, the known agricultural applications of nano chitin include growth promoters, fertilizers, insecticidal, antifungal, seed treatment, and antitranspirant or leaf sealants [12, 19].

2.2 *Characteristics of Chitin-Based Nanoparticles*

Chitin nanocrystal is the most common nanomaterial prepared from chitin powder. Generally, nanomaterials can be prepared either bottom-up or top-down [28]. For example, a study by [30] prepared chitin nanocrystals by the bottom-up method and found a chitin nanoparticle with an average size of 7 nm. They also found that the nanoparticles were stably dispersed in water, had higher heat resistance and lower viscosity, and easily redispersed in water after being concentrated by centrifugation. UV–Vis, FTIR, XRD, SEM, and TGA/DTA were used to characterize the nanoparticle. However, they did not use the DLS instrument to measure the zeta potential of the nanoparticle they synthesized. Similarly, a mechanical property study done by [4] found that the chitin nanoparticle had a rod-like structure, highly crystalline (60–94%) small particles with a high aspect ratio (~10 nm in width and 100–600 nm in length).

Generally, Chitin nanocrystal can be prepared from its powder by using the following methods: hydrochloric acid hydrolysis, TEMPO-mediated oxidation, partial deacetylation with NaOH by fibril surface cationization, ultrasonication, electrospinning, aqueous counter collision treatment, a simple grinding treatment and gelation with 1-allyl-3-methylimidazolium bromide [12]. Chitin nanocrystals are colloidal suspensions made from chitin, which are extracted from biological tissue and then dispersed in aqueous media to form the suspension. Since chitin is a renewable resource, its utilization creates a healthy ecosystem. Chitin nanocrystals exhibit the following properties: nano-sized dimensions, high surface area, high absorptivity, bio-degradability, nontoxicity, renewability, low density, and easy modification. Hence, it attracts many researchers from different fields [17]. However, generally, when the application of nanochitin is investigated, its mechanism of action comes from its high surface area, which makes nanochitin interact effectively with cells, enzymes, proteins, and other compounds. The other advantage of nano chitin is its surface is covered with a hydroxyl group. Hence, modification of chitin under mild conditions is possible, and this gives nano chitin a future candidate in many research and applications [17].

3 Applications in Agriculture

The advancement of agricultural technology is a crucial and necessary step to feed the ever-increasing world population. In this regard, while synthetic chemicals are used to improve agriculture yield, the soil and, generally, the ecosystem is polluted with these synthetic chemicals. Therefore, the use of renewable and degradable chemicals is recommended worldwide. Analogously, chitin and chitin nanocrystals are under deep study, and different researches indicate they are a good candidate for future agricultural advancement. This section summarizes some of the agricultural use and research findings of chitin and its derivative chitosan-based nanoparticle.

3.1 Pest and Disease Management

The annual loss of agriculture yield due to pests, weeds, and diseases is estimated to be 2000US billion dollars [9]. Chitin-based nanoparticles have been deeply studied since

2002 to alleviate such and other problems. For instance a study done by [21] indicates that a pesticide Spinosad and Permethrin loaded chitosan nanoparticle improved insecticidal activity of the pesticides against Drosophila melanogaster. This indicates that synthetic pesticides are reduced, and hence, the problem that comes with its use is also reduced. Similarly, a study done by [20] indicates that chitosan nanoparticles used to encapsulate the essential oil of *Piper nigrum* enhance the essential oil insecticidal activity against *Sitophilus oryzae* and *Tribolium castaneum.*

3.2 Crop Enhancement and Growth Promotion

Using chitosan nanoparticles as plant growth promoters is experimented with at different agricultural plants, including Robusta coffee, wheat, maize, and French beans. The result showed that the chitosan nanoparticle increased the uptake of nutrients (Ca, K, P, N, and so on.) when the nanoparticle was sprayed on the plant [7, 23]. As suggested by [7], the mechanism of action is positively charged rich protonated chitosan demonstrates enhanced affinity towards the cell membranes, resulting in increased reactivity in the plant system. Similarly, a study done by [10] indicates that when nanochitin is mixed with plant-promoting rhizobacteria and applied in Zea mays plant, it increases plant growing parameters like seed germination plant height, and leaf area. Their GC–MS study revealed plant metabolites, including alcohols, acid ester, and aldehyde compounds, were also increased after applying the mixture.

3.3 Soil Health Improvement

The health of the soil is a determinate factor for crop yield increment. In this regard, the application of chitosan nanoparticles in the soil improves the activity of enzymes found in the soil. For instance, a study done by [10] indicates that when chitosan nanoparticle is used, the activity of soil health indicator enzymes, dehydrogenase, fluorescein diacetate hydrolysis, and alkaline phosphatase increased by 2 to 3-fold. The other parameter of soil health is the level and availability of nutrients. So, since chitin and chitosan have high nitrogen content, they can be used as fertilizer by incorporating them with commercial fertilizer [16]. The third application is that although it contradicts with increased nutrient uptake by the plant, chitin and its nanoparticles are used to remove pollutant heavy metals [6]. Hence, the soil health is improved.

4 Challenges

Although nanochitin has such enormous applications in agriculture, its application is also hindered by various challenges and limitations. For example, chitosan nanoparticles synthesized via acid hydrolysis showed some phytotoxicity to plants, including a delay of seed germination or inhibition of plant growth. Mostly plant seed germination, root elongation, and calculations of growth vigour index are usually used as a significant phytotoxic tests. On the other hand, large-scale chitin-base nanoparticle production has yet to be practical [19, 27]. Therefore, this chapter will highlight the challenges and limitations of chitin-based nanoparticles in agriculture separately in Sects. 4 and 5, respectively.

4.1 Chitin Extraction and Nano Chitin Preparation Difficulties

There are two ways of extracting chitin from sea animal derbies. These are chemical and biological methods. The chemical method commonly employs the mineral acid hydrolysis (HCl or H_2SO_4) to prepare nanochitin. However, such an approach has faced various challenges, including transportation of concentrated acid, equipment corrosion, worker risk, high acid prices, and large water loads. To alleviate these problems nowadays, organic acids are in use, even though they claim to have low yield [14]. With all the challenges, the chemical method is still widely used for industrial-scale production. The chemical method uses two steps: first is removing the calcium carbonate from the exoskeletons of crustacean shells using acids (demineralization), then removing the protein by dissolving it in an alkaline solution (deproteinization). On the other hand, the biological method uses proteolytic enzymes to dissolve and remove the protein and the mineral. It gives good-quality chitin with a high degree of acetylation. However, this method does not completely remove the mineral and the proteins from the derbies of exoskelctons of crustacean shells [18, 26].

4.2 Nature and Resistance of Insect Pests

Although biopesticides are recommended over synthetic pesticides because of eutrophication and persistence problems, some biopesticides, like Abamectin, are not stable enough for storage and preparation because they are sensitive and need a short time duration for preparation [11]. Therefore, researchers are looking for alternatives and found that nanomaterials like cationic nano chitin are stable and show good pesticide activity [12]. The other problem is that plant pests are developing resistance against synthetic pesticides. Therefore, there is a high need for alternatives. Although there are different insect pests, only small types are studied against the chitin nanomaterial. The insect pests that have been studied so far are wheat aphids [12] and cotton leaf worms Noctuidae (*Lepidoptera larvae*) [1, 11]. Even if wide-scale and intensive research hasstill needed to be been done, the reported research findings confirm the strong potential of nanochitin against plant pests.

5 Limitations

The first limitation of nanochitin as an anti-bacterial application in agriculture depends on the degree of acetylation, molecular weight, concentration, and nature of the acid used. For example, higher chitin concentrations (above 0.8%) exhibited a phytotoxic effect on plants, whereas chitin at or above 0.4% reduced root galling caused by nematodes [19]. Similarly, chitosan, a chitin derivative with molecular weights in the range of $2.27 \times 10^5 - 5.97 \times 10^5$ g/mol, showed the highest efficacy in pest control [2]. A report also shows chitosan obtained by acid hydrolysis is somewhat toxic to plants [19]. The other limitation the author noticed is that the source of chitin and nanochitin provides good agricultural activity, which has not been clearly studied. There is also minimal research on chitin and nano chitin regarding agricultural use.

6 Future Directions and Potential Solutions

So far, chitin nanocrystals alone and loaded with agrochemicals shows a good result regarding solving agricultural problems. However, there are still problems like large-scale production, in vivo study at the molecular level, knowing the deep mechanism of action, and so on. In the future, more emphasis should be given to a molecular-level study on chitin nanocrystals like the chitin. For example, investigations of the molecular mechanism of Chitin perception and Chitin-induced immunity led to the discovery of a specific receptor, a plasma membrane LysM-containing protein [16].

7 Conclusion

Chitin is the second most abundant natural polymer next to cellulose. It is mainly found in sea animals as marine waste like crustaceans, the waste exoskeletons of crabs, shrimp, and arthropods, and microorganisms such as fungi. Due to water insolubility and adsorption of heavy metals, its application is limited in agriculture. However, its nano form, which is nanochitin, solves the above problems; hence, different research shows nanochitin has six applications in agriculture. These are growth promoters, fertilizers, insecticidal, antifungal, seed treatment, and antitranspirant or leaf sealant. However, some applications have yet to be applied in the real world. This is because of different challenges and limitations. These include chitin isolation and nano chitin preparation difficulties, the nature and resistance of insect pests, and the high concentration of chitin, which shows some phytotoxicity.

References

1. Ayoub HA, Khairy M, Rashwan FA, Abdel-Hafez HF (2021) Comparison of the insecticidal activity of calcium borate with mushroom-like structures and chitin synthesis inhibitor insecticide (Dimilin) against Egyptian cotton leaf worm. J Entomol Zool Stud 9:119–124
2. Badawy ME, El-Aswad AFJPPS (2012) Insecticidal activity of chitosans of different molecular weights and chitosan-metal complexes against cotton leafworm spodoptera littoralis and oleander aphid aphis nerii. Plant Prot Sci 48(3):131–141
3. Brzezinska MS, Jankiewicz U, Walczak M (2013) Biodegradation of chitinous substances and chitinase production by the soil actinomycete streptomyces rimosus. Int Biodeterior & Biodegrad 84:104–110
4. Colijn I, Fokkink R, Schroën K (2021) Quantification of energy input required for chitin nanocrystal aggregate size reduction through ultrasound. Sci Rep 11(1):17217
5. Elieh-Ali-Komi D, Hamblin MR (2016) Chitin and chitosan: production and application of versatile biomedical nanomaterials. Int J Adv Res 4(3):411
6. Gomaa EZ (2018) Iron nanoparticles α-chitin nanocomposite for enhanced antimicrobial, dyes degradation and heavy metals removal activities. J Polym Environ 26:3638–3654
7. Ingle PU, Shende SS, Shingote PR, Mishra SS, Sarda V, Wasule DL, Rajput VD, Minkina T, Rai M, Sushkova S, Mandzhieva S (2022) Chitosan nanoparticles (ChNPs): a versatile growth promoter in modern agricultural production. Heliyon 8(11):e11893
8. Jayakumar R, Prabaharan M, Nair S, Tamura H (2010) Novel chitin and chitosan nanofibers in biomedical applications. Biotechnol Adv 28(1):142–150
9. Kashyap PL, Xiang X, Heiden P (2015) Chitosan nanoparticle based delivery systems for sustainable agriculture. Int J Biol Macromol 77:36–51
10. Khati P, Chaudhary P, Gangola S, Bhatt P, Sharma A (2017) Nanochitosan supports growth of zea mays and also maintains soil health following growth. 3 Biotech 7:1–9
11. Li Z, Su L, Wang H, An S, Yin X (2020) Physicochemical and biological properties of nanochitin—abamectin conjugate for noctuidae insect pest control. J Nanobiotechnol 22:1–16
12. Li, Z, Wang, H, An, S, Yin, X. J. J. o. N (2021) Nanochitin whisker enhances insecticidal activity of chemical pesticide for pest insect control and toxicity. *19*, 1–13.
13. Liang R, Li X, Yuan W, Jin S, Hou S, Wang M (2018) Antifungal activity of nanochitin whisker against crown rot diseases of wheat. J Agric Food Chem 66(38):9907–9913
14. Liu L, Seta FT, An X, Yang J, Zhang W, Dai H, Cao H, Xu Q, Liu H (2020) Facile isolation of colloidal stable chitin nano-crystals from metapenaeus ensis shell via solid maleic acid hydrolysis and their application for synthesis of silver nanoparticles. Cellulose 27:9853–9875
15. Majeed A, Najar RA, Choudhary S, Rehman WU, Singh, A, Thakur S, Bhardwaj P (2017) Practical and plausible implications of chitin-and chitosan-based nanocomposites in agriculture, pp 409–430.
16. Malerba M, Cerana R (2019) Recent applications of chitin-and chitosan-based polymers in plants. Polymers 11(5):839
17. Mincea M, Negrulescu A, Ostafe V (2012) Preparation, modification, and applications of chitin nanowhiskers: a review. Rev Adv Mater Sci 30(3):225–242
18. Ngasotter S, Sampath L, Xavier KM (2022) Nanochitin: an update review on advances in preparation methods and food applications. Carbohydr Polym 291:119627
19. Ngasotter S, Xavier KM, Meitei MM, Waikhom D, Pathak J, Singh SK (2023) Crustacean shell waste derived chitin and chitin nanomaterials for application in agriculture, food, and health–a review. Carbohydr Polym Technol Appl 6:100349

20. Rajkumar V, Gunasekaran C, Dharmaraj J, Chinnaraj P, Paul CA, Kanithachristy I (2020) Structural characterization of chitosan nanoparticle loaded with piper nigrum essential oil for biological efficacy against the stored grain pest control. Pestic Biochem Physiol 166:104566
21. Sharma A, Sood K, Kaur J, Khatri M (2019) Agrochemical loaded biocompatible chitosan nanoparticles for insect pest management. Biocatal Agric Biotechnol 18:101079
22. Singh R, Upadhyay SK, Singh M, Sharma I, Sharma P, Kamboj P, Saini A, Voraha R, Sharma AK, Upadhyay TK, Chitin KF (2021) Chitin, chitinases and chitin derivatives in biopharmaceutical, agricultural and environmental perspective. Biointerface Res Appl Chem 11(3):9985–10005
23. Van SN, Minh HD, Anh DN (2013) Study on chitosan nanoparticles on biophysical characteristics and growth of robusta coffee in green house. Biocatal Agric Biotechnol 2(4):289–294
24. Yanat M, Schroën K (2023) Advances in chitin-based nanoparticle use in biodegradable polymers: a review. Carbohydr Polym 312:120789
25. Yihun FA (2022) Nanochitin preparation and its application in polymer nanocomposites: a review. Emergent Mater 5(6):2031–2060
26. Younes I, Rinaudo M (2015) Chitin and chitosan preparation from marine sources. Structure, properties and applications. Mar Drugs 13(3):1133–1174
27. Yu J, Wang D, Geetha N, Khawar KM, Jogaiah S, Mujtaba M (2021) Current trends and challenges in the synthesis and applications of chitosan-based nanocomposites for plants: a review. Carbohydr Polym 261:117904
28. Zebeaman, M, Bachheti, R. K, Bachheti, A, Pandey, D, Husen, A (2023) Green and cost-effective nanoparticles synthesis from some frequently used medicinal plants and their various applications. In: Secondary metabolites from medicinal plants. CRC Press, pp 287–304
29. Zeng J-B, He Y-S, Li S-L, Wang Y-Z (2012) Chitin whiskers: an overview. Biomacromol 13(1):1–11
30. Zewude DA, Noguchi T, Sato K, Izawa H, Ifuku S (2022) Production of chitin nanoparticles by bottom-up approach from alkaline chitin solution. Int J Biol Macromol 210:123–127

Future Perspectives and Opportunities for Chitin-Based Nanoparticles in Agriculture

Deepti Singh Vashishth, Rekha Goswami, Ummer Bashir Khoja, and Archana Bachheti

Abstract The overgrowing population is leading to challenges in global agriculture. Chitin nanoparticles (CNPs) fabricated from a polymer found in crustaceans and fungi came up to face these challenges and provide security to agriculture. The CNPs not only have potential to enhance the agricultural yield but can also help in minimising the environmental impact. A comprehensive study on the positive and negative aspects of CNPs, safe and responsible usage, and their impact on the environment is required to use these NPs in the future. This chapter will help in understanding various factors like the efficacy of CNPs in accelerating agricultural productivity, their impact on the ecosystem, and possible ways to rectify any negative consequences. By obtaining the knowledge of CNPs and their repercussions, the stakeholder can make better decisions concerning their incorporation into agricultural practices. Hence, this chapter gives a sound theoretical knowledge of how CNPs can help change farming methods of agriculture, leading them to be more sustainable, and what are the challenges and opportunities in this field.

Keywords Chitin Nanoparticles (CNPs) · Sustainable Agriculture · Environmental Impact · Crop Productivity Enhancement

---

D. S. Vashishth · A. Bachheti (✉)
Department of Environmental Science, Graphic Era (Deemed to be) University, Dehradun, Uttarakhand, India
e-mail: baccheti.archana@gmail.com

R. Goswami
Department of Environmental Science, Graphic Era Hill University, Dehradun, Uttarakhand, India

U. B. Khoja
Department of Physics, Graphic Era (Deemed to be) University, Dehradun, Uttarakhand, India

R. K. Bachheti et al. (eds.), *Chitin-Based Nanoparticles for the Agriculture Sectors*, Smart Nanomaterials Technology, https://doi.org/10.1007/978-981-96-0920-8_11

1 Introduction

Agriculture around the world demands new inventions inevitably to satisfy environmental concerns along with augmenting the population. In this aspect, chitin-based nanoparticles arise as a new contender as the new era of agriculture begins. Chitin nanoparticles (CNPs) are made up of naturally occurring polymers found in fungi and crustaceans' cell walls [28]. This chapter aims to look into the future uses of NPs based on chitin in agriculture, as NPs play a crucial role in changing farming practices and offering sustainable solutions (Fig. 1).

Providing food for the rapidly growing population without compromising the environment is difficult, but chitin-based NPs arise as a sustainable approach in this condition. The chitin is fragmented at a microscopic scale to make NPs, which range around 100 nm in size. These NPs possess good yield and immense surface area and are biocompatible as well as biodegradable in nature [53, 71]. Also, they serve as a medium for transporting the correct amount of nutrients and hormones when the plants need [61]. The protective coating of CNPs on plants protects pest and disease control, which can reduce the use of pesticides and aid environment-friendly

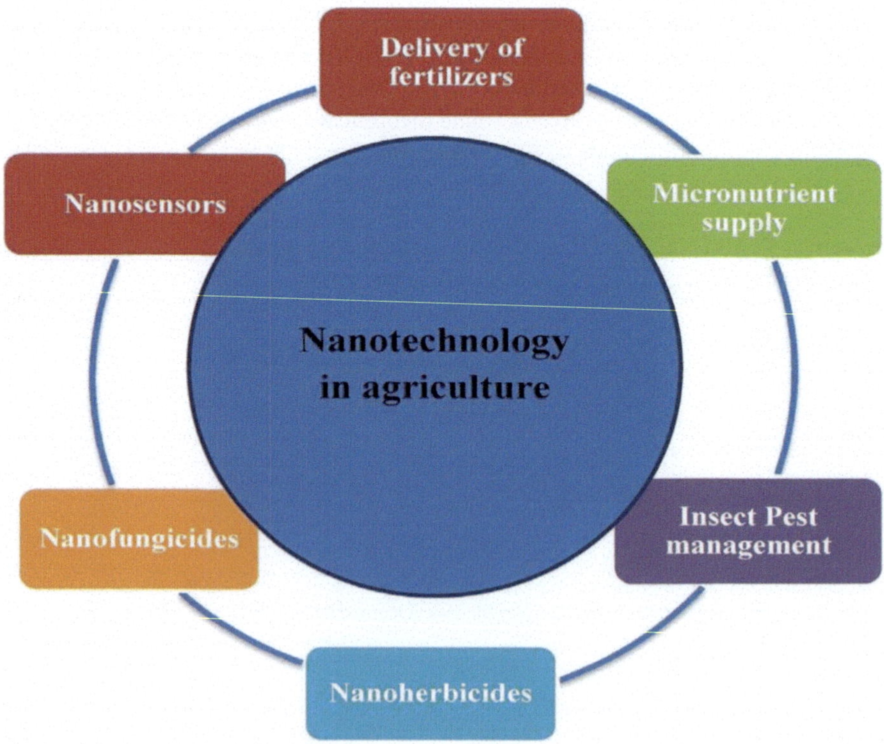

Fig. 1 Various applications of nanotechnology in agriculture

sustainable agriculture [43]. The CNPs serve as carbon source for soil microbes that helps enhance soil health and confine pollutants, healing the contaminated soil and decreasing the impact of pollutants on crops [72]. In the present era of finding alternatives for sustainable agriculture, CNPs are considered a good alternative to synthetic agricultural chemicals due to their biodegradable nature. The CNPs intensify the uptake efficiency of the plant nutrient, which not only minimizes the risk of water contamination for crop production but also saves the farmer's money [64]. The fertilizers containing these NPs are supplied to plant roots, which lessens the negative effects of synthetic chemicals and helps treat contaminated soil by clearing heavy metals and other toxic substances [5, 6, 41]. The NPs work in fighting drought resistance by increasing the water retaining power of soil [26, 34].

Furthermore, the NPs also assist in upholding the herbicides and insecticides to control weeds and pests, which lowers the quantity and effects of synthetic herbicides [63]. Similarly, nano-fungicides and nano-pesticides facilitate controlling disease and pests in plants and seeds with minimal chemical input. This might boost seed germination, yield, stress tolerance, and photosynthesis [30]. The NPs are also used for bioremediation techniques by supporting the growth and function of microbes [66]. Moreover, the plant health, soil, and environmental conditions can be monitored using nano-sensors, leading to help the farmers better understanding of using explicit agricultural practices like irrigation, fertilizer, and pest management [4, 49, 76].

Hence, Chitin-based nanoparticles have a great future in agriculture, as they have miscellaneous applications to protect crops such as pest, weed, and disease control. Counting the benefits of chitin-based NPs opens many research domains like developing pesticides and fungicides, seed coatings, chemical stimulants, biological diseases control, soil mitigation, nano-sensors, and monitoring. The chitin-based NPs can make sustainable, environmentally friendly farming practices more productive. As the research in this area progresses, more innovative applications of NPs will arise to refine sustainable agricultural practices. This chapter discusses the present and future uses and challenges of chitin-based NPs in agriculture.

2 Chitin-Based Nanoparticles: Basics

Nanotechnology is a fast-emerging field with a wide range of potential applications in materials engineering, environmental science, medicine, and agriculture [5, 6, 77–81, 83, 84]. The use of natural biopolymers, like chitin, to make chitin-based nanoparticles is one fascinating advancement in this field. Second only to cellulose in terms of global abundance, chitin is a naturally occurring biopolymer. Chitin found in fungus cell walls, arthropods, and crustacean shells [37, 47, 73, 74]. It is a linear polysaccharide consisting of N-acetylglucosamine (GlcNAc) units with β-(1 → 4) glycosidic linkages. Chitin is known for various nanoparticle formulations due to less toxicity, biodegradability, and biocompatibility [1, 29, 37, 73]. In various fields such as agriculture, biomedical, packaging, wastewater remediation, tissue engineering, and drug delivery, chitin is known as a suitable material for developing nano-based constituents [22, 54, 74]. Chitin has versatile properties such as small size, less toxicity, highly biodegradable in behavior, and high compatibility, due to which it is highly sustainable and environmentally friendly for the formulation of nanomaterials [2, 14, 36, 37, 68]. Due to its strong mechanical properties, chitin-based nanoparticles are also used as a reinforcing material, which results in improved mechanical performance, strong wear and tear tolerance, and overall lifecycle [37, 74]. This helps reduce the system's total cost, enhance safety measures, and reduce major environmental impacts [22, 54, 62].

2.1 Extraction and Preparation of Chitin-Based Nanoparticles

It is a multi-step process, From sourcing the natural material for extraction and preparation of chitin-based nanoparticles to completing the production for various applications.

Source Material Selection:

Considering some natural sources of chitin like exoskeletons of insects, fungi, and crustacean shells (shrimp, crab, and lobster), a specific requirement for desired nanoparticles, the availability and cost play important roles in selecting the source material [28, 35].

Chitin Extraction:

Basically, the deproteinization and demineralization processes are used to extract chitin from raw material [18, 35].

Chitin Deacetylation:

Normally, the chitin obtained from natural sources is partially acetylated, transforming it into chitosan, which is more soluble. Deacetylation involves treating chitin with an alkali solution that helps remove acetyl groups from the glucosamine units [18, 35].

Chitosan Nanoparticle Formation:

After deacetylation, different techniques, such as ionotropic gelation, emulsion cross-linking, and coacervation, are used to convert chitosan into nanoparticles [24].

Characterization and Purification:

To define the shape, size, surface chemistry, and zeta potential of chitosan nanoparticles, a number of techniques, such as fourier-transform infrared spectroscopy, dynamic light scattering, and electron microscopy, are employed. These techniques help in characterization, while purification helps remove contaminants [18, 54].

Functionalization (Optional):

According to the requirement, chitosan nanoparticles can be functionalized, for example, encapsulated with other nanoparticles, loaded with drugs, or coated with specific molecules to enhance their properties [3, 68].

Sterilization and Preservation (for Biomedical Applications):

Sterilization is required if the chitosan nanoparticles are to be used medicinally to guarantee that they are free of microbiological contamination. For stability, they can also need appropriate storage conditions.

3 Current Applications in Agriculture

The expeditious increase in population increases the food demand, and to provide food security, many agrochemicals are being used in agriculture, affecting human health and the environment. So, to achieve global agriculture security, many sustainable agricultural methods are being introduced agrochemicals have advantages, but they also have poisonous qualities that make them dangerous for living things in rivers and soil because of losses that occur from degradation, volatilization, photolysis, and leaching when they are applied. Agro-nanochemicals, also referred to as nano-enabled agrochemicals, have the potential to advance integrated disease and pest control significantly. These compounds allow active chemicals to be delivered precisely and released under controlled conditions, improving agricultural practises' efficacy and efficiency [23]. The Food and Drug Association of the United States (USA) states that chitosan is a recognised non-toxic, biodegradable, and biocompatible substance. Antimicrobial, antiviral, antifungal, antioxidant, anti-inflammatory, bio-adhesion, adsorption enhancer, and other intriguing qualities are among them [9]. In agriculture, chitosan nanoparticles on their own can stimulate plant development and function as strong antibacterial agents against harmful bacteria and fungi. The goal of nano-formulations is to minimize the negative effects of agrochemicals and chitosan while concurrently increasing their positive effects. Because of its amphiphilic qualities, chitosan encapsulation can improve the solubility of many agrochemicals in water, giving conventional agrochemicals an alternate application for inert substances and lowering their degree of toxicity in the process. The encapsulated agrochemicals are well protected by the bio-adhesive qualities of chitosan, which raises the stability and bioavailability of the plant [5, 11] (Fig. 2).

3.1 Chitin Nanoparticles for Crop Protection

Currently, plant diseases and changes in the global climate are the leading causes of significant losses in crop yields, as some phytosanitary outbreaks have increased. Chitosan has been shown to provide disease resistance in several plants, with pathogen and plant cultivar specificity based on the defensive gene induction activity.

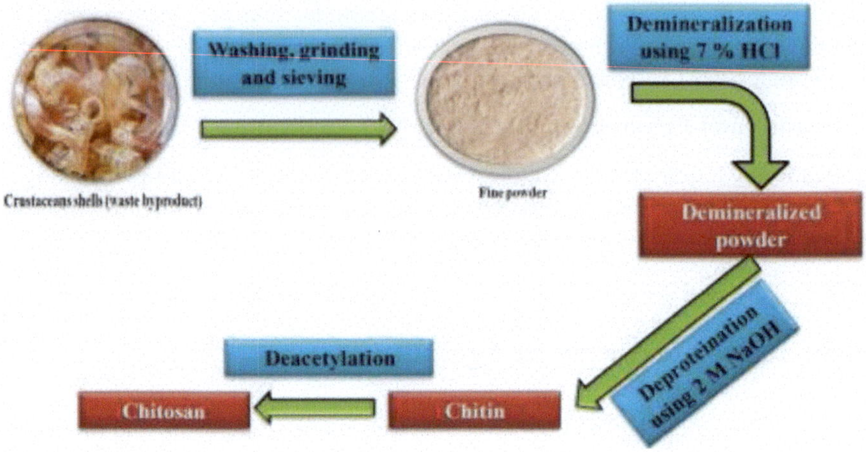

Fig. 2 Process of extraction of chitin/chitosan

Many studies have been conducted on using chitosan to control plant diseases, with varying degrees of efficacy depending on the pathosystem; the derivatives utilized, concentration, degree of deacylation, viscosity, and the applied formulation. For instance, the efficiency of five chitosan derivatives that have undergone chemical modification in limiting *Saprolegnia parasitica* growth. The growth of the fungus was inhibited by methylpyrrolidinonechitosan, N-phosphonomethylchitosan, and N-carboxymethylchitosan, but not by N-dicarboxymethylchitosan, according to [42]. Chitosan coating of rice seeds has been demonstrated by [52] to hasten germination and increase stress resistance. According to [19], using chitosan as a foliar spray on barley decreased the infection caused by the powdery mildew pathogen *Blumeria graminis* f.sp. *hordei* both locally and systemically. It has been demonstrated that adding chitosan to the soil can prevent Fusarium wilts in a variety of plant species. When applied at the ideal dosage, this biomaterial can cause a delay in the onset of disease, which reduces the amount of wilting that occurs in plants [50]. Increasing plant resilience to abiotic stress is also vital to ensure that the world's expanding population has access to safe, nutrient-dense, and healthful food, especially given the current state of global climate change. Abiotic stress tolerance can be conferred by chitin and chitosan. For instance, chitin added to the soil at the beginning of the experiment helps plants grown under regulated deficit irrigation, such as maize (*Zea mays*) acquire a tolerance to water stress [39], while the application of chitosan helps in reducing the effects of water deficit on wheat (*Triticum aestivum*) yield and quality [20] and nanoparticles of chitosan lessen the negative effects of drought stress [7].

Chitin and chitosan have varied operational methods to protect crops against abiotic stress, such as foliar application, direct application, and abscisic acid formation. In each method, their mode of action is different in the direct application as anti-transpirant coating, they cause stomatal closure and congregate stress-defending enzymes and metabolites, while foliar application provides protection in drought tolerance, and formation of abscisic acid works against heat stress [40].

3.2 Soil Health Improvement and Remediation

Chitin and chitosan also help in administering the soil quality and settling environmental issues. They reduce the mobility and bioavailability of the noxious metals, which minimizes the probability of these metals getting absorbed by plants and entering the food chain [33]. These NPs also assist the soil aggregation, which enhances the water infiltration in soil, capacity to reserve nutrients, and decreases erosion. They work as substrates, promoting the growth and activity of helpful bacteria [51].

4 Challenges and Hurdles

The prevailing motto in contemporary society is the concept of sustainable and environmentally friendly economies. However, it is disappointing that biopolymers, such as chitin, have not garnered the recognition they deserve. In light of the current emphasis on bio-based polymers, chitin is one of the most extensively studied materials.

From a safety standpoint, the compound has achieved the consent of Food and Drug Administration, United States (FDA), and has been designated as Generally Recognized as Safe (termed GRAS) [25]. However, in the realm of plants species, the full potential of chitin has yet to be fully investigated. The utilization of Ch-NCs as pesticides, fertilizers, and carriers for plant nutrients still needs to be improved. This can be attributed to the need for more attention and research on chitin-based products. Further exploration of chitin polymeric nanomaterials and nanogels is necessary to develop novel approaches for the controlled release of biomolecules and plant genome editing.

Chitin exhibits exceptional properties in forming films, but its applications in food coating and packaging at an industrial level are limited in the case of Ch-NCs or blends. It is necessary to shift the focus towards standardizing protocols for chitin synthesis and its industrial applications.

In addition, it is necessary to investigate the exploration of innovative and cost-effective sources for separating chitin. Presently, many studies are documenting the generation of chitin-based NCs on a smaller scale in laboratories. However, converting these from lab scale to large scale production and ultimately to the agricultural practitioners remains a biggest constraint. Research actions must be directed towards the formulation of economically viable and industrially suitable synthesis procedures for chitin and its derivatives. This path could facilitate the progression of chitin from academic laboratories to industries and subsequently to farmers [75].

4.1 Regulatory and Safety Considerations

Nanotechnology has emerged as a state-of-the-art technology with significant realized and potential consequences and influences. It constitutes a formidable instrument for ensuring the security of food and nutrition, the administration of abiotic and biotic tensions, the enhancement of input utilization effectiveness, the amplification of yield potential, and the enhancement of quality traits. Nevertheless, if the procedure and commodities are not adequately managed, they may present hazards to human well-being and the environment.

Therefore, it is imperative to utilize strategic research and policy alternatives and measures well-informed by rigorous scientific methods to cultivate this emerging technology for its application in agriculture appropriately. This will improve yield, viability, biosafety, and environmental sustainability.

4.2 Adoption of Strategic Research During Nanotechnology Usage

Enhance the effectiveness of the biological synthesis (commonly known as green synthesis) of nanoparticles with appropriate dimensions and morphology. Augment the knowledge regarding mycorrhizal fungi and root endophytes, which generate a significant quantity of unidentified and uncharacterized nanoparticles that play a crucial role in enhancing the productivity of plants. Additionally, it deepens the understanding of the mechanisms involved in the transportation of nanoparticles from the soil solution to plants and the intricate interactions between plants and nanoparticles. To evaluate comprehensively the unintended impacts of different nano formulations in the forms of nanoparticles and nano-gels on the ecological counterparts and the populace of advantageous soil-borne microorganisms, it is imperative to investigate diverse biological control mechanisms such as NPV formulations, But formulations, Trichoderma formulations, and EPN formulations. Additionally, an analysis of the adverse consequences of nanoparticles on the microbial ecology of soil, with particular emphasis on the structure and dynamics of microbial populations, is warranted. Develop an appropriate procedure for applying micronutrient nanoparticles in seed coating, aiming to produce tailored seeds for regions in the country that suffer from micronutrient deficiencies. Additionally, create sensor-based tools utilizing nanotechnology for testing soil, water, and pesticides. The economic feasibility of using nanotechnology through the implementation of statistically significant pilot trials, participatory demonstrations, and field trials is to be analysed. Furthermore, conduct scientific evaluations on public perception and ethical considerations related to nanotechnology, such as nano semantics, nano praxis, and nano pragmatism. Lastly, identify priority areas, particularly in the field of organics, for the application of nanotechnology.

4.3 Appropriate Policy and Actions

4.3.1 Human and Environmental Health Security

Identify, enhance, and establish (if necessary) organizations possessing pertinent knowledge and resources, and validate and establish connections with them to augment their efficacy in assessing the safety of nanoparticles. In accordance with the regulations set forth by the Food Safety and Standards Authority of India (FSSAI) and the Food and Agriculture Organization/World Health Organization (FAO/WHO), develop explicit principles about the criteria and approach for surveillance and appraisal. In order to ensure the safety of specific nano food/products before their commercial release, it is crucial to conduct a thorough evaluation of their food and environmental safety. Furthermore, we must enhance our capacity to handle a substantial number of samples, considering the growing number of new nano products. Given the potential risks associated with the exposure of the environment to nanotechnology-based materials and products, it becomes increasingly important to address the risks they pose to human health. This is particularly significant due to the vulnerability of water resources to both direct and indirect contamination with nano-materials. Hence, evaluating the potential toxicity and environmental consequences of nano-materials on aquatic organisms is essential. Establishing a network comprising of proficient and accredited dealers, well-trained.

Applicators and other personnel are crucial in advocating for high-quality products and mitigating issues at the operational level. Adherence to the prescribed timetables and guidelines about the handling, storage, packaging, transportation, and other related aspects of the products should be strictly enforced from the moment of their production until the field impact assessment while implementing effective measures to deter any substandard approaches.

4.3.2 Build Necessary Human Resources, Investment and Institutions

Initiate disseminating educational guidance and curricula regarding agricultural nanotechnology at designated agricultural universities and establish connections between the relevant institutions. To commence this endeavor, it is advisable to financially support at least five students pursuing a Ph.D. in agricultural nanotechnology. The Indian Council of Agricultural Research (ICAR) may establish Memorandums of Understanding (MoUs) with esteemed institutes such as the Bhabha Atomic Research Centre (BARC), the Saha Institute of Nuclear Physics, and the Inter-University Accelerator Centre (IUAC), among others, in order to facilitate the sharing of expensive equipment, including ion beams, to functionalize, manipulate, and fabricating nano-materials. Given that research and development activities in nanotechnology necessitate advanced equipment facilities in addition to infrastructure, it is imperative to intensify investments in Agricultural Nanotechnology. Furthermore, the private sector should supplement the fundamental and strategic research conducted by the public sector through appropriate funding and resource allocation. It is recommended that leading experts in fields such as physics, chemistry, nanotechnology, nanobiotechnology electronics at strategic agricultural institutions be employed to gain valuable insights into products that can be swiftly integrated into the Indian agricultural sector.

Enhance the public's level of awareness regarding the extent and role of nanotechnology in the management of sustainable natural resources, control of pests, and other related areas by utilizing the information media. It is necessary to educate the extension personnel on the practicality of nanotechnology. The Krishi Vigyan Kendras (KVKs) should be equipped to carry out this endeavor. Evaluate the technological application of nano-products through the regulatory processes, taking into consideration the social, political, and cultural implications. This evaluation should be a collective effort involving various subject matter specialists, planners, and stakeholders.

Due to our current lack of preparedness in Agricultural Nanotechnology, specifically in terms of risk assessment research, which hinders our ability to provide the necessary scientific support for the regulatory process, it is essential to establish a "National Institute of Agricultural Nanotechnology and Nano-Safety" equipped with state-of-the-art infrastructure, human resources, and research programs. This institute will conduct cutting-edge research, enhance expertise in this field, and offer government policy support and technical guidance.

4.4 Management of Occupational Hazards During Usage of Nanotechnology in Agriculture

4.4.1 Risk Assessment

Evaluating the potential harm of agricultural chemicals on human health is challenging due to the widespread series of elements utilized, mixtures employed in the field, variations in the amount of exposure, and the geographical and meteorological attributes of the agronomic extents where agro-based chemicals are utilized [10, 13, 46].

Suppose the nano-constituent merely serves to safeguard the active element against degradation. In that case, the destiny and conduct of the nano-component might be identical to that of traditional pesticide formulation. In instances where inorganic, engineered NPs, and pristine are directly utilized as biologically efficient fertilizers, pesticides, or for soil or water remediation, the perilous conduct should be meticulously examined from an environmental life-cycle perspective, spanning from the inception of implementation in application fields to the elimination of residual materials [65].

Considering the substantial number of potential nano-formulations used in agricultural practices and the uncertainties surrounding their interactions with various environmental factors, the issue becomes even more pressing. The uncertainties of naturally occurring ultrafine particles on nano-agrochemicals are unknown. The uncertainties of aging, soil, and water characteristics are also unknown. It is challenging to quantify these variables to conduct a comprehensive risk assessment. Nevertheless, it is crucial to carefully consider these aspects due to their potential influence on the physicochemical properties of nanomaterials; the alteration of chemical properties can subsequently result in the modification of their toxicological characteristics, thereby potentially giving rise to occupational hazards. Studies examining the detrimental characteristics of nano-chemicals used in agriculture should take into account the potential relations among nanosized elements and the various stressors present in agroecosystems. These interactions may lead to incompatible, coactive, and additive effects, resulting in a combination of outcomes that modify the toxicities caused by individual substances and potentially affect susceptibility to environmental and health-related adverse effects [32, 48, 60].

The evaluation of the environmental and human health effects of purposely created nano-solutions for agriculture becomes more intricate due to the added risks connected with biosolid-induced "nano-emerging" contamination. Biosolids, which contain nanosized metals originating from consumer products such as Ag, TiO2, and ZnO-NPs, are used as land fertilizers, disposed of through landfilling and incineration, and can contaminate surface water through irrigation with wastewater. Therefore, these biosolids represent a potential substitute for nanomaterials in the environment [12, 31].

In this initial stage of understanding, the identification of potential target organs affected by the actions of nanomaterials and the establishment of biomarkers indicating early effects appear to be valuable approaches for characterizing the toxicological profile of these foreign substances and, ultimately, implementing appropriate preventive and protective measures. In general, the assessment of internal exposure to nanomaterials of a nano-scale and the identification of early biological changes induced by such exposure may be crucial in elucidating the relationship between the properties of nanomaterials and their toxicity and are deemed essential in facilitating a thorough risk evaluation process.

4.4.2 Risk Management Process

The primary objective is offering quantifiable forecasts of specified risks, facilitating evidence-based management. However, in the emerging field of nano-agriculture, a considerable amount of uncertainty exists regarding hazards, exposures, and risks [55]. Hence, it is imperative to adopt a dynamic and cautious approach to administration before the availability of complete evidence concerning the deliberate and planned utilization of nanomaterials in the surroundings. This implies that there will be alterations and ongoing assessment, enhancement, and authentication of strategies and guidance for managing risks as health and safety assessments and information regarding risks become more substantial. In order to effectively address potential risks related to agricultural nanotechnology, it is of utmost importance to prioritize implementing an appropriate management plan that encompasses a hierarchy of controls [27, 58].

The most efficient methodology, situated at the pinnacle of the control framework, involves the eradication or strategic elimination of dangers from nanomaterials by the longstanding ideologies of "prevention through design" and "safety by design" [21, 56]. The fundamental concept underlying green engineering posits that by implementing strategies to safeguard the well-being of human beings and the surrounding ecosystem, one can attain the most significant influence and economic effectiveness when executed at the initial stages of the conception and creation of a procedure or item. Green engineering considers the entire existence of a product, encompassing the acquisition of resources, the production process, the utilization of the item, and its ultimate disposal [8].

Green nanotechnology, with its emphasis on the entire life-cycle, enables users better to minimize the creation of new hazards through unintended consequences. Alongside the development of "safer" nanomaterials, effective containment measures that minimize or eliminate exposure are crucial in the occupational setting. Administrative controls should involve identifying workers who may be exposed and determining which tasks pose the highest risks of exposure to allocate and organize the workforce in the most efficient, healthy, and safe manner in terms of both numbers and expertise. Training programs should be established for companies to inform workers about workplace exposure to nanomaterials, risks involved, job procedures, preventive and protective measures, and implemented policies. In training and risk communication, addressing insufficient information on nanomaterials and chemical products containing nanomaterials may be essential.

Industry may have a significant role to play, for instance, by providing essential data and product information and sharing technical, scientific, and policy expertise [69, 70]. It is important to review product labels and safety data sheets (SDS) in order to obtain information regarding the quantity, quality, traceability, proper application, as well as health and safety aspects related to nanomaterials, whether they are active nano-ingredients or more complex nano-formulations [17]. In the risk management policies, properly communicating risk-related issues is utmost important. During the risk management plan, various information disseminates to inhabitants, such as technical, medical, and regulatory knowledge. For the safety purpose of outdoor workers, the usage of personal protective equipment (PPE) is mandatory, in addition to preventing infection in dermal layers of the worker's usage of gloves, coats, eye and respiratory protection. To determine the most suitable protection measures for precise occupational activities, it is important to collect relevant nanomaterials such as quantity, physical state, degree of dispersion, and the time interval during agricultural use [45]. With each increase in these aforementioned variables, the likelihood of exposure also increases. Consequently, it becomes imperative to employ more efficient exposure control measures.

In addition, occupational health surveillance programs can serve as valuable components of a risk management plan for nanomaterials, particularly in secondary prevention. These programs may encompass various aspects of surveillance, such as monitoring health outcomes or biological changes, as well as examining effects at both group and individual levels. The literature by [44, 57, 59] highlights the inclusion of hazard and medical surveillance in these programs.

Crucially, these surveillance programs should involve several key steps: initial, periodic, and post-incident medical examinations accompanied by specific medical screening tests. Additionally, worker training is vital to enable the recognition and reporting of symptoms associated with exposure to specific hazards. Finally, employers must take appropriate action upon detecting potential hazards and risks to health, as emphasized by [67].

4.4.3 Risk Governance

At the societal level, the expeditious expansion of nanotechnology necessitates alterations in regulations that can influence the utilization of nanomaterials in agriculture. Consequently, this can contribute to the thorough assessment and prudent management of potential risks. This is contingent on the notion that the outcomes and advantages of innovative technologies may, to some extent, rely on the existing regulatory systems and policy environment [69, 70].

Therefore, the academic, industry, governmental bodies, and stakeholders must collaborate to understand nanomaterials' characteristics and toxicological behaviour comprehensively. This will also involve determining the current applications of nanomaterials in the field of nano-agriculture, as well as their potential intended uses. The primary focus should be developing new policies specifically tailored to nanomaterials or effectively implementing existing policies for their application in the nano-agricultural context. Instead of reacting to unsafe conditions after they have been defined, it is crucial to establish relevant occupational safety practices and policies right from the outset of any nanomaterials activity. This proactive approach is necessary due to the safety concerns of certain nanomaterials and the risk of inappropriate generalizations arising from the vast range of nanotechnological applications. Addressing this regulatory gap requires utilizing the insights gained from ongoing projects in toxicity testing, decision-making regarding nanomaterial characterization and testing protocols, and data on exposure and precautionary management. An alternative approach to governing nanotechnology, which combines anticipatory ethics, future-oriented responsibility, upstream public engagement, and theories of justice, is beneficial.

5 Future Opportunities

Current progressions in technology are underway to form nanodevices with the competence to substitute diverse cellular apparatus resourcefully. The employment of nanotechnology has the potential to advance improvement in agricultural research, more precisely in reproductive science and technology. This results in producing rich, diverse seeds and fruit varieties unaffected by seasonal variations. Moreover, nanotechnology usage in agricultural sectors helps detect early disease stages, stress, and its effects. To enhance the working potential, various smart sensors and smart delivery systems were implemented, which help in combating diseases in plants, viruses, and other crop pathogens in the agricultural field. Although nanotechnology has many benefits in agriculture, it is also very important to consider and study the associated risks to human life and the ecosystem. Considering nanotechnologies' safety measures towards human health concerns, it remains unclear how much is safe and harmful during its usage. While the commercialization of nanotechnology, some major challenges were observed, and they should be addressed to improve the working efficiency, including a lack of research for scaling up the work, high operation and maintenance costs, and safety measures regarding environmental and human health issues. Before widespread usage and commercialization of these nanomaterials, the government should establish strict monitoring regulations, standards, and protocols. To date, nanotechnology has become one of the innovative techniques in various agricultural practices, which involve enhanced crop production, good food quality, increased soil fertility rate, pest control, and management. The successful utilization of nanotechnology in agriculture shows nano-fertilizers, nanosensors, food packaging nanofilms, and disease detection nanodevices. To investigate the appropriate mechanism behind nanotechnology-based techniques in agricultural practices, continuous and rigorous investigation is required to combat all the issues such as ecological and economic consequences.

6 Conclusion

Chitin-based nanoparticles are a promising frontier in modern agriculture, with numerous advantages ranging from higher crop yields to reduced environmental impact. The importance of chitin-based nanoparticles cannot be stressed as the agriculture industry seeks sustainable and environmentally friendly solutions to fulfill the demands of a growing global population. Their use in crop productivity, pest management, soil health improvement, and environmental conservation can alter agriculture's future, assuring food security while protecting our planet for future generations. Despite the potential benefits, using nanoparticles in agriculture raises concerns about environmental and health risks. To ensure appropriate and sustainable usage, extensive study is required on the safety and environmental effects of employing nanoparticles in farming.

References

1. NIOSH National Institute for Occupational Safety and Health (2009) Approaches to safe nanotechnology managing the health and safety concerns associated with engineered nanomaterials publication no 2009-125. US Department of Health and Human Services Public Health Service Centers for Disease Control National Institute of Occupational Safety and Health DHHS (NIOSH), Cincinnati Ohio USA. http://www.cdc.gov/niosh/docs/2009-125/pdfs/2009-125.pdf. Accessed 07 October 2016
2. Agnihotri SA, Mallikarjuna NN, Aminabhavi TM (2004) Recent advances on chitosan-based micro-and nanoparticles in drug delivery. Journal Control Release 100(1):5–28
3. Ahmad M, Manzoor K, Ahmad S, Akram N, Ikram S (2019) Chitosan-based nanocomposites for cardiac liver and wound healing applications In: Applications of nanocomposite materials in orthopedics. Woodhead Publishing, pp 253–262
4. Aranaz I, Mengíbar M, Harris R, Paños I, Miralles B, Acosta N, Heras Á (2009) Functional characterization of chitin and chitosan. Curr Chem Biol 3(2):203–230
5. Ashraf SA, Siddiqui AJ, Abd Elmoneim OE, Khan MI, Patel M, Alreshidi M, Adnan M (2021) Innovations in nanoscience for the sustainable development of food and agriculture with implications on health and environment. STOTEN 768:144990
6. Bachheti A, Bachheti RK, Husen A (2024b) Carbon-based smart nanomaterials (synthesis, agricultural, biomedical and environmental interventions). Springer Nature Singapore Pte Ltd., Singapore. https://doi.org/10.1007/978-981-97-0240-4, ISBN: 978-981-97-0239-8
7. Bachheti RK, Bachheti A, Husen A (2024) Metal and metal-oxide based nanomaterials (Synthesis, agricultural, biomedical and environmental interventions). Springer Nature Singapore Pte Ltd., Singapore
8. Bachheti RK, Bachheti A, Husen A (eds) (2023) Nanomaterials for environmental and agricultural sectors. Springer, Singapore. https://doi.org/10.1007/978-981-99-2874-3
9. Bachheti RK, Bachheti A, Husen A (eds) (2024). Metal and metal-oxide based nanomaterials: In: Synthesis, agricultural, biomedical and environmental interventions. Springer Nature Singapore, Singapore. https://doi.org/10.1007/978-981-99-7673-7
10. Behboudi F, Tahmasebi-Sarvestani Z, Kassaee MZ, Modarres-Sanavy SAM, Sorooshzadeh A, Mokhtassi-Bidgoli A (2019) Evaluation of chitosan nanoparticles effects with two application methods on wheat under drought stress. J Plant Nutri 42(13):1439–1451
11. Bergeson LL (2013) Sustainable nanomaterials: emerging governance systems.
12. Bernkop-Schnürch A, Dünnhaupt S (2012) Chitosan-based drug delivery systems. Eur J Pharm Biopharm 81:463–469
13. Bolognesi C (2003) Genotoxicity of pesticides: a review of human biomonitoring studies. Mutat Res/Rev Mutat Res 543(3):251–272
14. Campos EVR, de Oliveira JL, Fraceto LF, Singh B (2015) Polysaccharides as safer release systems for agrochemicals. Agron Sustain Dev 35:47–66
15. Chen C, Unrine JM, Judy JD, Lewis RW, Guo J, McNear DH Jr, Tsyusko OV (2015) Toxicogenomic responses of the model legume Medicago truncatula to aged biosolids containing a mixture of nanomaterials (TiO_2 Ag and ZnO) from a pilot wastewater treatment plant. Environ Sci Technol 49(14):8759–8768
16. Damalas CA, Eleftherohorinos IG (2011) Pesticide exposure safety issues and risk assessment indicators. Int J Environ Res Public Heal 8(5):1402–1419
17. Duan B, Huang Y, Lu A, Zhang L (2018) Recent advances in chitin based materials constructed via physical methods. Prog Polym Sci 82:1–33
18. Dudhani AR, Kosaraju SL (2010) Bioadhesive chitosan nanoparticles: preparation and characterization. Carbohydr Polym 81:243–251
19. Duhan JS, Kumar R, Kumar N, Kaur P, Nehra K, Duhan S (2017) Nanotechnology: the new perspective in precision agriculture. Biotechnol Rep 15:11–23
20. Eastlake A, Hodson L, Geraci C, Crawford C (2012) A critical evaluation of material safety data sheets (MSDSs) for engineered nanomaterials. J Chem Health Saf 19(5):1–8

21. El-Knidri H, Belaabed R, Addaou A, Laajeb A, Lahsini A (2018) Extraction chemical modification and characterization of chitin and chitosan. Int J Biol Macromol 120:1181–1189
22. Faoro F, Maffi D, Cantu D, Iriti M (2008) Chemical-induced resistance against powdery mildew in barley: the effects of chitosan and benzothiadiazole. Biocontrol 53:387–401
23. Farouk S, El-Metwally IM (2019) Synergistic responses of drip-irrigated wheat crop to chitosan and/or silicon under different irrigation regimes. Agric Water Manag 226:105807
24. Geraci C, Heidel D, Sayes C, Hodson L, Schulte P, Eastlake A, Brenner S (2015) Perspectives on the design of safer nanomaterials and manufacturing processes. J Nano Res 17:1–13
25. Gholap AD, Rojekar S, Kapare HS, Vishwakarma N, Raikwar S, Garkal A, Annapure U (2023) Chitosan scaffolds: expanding horizons in biomedical applications. Carbohydr Polym 121394
26. Ghormade V, Deshpande MV, Paknikar KM (2011) Perspective nanobiotechnology enabled protection and nutrition of plants. Biotechnol Adv 29(6):792–803
27. Grenha A (2012) Chitosan nanoparticles: a survey of preparation methods. J Drug Target 20(4):291–300
28. Gutiérrez TJ (2017) Chitosan applications for the food industry. In: Chitosan: derivatives composites and applications, pp 183–232
29. Husen A (2023) Secondary metabolites based green synthesis of nanomaterials and their applications. Springer Nature Singapore Pte Ltd., Singapore https://doi.org/10.1007/978-981-99-0927-8
30. Husen A (2022) Engineered nanomaterials for sustainable agricultural production, soil improvement and stress management. Elsevier Inc., Cambridge, MA. https://doi.org/10.1016/C2021-0-00054-7
31. Husen A (2023) Nanomaterials from agricultural and horticultural products. Springer Nature Singapore Pte Ltd., Singapore. https://link.springer.com/book/9789819934348
32. Husen A (2023) Nanomaterials and nanocomposites exposures to plants (Response, interaction, phytotoxicity and defense mechanisms). Springer Nature Singapore Pte Ltd., Singapore. https://doi.org/10.1007/978-981-99-2419-6
33. Husen A (2020) Introduction and techniques in nanomaterials formulation, In: Husen A, Jawaid M (eds) Nanomaterials for agriculture and forestry applications. Elsevier Inc., Cambridge, MA, pp 1–14. https://doi.org/10.1016/B978-0-12-817852-2.00001-9
34. Husen A, Bachheti RK, Bachheti A (eds) (2023). Current trends in green nano-emulsions: food, agriculture and biomedical sectors. Springer Nature Singapore, Singapore. https://doi.org/10.1007/978-981-99-5398-1
35. Iavicoli I, Leso V, Ricciardi W, Hodson LL, Hoover MD (2014) Opportunities and challenges of nanotechnology in the green economy. Environ Health 13(1):1–11
36. Iber BT, Kasan NA, Torsabo D, Omuwa JW (2022) A review of various sources of chitin and chitosan in nature. J Renew Mat 10(4):1097
37. Islam S, Bhuiyan MR, Islam MN (2017) Chitin and chitosan: structure properties and applications in biomedical engineering. J Poly Environ 25:854–866
38. Jiang M, Song Y, Kanwar MK, Ahammed GJ, Shao S, Zhou J (2021) Phytonanotechnology applications in modern agriculture. J Nanobiotechnol 19(1):1–20
39. Judy JD, McNear DH Jr, Chen C, Lewis RW, Tsyusko OV, Bertsch PM, Unrine JM (2015) Nanomaterials in biosolids inhibit nodulation shift microbial community composition and result in increased metal uptake relative to bulk/dissolved metals. Environ Sci Technol 49(14):8751–8758
40. Kahru A, Savolainen K (2010) Potential hazard of nanoparticles: from properties to biological and environmental effects. Toxicology 2(269):89–91
41. Kamari A, Pulford ID, Hargreaves JSJ (2011) Binding of heavy metal contaminants onto chitosans–an evaluation for remediation of metal contaminated soil and water. J Environ Manag 92(10):2675–2682
42. Komendova R, Žídek J, Berka M, Jemelkova M, Řezáčová V, Conte P, Kučerík J (2019) Small-sized platinum nanoparticles in soil organic matter: influence on water holding capacity evaporation and structural rigidity. STOTEN 694:133822

43. Kou SG, Peters LM, Mucalo MR (2021) Chitosan: a review of sources and preparation methods. Int J Biol Macromol 169:85–94
44. Kumari S, Kishor R (2020) Chitin and chitosan: origin properties and applications In: Handbook of chitin and chitosan. Elsevier, pp 1–33
45. Kumari A, Yadav SK, Yadav SC (2010) Biodegradable polymeric nanoparticles based drug delivery systems. Colloids Surf B: Biointerfaces 75(1):1–18
46. Liaqat A, Ihsan MZ, Rizwan MS, Mehmood A, Ijaz M, Alam M, Yaqub MS (2019) Inducing effect of chitosan on the physiological and biochemical indices of eggplant (solanum melongena l) Genotypes under heat and high irradiance. Appl Ecol Environ Res 17(5)
47. Lin KH, Lin FW, Wu CW, Chang YS (2019) Biostimulation of maize (*Zea mays*) and irrigation management improved crop growth and water use under controlled environment. Agronomy 9(9):559
48. Madhusudhan A, Purohit SD, Prasad R, Husen A (2024) Functional smart nanomaterials and their theranostics approaches. Springer Nature Singapore Pte Ltd., Singapore
49. Malerba M, Cerana R (2020) Chitin-and chitosan-based derivatives in plant protection against biotic and abiotic stresses and in recovery of contaminated soil and water. Polysaccharides 1(1):21–30
50. McGee CF, Storey S, Clipson N, Doyle E (2017) Soil microbial community responses to contamination with silver aluminium oxide and silicon dioxide nanoparticles. Ecotoxicology 26:449–458
51. Muzzarelli RA, Muzzarelli C, Tarsi R, Miliani M, Gabbanelli F, Cartolari M (2001) Fungistatic Activity of Modified Chitosans against *Saprolegnia parasitica*. Biomacromol 2(1):165–169
52. Nagpure A, Choudhary B, Gupta RK (2014) Chitinases: in agriculture and human healthcare. Crit Rev Biotechnol 34(3):215–232
53. Nasterlack M, Zober A, Oberlinner C (2008) Considerations on occupational medical surveillance in employees handling nanoparticles. Int Arch Occup Environ Heal 81:721–726
54. Pastor S, Creus A, Parrón T, Cebulska-Wasilewska A, Siffel C, Piperakis S, Marcos R (2003) Biomonitoring of four European populations occupationally exposed to pesticides: use of micronuclei as biomarkers. Mutagenesis 18(3):249–258
55. Peniche C, Argüelles-Monal W, Goycoolea FM (2008) Chitin and chitosan: major sources properties and applications In: Monomers polymers and composites from renewable resources. Elsevier, pp 517–542
56. Pérez S, la Farré M, Barceló D (2009) Analysis behavior and ecotoxicity of carbon-based nanomaterials in the aquatic environment. TrAC 28(6):820–832
57. Pramanik P, Krishnan P, Maity A, Mridha N, Mukherjee A, Rai V (2020) Application of nanotechnology in agriculture. Environ Nanotechnol 4:317–348
58. Rabea EI, Badawy MET, Stevens CV, Smagghe G, Steurbaut W (2003) Chitosan as antimicrobial agent: applications and mode of action. Biomacromol 4(6):1457–1465
59. Rath SK, Dwivedi V, Husen A, Akhtar N (2024) Smart nanomaterials for drug delivery and neurological diseases management. Springer Nature Singapore Pte Ltd., Singapore. https://doi.org/10.1007/978-981-97-0308-1, ISBN: 978-981-97-0307-4
60. Riseh RS (2023) Advancing agriculture through bioresource technology: the role of cellulose-based biodegradable mulches. Int J Biol Macromole 128006
61. Ruan SL, Xue QZ (2002) Effects of chitosan coating on seed germination and salt-tolerance of seedlings in hybrid rice (*Oryza sativa* L). Acta Agron Sin 28(6):803–808
62. Sahraee S, Milani JM, Ghanbarzadeh B, Hamishehkar H (2017) Physicochemical and antifungal properties of bio-nanocomposite film based on gelatin-chitin nanoparticles. Int J Biol Macromol 97:373–381
63. Sathiyaseelan A, Naveen KV, Zhang X, Han K, Wang MH (2023) Research progress on chitosan-zinc oxide nanocomposites fabrication characterization biomedical and environmental applications. Coord Chem Rev 496:215398
64. Savolainen K, Alenius H, Norppa H, Pylkkänen L, Tuomi T, Kasper G (2010) Risk assessment of engineered nanomaterials and nanotechnologies—a review. Toxicology 269(2–3):92–104

65. Schulte PA, Trout DB (2011) Nanomaterials and worker health: medical surveillance exposure registries and epidemiologic research. J Occup Environ Med S3–S7
66. Schulte PA, Trout D, Zumwalde RD, Kuempel E, Geraci CL, Castranova V, Halperin WE (2008) Options for occupational health surveillance of workers potentially exposed to engineered nanoparticles: state of the science. J Occup Environ Med 517–526
67. Schulte P, Geraci C, Zumwalde R, Hoover M, Kuempel E (2008) Occupational risk management of engineered nanoparticles. J Occu Environ Hyg 5(4):239–249
68. Schulte PA, Geraci CL, Hodson LL, Zumwalde RD, Kuempel ED, Murashov V, Heidel DS (2013) Overview of risk management for engineered nanomaterials. J Phys: Conf Ser 429(1):012062 IOP Publishing
69. Scown TM, Santos EM, Johnston BD, Gaiser B, Baalousha M, Mitov S, Tyler CR (2010) Effects of aqueous exposure to silver nanoparticles of different sizes in rainbow trout. Toxicol Sci 115(2):521–534
70. Shahrajabian MH, Chaski C, Polyzos N, Tzortzakis N, Petropoulos SA (2021) Sustainable agriculture systems in vegetable production using chitin and chitosan as plant biostimulants. Biomolecules 11(6):819
71. Shamshina JL, Berton P, Rogers RD (2019) Advances in functional chitin materials: a review. ACS Sustain Chem Eng 7(7):6444–6457
72. Shang Y, Hasan MK, Ahammed GJ, Li M, Yin H, Zhou J (2019) Applications of nanotechnology in plant growth and crop protection: a review. Molecules 24(14):2558
73. Sharp RG (2013) A review of the applications of chitin and its derivatives in agriculture to modify plant-microbial interactions and improve crop yields. Agronomy 3(4):757–793
74. Shatkin JA, Kim B (2015) Cellulose nanomaterials: life cycle risk assessment and environmental health and safety roadmap. Environ Sci Nano 2(5):477–499
75. Shilpa BM, Rashmi R, Manjula NG, Sreekantha A (2022) Bioremediation of heavy metal contaminated sites using phytogenic nanoparticles In: Phytonanotechnology. Springer Nature Singapore, Singapore, pp 227–253
76. Trout DB, Schulte PA (2010) Medical surveillance exposure registries and epidemiologic research for workers exposed to nanomaterials. Toxicology 269(2–3):128–135
77. Verma D, Okhawilai M, Goh KL, Thakur VK, Senthilkumar N, Sharma M, Uyama H (2023) Sustainable functionalized chitosan based nano-composites for wound dressings applications: a review. Environ Res 116580
78. Watson SB, Gergely A, Janus ER (2011) Where is agro-nanotechnology heading in the United States and European Union. Nat Resour Environ 26:8
79. Watson SB, Gergely A, Janus ER (2011) Where is agronano-technolgoy heading in the United States and European Union. Nat Resour Environ 26:8
80. Worku LA, Tadesse MG, Bachheti A, Pandey DP, Chandel AK, Ewuntu AW, Bachheti RK (2024) Experimental investigations on PVA/chitosan and PVA/chitin films for active food packaging using *Oxytenanthera abyssinica* lignin nanoparticles and its UV-shielding, antimicrobial, and antiradical effects. Int J Biol Macromol 254:127644. https://doi.org/10.1016/j.ijbiomac.2023.127644
81. Wu D, Wang W, Yao Y, Li H, Wang Q, Niu B (2023) Microbial interactions within beneficial consortia promote soil health. STOTEN 165801
82. Yanat M, Schroën K (2023) Advances in chitin-based nanoparticle use in biodegradable polymers: a review. Carbohydr Polym 120789
83. Younes I, Rinaudo M (2015) Chitin and chitosan preparation from marine sources Structure properties and applications. Mar Drugs 13(3):1133–1174
84. Yu J, Wang D, Geetha N, Khawar KM, Jogaiah S, Mujtaba M (2021) Current trends and challenges in the synthesis and applications of chitosan-based nanocomposites for plants: a review. Carbohydr Polym 261:117904
85. Zhang Q, Ying Y, Ping J (2022) Recent advances in plant nanoscience. Adv Sci 9(2):2103414

Chitin and Chitosan-Based Hydrogel and Their Use in the Agriculture

Limenew Abate Worku, Marie Yayinie, Rakesh Kumar Bachheti, and Archana Bachheti

Abstract All over the world, biopolymeric hydrogels are used in a variety of applications. These hydrogels are primarily used in agriculture, where they are employed as nutrient and pesticide release agents and water reservoirs. Because chitosan-based hydrogels can operate as a carrier of nutrients and other agrochemicals and improve soil's capacity to retain water, they can be employed as soil conditioning agents. Irrigation methods and water-saving measures influence crop growth, productivity, and survival. This paper addresses the possible usage of biodegradable hydrogels based on chitosan in horticulture and agriculture. This article describes how to create hydrogels by chemical or physical crosslinking and the subsequent functional characteristics of recently described hydrogels, like the release of active substances and water retention.

Keywords Chitosan and chitin · Hydrogel · Agricultural use

1 Introduction

Biopolymers are a significant class of polymeric materials of many varieties. Biodegradable polymeric materials have been the subject of extensive investigation, primarily because most of them come from renewable sources [109]. Some of them

L. A. Worku (✉) · M. Yayinie · R. K. Bachheti (✉)
Department of Chemistry, College of Natural and Computational Science, Debre Tabor University, Debre Tabor, Ethiopia
e-mail: limenewabate@gmail.com

R. K. Bachheti
e-mail: rkbachheti@gmail.com

A. Bachheti
Department of Environment Science, Graphic Era University, Dehradun 248002, Uttarakhand, India

R. K. Bachheti
Department of Industrial Chemistry, College of Natural and Applied Sciences, Addis Ababa Science and Technology University, Addis Ababa, Ethiopia

R. K. Bachheti et al. (eds.), *Chitin-Based Nanoparticles for the Agriculture Sectors*, Smart Nanomaterials Technology, https://doi.org/10.1007/978-981-96-0920-8_12

are inexpensive and widely available in the natural world. They frequently display low toxicity, controlled reactivity, biocompatibility, biodegradability, and filmogenic qualities [24]. Environmental concerns, including preserving natural resources and reusing or reducing organic waste, have drawn attention to these materials [12].

Many different types of polymexistound in nature, including those found in bacteria, proteins, polypeptides, animal-derived polymers, and members of the polyester claPolymersThey are a simple way to generate micro- or nanoscale fibers or particles for various intriguing therapeutic uses (Balaji et al, 201Polysaccharidesides are harmless, biodegradable, and somewhat abundant in nature [48].

Because they include hydrophilic functional groups that readily undergo crosslinking and grafting copolymerization reactions utilizing well-established chemical and physical techniques, polysaccharides are frequently utilized in the creation of hydrogels [38, 63]. Chitosan (CS) is one of the most important polysaccharides added to hydrogels that have the potential to be used in many different agricultural applications [17], cellulose (Cell) [22], starch (ST) [47], and alginates (Alg) [1]. For these kinds of applications, CS is the most popular polysaccharide-based raw material since, in addition to being biocompatible and biodegradable, it also possesses antibacterial qualities. It can be generated commercially from naturally renewing resources (such as leftover seafood shells) [93].

Chitosan hydrogels can be produced in several methods, such as by joining different formulations and geometries chemically or physically [35]. The traditional sol–gel method of making chitosan hydrogels involves dissolving chitosan polymers in diluted acid to produce an aqueous solution. Crosslinkers are introduced once the polymer chains dissolve entirely to form the hydrogel networks by covalent bonding [86]. The crosslinks created by the polymer chains and crosslinking agent dictate the transition solution/gel to form a network; in highly diluted solutions, these crosslinks do not occur in sufficient quantities [94].

Furthermore, as chitosan and anionic molecules interact electrostatically to drive chitosan-based encapsulation mechanisms, the cationic property of chitosan in acidic environments can produce capsules to encapsulate nutrients. The potential applications of chitosan hydrogel beads in agriculture and their availability, affordability, and abundance of hydroxyl and amino groups have made them increasingly popular in recent years. Apart from its use as an adsorbent in wastewater treatment, chitosan can also produce micronutrients, fertilizers, and herbicides, with proven results in the past [83].

Even though several techniques for creating and using CS-based hydrogels for diverse biomedical applications have been established [79, 80]. The number of formulations that satisfy agricultural needs is small. The capacity to capture and release active chemicals from the hydrogel network and its water absorption capabilities are the primary points of emphasis in this evaluation of CS-based hydrogels for agricultural usage. Table 1 presents examples of CS-based hydrogels and their performance; a detailed discussion of these hydrogels is provided in Sect. 3. The review supplements existing reviews on the broad usage of CS hydrogels, not just in the agricultural sector, and encompasses the literature from the past 10 years on CS-based hydrogel uses in agriculture [16, 17]. When compared to other absorbent materials like paper, cellulose, and cotton, hydrogels have the following advantages: they can absorb several times their weight, they are resistant to pressure, and 90% of their constituents may disintegrate, making them environmentally benign [3, 36].

2 Chitin and Chitosan: Fundamental Concepts

2.1 Definition and Structure of Chitin and Chitosan

Chitin is a long-chain polymer of repeated units of N-acetylglucosamine, a glucose derivative. It is a crucial structural element present in the radula (eating organs) of mollusks, the cell walls of fungi, and the exoskeletons of arthropods (such as insects and crustaceans) [108].

Chitin shares a structure similar to cellulose, wherein β-1,4 glycosidic linkages bind every N-acetylglucosamine unit. This arrangement results in chitin's unique stiffness and strength [30]. Chitin is a strong and adaptable material because its molecular structure gives it qualities like resistance to enzyme breakdown and insolubility in water [30].

Moreover, chitosan, which has distinct qualities and uses, can be created chemically from chitin [27]. Deacetylating chitin produces a polymer with positively charged amino groups; chitosan has unique qualities like biocompatibility, antibacterial activity, and the capacity to bind to specific metals and colors [6].

In general, chitin's distinct composition and characteristics render it a valuable biomaterial with a wide range of uses in industries like biotechnology, healthcare, agriculture, and environmental restoration [20].

Chitosan is a byproduct of chitin, the second most prevalent natural polymer after cellulose. "Chitosan" refers to the entire family of chitin-derived polymers with variable levels of deacetylation, starting at 50% [60]. D-glucosamine and N-acetyl-D-glucosamine are the two subunits comprising chitosan, a linear polymer connected by 1,4-glycosidic linkages. The chitosan molecule's general structure comprises three rings [23]. Its three functional groups—primary and secondary hydroxyl groups and amino groups—allow for simple chemical modification. These functional groups also affect the mechanical and solubility properties of chitosan. Moreover, β-1,4 glycosidic bonds are present in chitosan[68]. Chitosan is more soluble in acidic aqueous conditions than chitin. The leading cause of chitosan's solubility is the protonation of -NH2 at the D-glucosamine repeat unit's C-1 position, which, in an acidic environment, converts the polysaccharide into a polyelectrolyte [8]. Chitosan has a solubility that makes it useful in almost every aspect of human life and health, including industries, agriculture, medicine, and process engineering [88]. Chitosan can be created by extracting chitin and then undergoing a deacetylation process, or it can be extracted directly from various fungi. Chitin is a polymer produced more frequently than chitosan by zygomycetes, ascomycetes, basidiomycetes, Deuteromycetes, and Phycomycetes, which are members of the fungus kingdom.

Table 1 Chitosan hydrogels prepared by physical and chemical methods

Crosslinking method	Primary materials	Hydrogel name	Crosslinking agent	References
Schiff base reaction	Chitosan, sodium periodate, hyaluronic acid	API-CS-oxCS/oxHA	oxCS, oxHA	Tatarusanu et al. [102]
Schiff base reaction)	Chitosan, potassium hydroxide, sodium chloride, aldehyde-4-arm polyethylene glycol, and amino-4-arm polyethylene glycol	Cn-Nm hydrogel	4r-PEG-CHO	Lu et al. [58]
Chemical cross-linking	Glutaraldehyde, polyvinyl alcohol, chitosan, dextran	PVA/dextran/chitosan hydrogel	Glutaraldehyde	Lin et al. (2019)
Chemical cross-linking	Br12 dye, chitosan, isothiocyanatotriphthalic anhydride	Chitosan-based isothiocyanatotrimellitic anhydride cross-linked hydrogels	Isothiocyanatotrityltriphthalic anhydride	Mohamed et al. [64]
Amidation reaction	Triethylamine (TEA), succinic anhydride, carboxylated polyvinyl alcohol (PC), and chitosan hydrochloride	PC/CS (PCCS) hydrogels	Carboxy polyvinyl alcohol (PC)	Liu et al. [57]
Chemical cross-linking	Genipin, chitosan, bacterial cellulose (BC)	BC-Ch hydrogel	Genipin	Arikibe et al. [11]
Hydrogen bonding interactions, electrostatic interactions)	Chitosan, carboxymethyl cellulose sodium salt,	Frozen-Thawed CMC-CS-CA hydrogels	Citrate	Bósquez-Cáceres et al. [14]
Chemical cross-linking	Chitosan powder, LiOH, epichlorohydrin (ECH)	Hierarchical microCS hydrogel scaffold	Epichlorohydrin	Wang et al. [111]
Physical cross-linking (ionic interactions)	EDC, sodium bicarbonate, chitosan	Cat-CH/Hydrogel	Sodium bicarbonate	Guyot et al. [40]

(continued)

Table 1 (continued)

Crosslinking method	Primary materials	Hydrogel name	Crosslinking agent	References
Physical cross-linking (electrostatic interactions)	Carboxymethyl chitosan, alginate, calcium chloride, EGF powder	Dual-crosslinked CMC-ALG hydrogels		Hu et al. [44]
Physical cross-linking (electrostatic interactions)				
Physical cross-linking (interactions between ions and hydrogen bonds)	Tetraacetylethylenediamine (TEPA), epichlorohydrin (ECH), low chitosan	TEPA-COS hydrogel	TEPA	Mei et al. [61]
Physical cross-linking: interactions involving hydrogen bonds	Chitosan, tripolyphosphate (TPP), SQR22 dye, dimethyl sulfoxide (DMSO)	CS-TPP hydrogel	TPP	Lim et al. [56]
Physical cross-linking (electrostatic interactions)	Powdered chitosan, sulfate (SP), and two-component silicone elastomer (ACEO)	CS/SP printing inks	SP	Zhang et al. [117]

On the other hand, chitosan is only present in the cell walls of a few fungal species, zygomycetes in particular. The ability to extract chitosan from fungal biomass whenever desired and its resistance to seasonal variations make it a highly advantageous process. The method used to remove chitin from fungus, whose cell walls are their only source, is comparable to that used to extract chitosan from crustaceans. Therefore, It is more economical to use just fungi that can produce the desired output [98]. An additional significant and plentiful source of chitin is found in the exoskeletons of crustaceans, primarily from a range of marine species like lobsters, crabs, and shrimp. Currently, exoskeletons from the shrimp industry—where chitin and calcium are obtained—are the primary source of chitin waste. The chemical process of N-deacetylation is used to produce chitosan. The cationic nature of chitosan is due to the free amino group that remains after part of the acetyl group of chitin is removed [25] (Fig. 1).

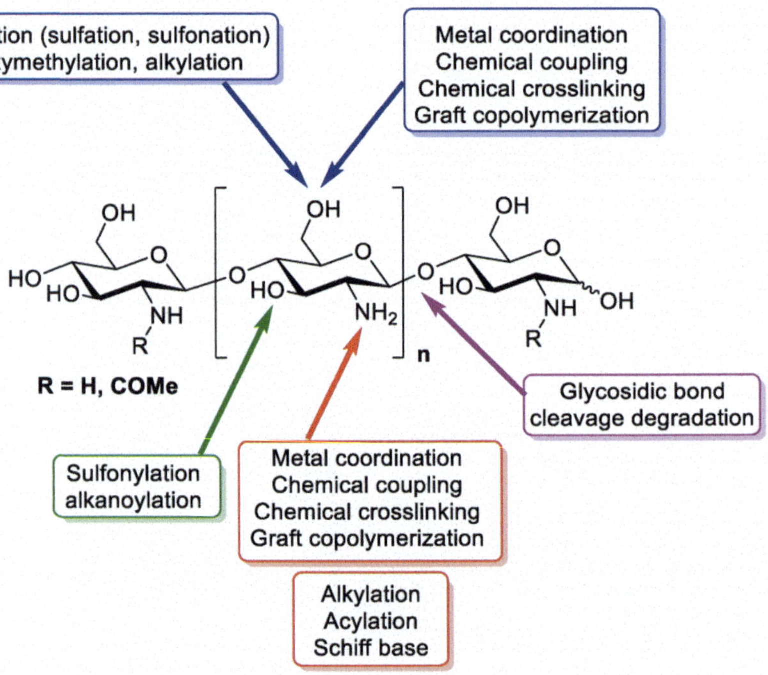

Fig. 1 Structure, chemistry, and modifiable functional groups in the design of chitin and chitosan [9]

2.2 *Extraction Methods for Chitin and Chitosan*

Natural sources include fungal cell walls, insect exoskeletons, and crustacean shells (shrimp, crab) that can extract chitin and chitosan [45]. In order to achieve pure polymers, the extraction processes for chitin and chitosan usually require many steps to eliminate contaminants [29]. Here are some common extraction methods for chitin and chitosan:

2.2.1 Chemical Extraction

The most widely used raw material for chitin is found in the stiff shells of crabs and shrimp, which are made of deposits of calcium salts and a complex network of chitin with proteins [116]. Deproteinization and demineralization procedures must thus be used to eliminate those components in order to separate chitin from the shells. Both can be achieved through several techniques, including chemical and biological extraction, with the latter being preferred in an environmentally benign manner [95].

2.2.2 Chemical Deproteinization

Breaking the chemical bonds that bind chitin to proteins makes the deproteinization phase harder. The biopolymer is also depolymerized by the chemicals used in this heterogeneous process. Since some of the human population is allergic to shellfish, and the protein component is the main cause of this allergy, biomedical applications must remove all proteins [116]. The first strategy for deproteinization was chemical. Many substances, such as NaOH, Na_2CO_3, $NaHCO_3$, KOH, K_2CO_3, $Ca(OH)_2$, Na_2SO_3, $NaHSO_3$, $CaHSO_3$, Na3PO4, and Na2S, have been studied as deproteinization reagents. Conditions for reactions differ significantly between studies. The preferred reagent, NaOH, is used at concentrations between 0.125 and 5.0 M, with treatment times ranging from a few minutes to many days and temperatures up to 160 °C. Using NaOH always causes deproteinization as well as partial acetylation of chitin and hydrolysis of the biopolymer, which lowers the molecular weight of the polymer [116].

Chemical Demineralization

The process of demineralization involves removing minerals, primarily calcium carbonate. Acid treatment with HCl, HNO3, H2SO4, CH3COOH, and HCOOH is commonly employed to achieve demineralization [71, 81]. Diluted hydrochloric acid is the ideal reagent for these acids. Demineralization is an easy process to complete since it involves breaking down calcium carbonate into calcium salts that are soluble in water and releasing carbon dioxide, as shown by the following equation:

$$2HCl + CaCO_3 \rightarrow CaCl_2 + H_2O + CO_2 \uparrow$$

Most other minerals in the clam cuticle react similarly to acid, creating soluble salts. The solid phase of chitin can then be filtered, and the salts can be easily removed by washing with deionized water. Demineralization treatments are often empirical and depend on several factors, including the degree of mineralization of each shell, the length of the extraction process, temperature, particle size, acid content, and the solute/solvent ratio. Calcium chloride depends on the acid content since it requires two molecules of HCl to convert one calcium carbonate molecule into calcium chloride. For a complete reaction, the acid intake must be equal to the stoichiometric amount of minerals [96]. Removing every mineral from the material is difficult because of its heterogeneity; larger volumes or more concentrated acid solutions are consequently required. Demineralization can be tracked via acidimetric titration; the medium's continuous acidity indicates the termination of the process, while the pH's movement towards neutrality indicates the consumption of acid [105].

There have been several demineralization procedures employed in the past with different reaction circumstances. Traditionally, varying doses of diluted hydrochloric acid (up to 10% w/v) at room temperature and for varying incubation times are used to achieve demineralization. These techniques include those of Anderson et al. [70], Hackman [41], Anderson et al. [7].

2.2.3 Enzymatic Extraction

Utilizing particular enzymes or microbes, a biological method of chitin extraction is thought to be quicker, safer, and less harmful to the physicochemical characteristics of native chitin [65]. The demineralization and deproteinization processes include removing certain pigments and lipids. To get a superior, colorless product, extra decolorization and purification are done as needed [75]. Using certain enzymes, such as proteases, to break down proteins and other organic components in the raw material is known as enzymatic extraction. This technique is frequently employed as a greener substitute for chemical extraction. Alternatively, deacetylation mediated by enzymes can be employed [59]. Chitin deacetylase, an enzyme in fungi and insects, catalyzes the hydrolysis of N acetamido linkages in the chitin substrate, which is the basis of this technique [118]. When compared to a traditional chemical method, the primary benefit of the biotransformation of chitin to chitosan is the improved molecular structure and larger molar mass of the end product [116]. The original polymer structure can be preserved through enzymatic treatment, which is much less harsh. However, achieving a high degree of deacetylation is challenging due to the crystalline structure of chitin chains and the restricted accessibility of N-acetyl D-glucosamine units. As a result, certain chitin alterations are required, such as those that attempt to open up the structure and reduce crystallinity [97].

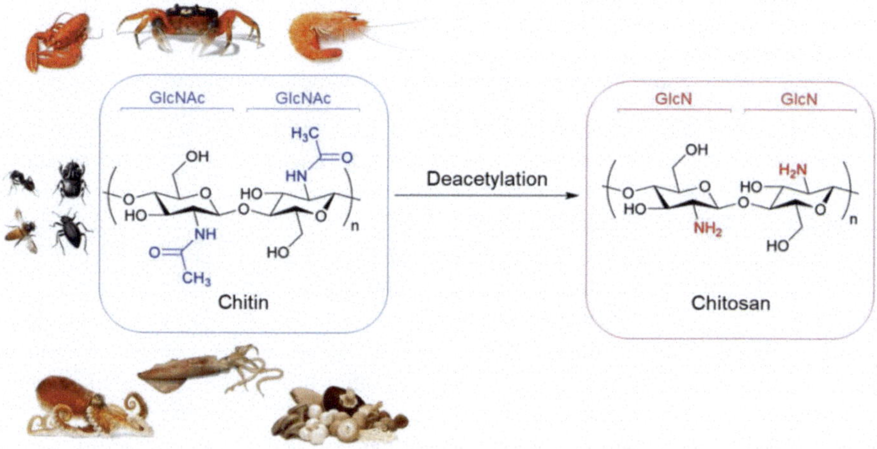

Fig. 2 The structure of chitosan, which is made up of D-glucosamine (GlcN) residues and amino acetyl groups of chitin (N-acetyl-D-glucosamine (GlcNAc) residues), and the origin of chitin [31]

2.2.4 Deacetylation

Chitin undergoes a procedure known as deacetylation, which entails treating it with an alkaline solution to eliminate acetyl groups and transform it into chitosan. Different characteristics of chitosan can be obtained by varying the degree of deacetylation [72]. Despite being low-cost and appropriate for large-scale production, chemical deacetylation has certain disadvantages, such as the potential to cause harm to the polymer backbone and adverse effects on the environment because of its high energy consumption, the need for large quantities of concentrated alkalis and the variety of soluble and insoluble reaction byproducts it produces [72] (Fig. 2).

2.2.5 Microbial Fermentation

Additionally, chitin can be obtained from fungal sources through microbial fermentation techniques. As a primary component of their cell walls, fungi like Aspergillus and Mucorales produce chitin, which can be separated via fermentation and further processing [2]. Following extraction, chitin and chitosan can undergo additional processing stages, such as milling, drying, and purifying, to get the appropriate forms for different uses [82]. The source material, required purity, environmental variables, and the planned uses of the chitin and chitosan products all influence the extraction process selection [49].

2.3 Properties of Chitin and Chitosan Relevant to Hydrogel Formation

One of the main differences between chitosan and chitin is the degree of deacetylation (DD) [21]. The DD value displays the molar proportion of monomeric units, including amino groups, or the ratio of D-glucosamine to the sum of D-glucosamine and N-acetyl D-glucosamine. According to [21, 26]. The chitosan's deacetylated chitin has a DD of 50–100%. Chitosan typically has a molar mass between 50 and 2000 kDa. The physicochemical (such as solubility and chain conformation) and biological properties of chitosan are directly impacted by the degree of deacetylation in conjunction with molar mass [26]. Deacetylation degree is closely correlated with solubility, viscosity, biocompatibility, and other biological features like hemostatic, analgesic, mucoadhesive, and antibacterial activity. Chitosan crystallinity and biodegradability both decrease in tandem with a drop in DD. Biodegradability and antioxidant activity show inverse proportionality with molar mass [19].

The main source of chitosan's unique qualities is the amino groups that are present in its structure. Water, aqueous bases, and organic solvents cannot dissolve chitosan; nevertheless, it can be converted into a cationic polyelectrolyte by protonating its amino groups in diluted acidic solutions ($pH < 6$). The difference between the positively charged and uncharged (soluble and insoluble) chitosan states happens in the pH range of 6–6.5 at that specific pKa value. The degree of protonation determines the properties of the biomaterial and is also influenced by the acid's pKa when used as a solvent [90, 110]. Therefore, the ability to adjust the "charged state" and the characteristics of chitosan is made possible by the presence of amino groups [19].

Chitosan chains include hydroxyl groups in addition to amino groups. Because they both facilitate covalent bonding, they expand the range of potential uses by adding different species throughout the polysaccharide backbone. While the amino groups can undergo reductive amination to have aldehyde activities, the alcohol groups permit nonspecific processes like etherification and esterification [18, 89].

Chitosan is a very intriguing natural polymer with a range of beneficial qualities when viewed through biomaterials. Among its excellent biological traits are antimicrobial properties. Kong et al. [55], antifungal [115], mucoadhesive [100], analgesic [39, 73], and hemostatic [67, 112] properties, as well as biodegradability [53] and biocompatibility [107].

The biodegradability of chitosan is mainly determined by its degree of deacetylation. However, the molar mass, dispersity, and distribution of the chitosan's N-acetyl D glucosamine residues also have an effect. The human body's lysozyme, acid, gastrointestinal enzymes, or colon bacteria can all break down chitosan enzymatically [53]. Glycosidic connections break to produce different-length oligosaccharides, safe substances that can either be absorbed into metabolic pathways or removed by renal filtration. The amino groups' protonation may cause the analgesic and anti-inflammatory properties of chitosan due to absorption of proton ions generated in the inflammatory area [53].

3 Hydrogel Formation

3.1 *Basics of Hydrogel Formation*

Hydrogels are three-dimensional (3D) polymer networks characterized by a high water content and, therefore, a great deal of flexibility. Because of this, their softness and malleability naturally mirror many human bodily tissues [99]. Soft tissues and hydrogels have similar mechanical properties, in particular. Their viscoelastic nature reduces adverse effects on the surrounding host tissue after implantation, along with other desirable features, including biocompatibility, functioning, and reversibility, making them an intriguing and significant member of the biomaterial group [19, 106].

The hydrophilic groups in the hydrogel polymer network enable considerable liquid water absorption. The more water there is in the hydrogel, the larger it gets. Three states of water molecules are present in a hydrogel structure: free, intermediate, and bound [4]. The amount of free water molecules in a hydrogel is determined by its structure; the more compact the structure, the fewer free water molecules it contains. Weak interactions between the functional groups of polymer chains and water molecules create intermediate water. In contrast, bound water results from strong hydrogen bonding between the polymer and water molecules.

Crosslinking generally enhances mechanical characteristics and increases hydrogel stability through the interconnection of molecules. In addition, it affects polymer degradability, alters rheological properties that may lead to processing challenges, and narrows the pool of functional groups that can be modified [87]. Chitosan-based hydrogels can be made in a variety of ways, but in order to guarantee hydrogel network stability, chitosan hydrogels must be chemically or physically crosslinked. Crosslinking determines the hydrogel system's stability and reversibility [13].

3.2 *Methods for Developing Chitin and Chitosan-Based Hydrogels*

Chitosan-based hydrogels can be prepared using several methods, but they need to be chemically or physically crosslinked to ensure hydrogel network stability. The crosslinking process establishes the stability and reversibility of the hydrogel system.

3.2.1 Chitosan-Based Hydrogels Fabricated by Chemical (Covalent) Crosslinking

The hydrogel network is permanently linked by chemical crosslinking created by photopolymerization, radical polymerization, enzymatic activities, and covalent bonding with substances like aldehydes. Chemical reactions that change the arrangement of molecules give chemical hydrogels their permanent and irreversible nature. Chemically crosslinked chitosan hydrogels can form semi-interpenetrating polymer networks (semi-IPN), hybrid polymer networks (HPN), and interpenetrating polymer networks (IPN) [4]. Crosslinking agents used for chitosan hydrogel preparation include glutaraldehyde (GA) [66], N, N-methylene-bisacrylamide (MBA) [85], glyoxal [114], and genipin [51].

The chitosan chain's network of covalently bonded amino and hydroxyl functional groups enables the chemical gel to engage with the cross-linking agent to form a stable three-dimensional gel structure. Covalent bond formation increases the gel structure's stability and rigidity [104]. True gels, or chemically cross-linked hydrogels, are made when a covalent bond joins two polymer molecules. Covalent units are produced between polymer chains by crosslinked molecules. Ghorpade [34], producing a hydrogel that is frequently durable, persistent, and irreversible. The hydroxyl and amino groups on the chitosan chain, which are very reactive, cause polymer molecules to cross-link readily. Moreover, derivatized chitosan's active groups can go through fresh chemical crosslinking, which enhances its mechanical properties, structural stability, bacteriostatic properties, and biocompatibility [10].

Hydrogels can be chemically crosslinked to generate stable gels with better mechanical properties by covalently joining polymer chains. These chemically crosslinked gels, as opposed to physical hydrogels, show improved resistance to outside environmental elements that could jeopardize the integrity of the gel structure [28]. As such, hydrogels can exhibit markedly enhanced mechanical properties.

Chemical (covalent) crosslinking, which includes creating strong links between chitosan molecules to generate a three-dimensional network structure, creates chitosan-based hydrogels [32]. The procedure of crosslinking improves the hydrogel's mechanical strength, stability, and biocompatibility, which makes it appropriate for a range of biomedical uses, including tissue engineering, drug delivery, wound healing, and regenerative medicine [54]. The covalent crosslinking process guarantees long-term preservation of the hydrogel's structural integrity and functionality [33].

As CS hydrogels generated by covalent crosslinking may be formed and applied without being pH-limited, controlling their characteristics is comparatively easier. However, they need more work to manufacture, and there could be a risk of contamination from hazardous reagents [62].

3.2.2 Based Hydrogels Fabricated by Physical Crosslinking

Physical crosslinking techniques, which produce reversible linkages between chitosan molecules to build a network structure, can also be used to create chitosan hydrogels [32].

Incorporating bioactive molecules or cells into the hydrogel matrix without compromising their activity is possible since physical crosslinking techniques are typically kinder and gentler than chemical crosslinking techniques. Compared to chemically crosslinked hydrogels, physical crosslinked hydrogels could have less stability and mechanical strength [113]. For chitosan-based hydrogels, temperature-induced gelation, ionotropic gelation with multivalent cations like calcium ions, and self-assembly of chitosan molecules through interactions with other polymers or nanoparticles are a few examples of physical crosslinking techniques [42].

Generally speaking, physical crosslinking techniques provide adaptability and flexibility in creating hydrogels based on chitosan for a range of biomedical applications. They utilize the special qualities of chitosan and allow for the customization of hydrogel properties according to particular needs [84].

3.3 Factors Influencing Hydrogel Properties

The characteristics of hydrogels, particularly those based on chitosan, can be influenced by a number of variables. The following are some important variables that may have an impact on hydrogel properties:

3.3.1 Composition

The kind and concentration of the polymer, the crosslinking agent, any additives, and any bioactive compounds included in the hydrogel's composition can all greatly impact its characteristics [50]. The type and concentration of crosslinking agents and the molecular weight and degree of acetylation of the chitosan can affect the hydrogel's mechanical strength, swelling behavior, and biocompatibility [35].

3.3.2 Crosslinking Density

The mechanical characteristics, swelling behavior, and rate of degradation of the hydrogel network can all be impacted by the density of crosslinks within it. Hydrogels having a higher crosslinking density tend to be stiffer, less able to swell, and more stable. The process and degree of physical or chemical crosslinking can also affect the hydrogel's characteristics [43].

3.3.3 pH and Temperature

The mechanical characteristics, rate of deterioration, and swelling behavior of hydrogels can all be impacted by environmental factors like temperature and pH. The protonation state of chitosan molecules in hydrogels based on chitosan can be influenced by the pH of the surrounding media, which can result in modifications to the hydrogel's mechanical characteristics and swelling behavior. Temperature can also affect the hydrogel's stability and gelation process [4].

3.3.4 Molecular Weight and Structure of Polymers

The molecular weight and structure of the polymers utilized in the formulation can influence the mesh size, porosity, and mechanical characteristics of the hydrogel network. Changes in the molecular weight and degree of acetylation of chitosan can affect how the hydrogel interacts with other polymers or crosslinking agents, altering the hydrogel's characteristics [42].

3.3.5 Processing Methods

The mixing, crosslinking, and drying procedures used during the hydrogel synthesis can affect the finished hydrogel's characteristics. Variations in the hydrogel's mechanical strength, pore size distribution, and network topology can be caused by different processing techniques [5]. Researchers can customize the properties of chitosan-based hydrogels to meet specific needs for a range of biomedical applications, including drug delivery, tissue engineering, wound healing, and regenerative medicine, by carefully controlling these factors during the design and fabrication process [15].

4 Applications of Chitin and Chitosan-Based Hydrogel in Agriculture

The combination of chitosan's structural and biological features has made it one of the most promising polymers for efficiently delivering agrochemicals and micronutrients in nanoparticle form [52]. Applications of chitosan-based nanoparticle delivery systems in agriculture, including the delivery of micronutrients, fertilizers, pesticides, and herbicides, have been examined [69]. Chitosan has several advantages over other biosorbents and activated carbon, including being less expensive, having a high affinity for various contaminants, being chemically stable, having high reactivity, and being selective regarding pollution [74].

4.1 Soil Conditioning

Hydrogels based on chitosan can also be used as soil conditioners to improve the structure, water-holding capacity, and soil nutrient availability. Crustacean exoskeletons contain chitin, which is the source of chitosan, a biocompatible and biodegradable polymer [94]. The study describes new soil conditioner systems that address soil water retention and fertilization. The systems are made by hydrogelating chitosan with salicylaldehyde in situ when urea fertilizer is present [46]. The formulations exhibited a high water absorbency of 68 g/g, induced a 154% increase in soil water holding capacity, and nearly doubled the amount of nitrogen in the soil, resulting in over 70% more plant growth than in the reference soil [46]. Chitosan has several advantages when applied to soil conditioning through hydrogels, including water retention. Hydrogels based on chitosan can absorb and hold onto water, which can help raise soil moisture content and lower evaporation loss. This can be especially helpful in dry or drought-prone areas when little water is available [101]. Chitosan-based hydrogels can improve soil structure by encouraging particle aggregation, which raises soil porosity, aeration, and drainage. This may result from better root development, nutrient uptake, and general plant health [78]. In situ hydrogelation of CS with salicylaldehyde as a crosslinker and urea as fertilizer was used to create new soil conditioner systems [46]. The arranging of produced imine units into supramolecular clusters, which serve as crosslinking nodes, has been proposed as the cause of hydrogelation. The improved composition increased the soil's ability to retain water by up to 154% and showed strong water absorption (68 g/g). Apart from the aforementioned, soil conditioning is employed to enhance nutrient availability. Hydrogels based on chitosan can keep nutrients in the soil longer by reducing leaching and increasing plant availability. This may lead to higher agricultural yields and better plant uptake of nutrients [101]. Hydrogels based on chitosan offer a sustainable alternative for soil conditioning because they are both environmentally benign and biodegradable. Chitosan oligomers, which have been demonstrated to positively impact plant development and soil health, are released as the hydrogel breaks down [62]. Because chitosan-based hydrogels have antibacterial, antifungal, and insecticidal qualities, they can also protect crops. Chitosan is a natural and environmentally beneficial substitute for synthetic pesticides because of its well-known capacity to impede the growth of many diseases and pests. Here are a few applications for chitosan-based hydrogels in crop protection [94].

4.1.1 Wound Healing

Chitosan-based hydrogels stimulate plant defense mechanisms and form a protective coating over injured tissues, aiding in the healing of wounds in plants. This can speed up a plant's recovery from disease or physical harm [15].

4.1.2 Biodegradability

Compared to synthetic pesticides, hydrogels based on chitosan are less environmentally hazardous and biodegradable. Over time, they decompose into non-toxic chemicals, preventing the buildup of toxic residues in soil and water [4].

Chitosan-based hydrogels provide an eco-friendly and sustainable method of crop protection by utilizing chitosan's inherent qualities to strengthen plant resistance to pests and diseases. Farmers can lessen their reliance on chemical pesticides, safeguard the long-term health of agricultural ecosystems, and encourage biodiversity by implementing chitosan-based hydrogels into integrated pest management programs [91].

4.2 Controlled-Release Systems

Hydrogels based on chitosan can also be used as controlled-release fertilizer solutions in agriculture. A slow-release fertilizer that progressively delivers vital nutrients to plants over an extended period of time can be made by adding nutrients to chitosan hydrogels. Numerous advantages are provided by this controlled-release system for crop growth and soil health [62]. Hydrogels based on chitosan can increase plants' nitrogen absorption efficiency. Leaching and runoff are decreased because of the hydrogel's slow-release characteristics, which guarantee that nutrients are delivered gradually. Plants can better utilize nutrients as a result, and environmental contamination is reduced [76].

Plants may access nutrients for a longer period of time because of the controlled-release properties of chitosan-based hydrogels, which give them a steady supply of vital components for development and growth. Throughout the growing season, this can encourage healthy plant growth and assist in keeping the soil's nutrient levels at their ideal levels [62].

Farmers might lessen how often they apply fertilizer by utilizing chitosan-based hydrogels as a carrier. This helps reduce the possibility of overfertilization, which can result in nutrient imbalances and environmental harm, in addition to saving time and work [92]. Hydrogels based on chitosan may also help strengthen the soil's structure and its ability to hold water. By improving soil aeration, decreasing compaction, and retaining moisture in the soil, the hydrogel matrix can improve the conditions that allow plant roots to absorb nutrients and water [62].

Hydrogel formulations based on chitosan present a sustainable substitute for traditional fertilizers due to their biodegradable nature and environmental friendliness. By reducing nutrient losses and enhancing soil health, using chitosan-based controlled-release systems can help lessen the adverse effects of agriculture on the environment [92].

Chitosan-based hydrogels present a viable way to create controlled-release fertilizer systems that can boost crop yield, encourage sustainable farming methods, and better manage nutrient availability in agriculture. By utilizing the special qualities of chitosan as a natural polymer, farmers can maximize nutrient delivery to plants while reducing environmental dangers associated with typical fertilizer applications [92].

4.3 Seed Coating

Hydrogels based on chitosan have potential applications in agriculture as materials for seed coatings. In modern agriculture, seed coating is a widespread process wherein seeds are coated with several compounds to boost seedling growth, enhance germination, and protect against diseases. Hydrogels based on chitosan provide several benefits as seed coating materials, including [77].

Hydrogels based on chitosan can increase the rate at which seeds germinate by enclosing the seed in a barrier that keeps moisture and nutrients. This may produce the ideal conditions for the growth and development of seedlings, resulting in quicker and more consistent germination [37]. Because of its antibacterial qualities, chitosan helps shield seeds from bacterial and fungal infections. Applying hydrogels based on chitosan to seeds can lower the risk of infections that spread from seeds and enhance the health of seedlings [94]. Hydrogels based on chitosan can provide growth stimulants, nutrients, or advantageous microbes to the seed covering. This facilitates the gradual and regulated flow of nutrients to the budding seedling, there by encouraging sound growth and development [94]. Chitosan-based seed coatings are a viable substitute for synthetic seed coatings since they are environmentally benign and biodegradable. Utilizing materials based on chitosan might lessen agriculture's negative environmental effects and encourage environmentally friendly farming methods [94]. Because of their strong adhesive qualities, chitosan-based hydrogels help the seed covering stick firmly to the seed surface. This guarantees that the seeds receive consistent covering and an efficient supply of nutrients and protective chemicals [15].

All things considered, hydrogels based on chitosan provide an adaptable and sustainable solution for agricultural seed coating applications. Farmers can lessen the adverse effects of agricultural activities on the environment while enhancing seed germination, shielding seeds from diseases, and encouraging healthy seedling growth by utilizing the special qualities of chitosan as a natural polymer [94] (Fig. 3).

5 Benefits

5.1 Environmental Sustainability

Hydrogels based on chitosan can also be used as controlled-release fertilizer solutions By adding nutrients to chitosan hydrogels, as in agriculture. A slow-release fertilizer that progressively delivers vital nutrients to plants over an extended period of time can be. This controlled-release system provides numerous advantagesstem for crop growth and soil health [62].

Hydrogels based on chitosan can increase plants' nitrogen absorption efficiency. Leaching and runoff are decreased because of the hydrogel's slow-release characteristics, which guarantee that nutrients are delivered gradually. Plants can better utilize nutrients as a result, and environmental contamination is reduced [62]. Plants may

Fig. 3 Agricultural application of hydrogel for water-storing purpose

access nutrients for a more extended period of time because of the controlled-release properties of chitosan-based hydrogels, which give them a steady supply of vital components for development and growth. Throughout the growing season, this can encourage healthy plant growth and assist in keeping the soil's nutrient levels at their ideal levels [78].

Farmers might lessen how often they apply fertilizer by utilizing chitosan-based hydrogels as a carrier. This helps reduce the possibility of overfertilization, which can result in nutrient imbalances and environmental harm, in addition to saving time and work [92]. Hydrogels based on chitosan may also help strengthen the soil's structure and its ability to hold water. By improving soil aeration, decreasing compaction, and retaining moisture in the soil, the hydrogel matrix can improve the conditions that allow plant roots to absorb nutrients and water [101]. Hydrogel formulations based on chitosan present a sustainable substitute for traditional fertilizers due to their biodegradable nature and environmental friendliness. By reducing nutrient losses and enhancing soil health, using chitosan-based controlled-release systems can help lessen the negative effects of agriculture on the environment [62].

All things considered, chitosan-based hydrogels present a viable way to create controlled-release fertilizer systems that can boost crop yield, encourage sustainable farming methods, and better manage nutrient availability in agriculture. By utilizing the special qualities of chitosan as a natural polymer, farmers can maximize nutrient delivery to plants while reducing environmental dangers associated with typical fertilizer applications [92].

5.2 As Seed Coating Materials in Agriculture

In modern agriculture, seed coating is a widespread process wherein seeds are coated with several compounds to boost seedling growth, enhance germination, and protect against diseases. Hydrogels based on chitosan have several benefits when used as seed-coating materials [77].

5.2.1 Enhanced Seed Germination

Hydrogels based on chitosan can increase the rate at which seeds germinate by enclosing the seed in a barrier that keeps moisture and nutrients. This may produce the ideal conditions for the growth and development of seedlings, resulting in quicker and more consistent germination [37].

5.2.2 Pathogen Protection

Because of its antibacterial qualities, chitosan helps shield seeds from bacterial and fungal infections. Applying hydrogels based on chitosan to seeds can lower the risk of infections that spread from seeds and enhance the health of seedlings [94].

5.2.3 Nutrient Delivery

Hydrogels based on chitosan can provide growth stimulants, nutrients, or advantageous microbes to the seed covering. This facilitates the gradual and regulated flow of nutrients to the budding seedling by encouraging sound growth and development [62].

5.2.4 Environmental Benefits

Chitosan-based seed coatings are a viable substitute for synthetic seed coatings since they are environmentally benign and biodegradable. Utilizing materials based on chitosan might lessen agriculture's adverse environmental effects and encourage environmentally friendly farming methods [94].

5.2.5 Adhesion Properties

Seed coatings stick effectively to the seed surface thanks to the excellent adhesion properties of hydrogels based on chitosan. By doing this, the seeds are guaranteed consistent covering and efficient delivery of nutrients and barrier agents.

Hydrogels based on chitosan provide an adaptable and sustainable solution for agricultural seed coating applications. By utilizing the special qualities of chitosan as a natural polymer, farmers can lessen the negative effects of agricultural activities on the environment while enhancing seed germination, shielding seeds from diseases, and encouraging healthy seedling growth [94].

5.3 Environmentally Sustainable

The natural polymer chitin, which is present in the exoskeletons of crustaceans like shrimp and crabs, is the source of chitosan, a biodegradable and biocompatible substance. Chitosan-based hydrogels have a lower environmental impact since they decompose into non-toxic substances [30].

Chitosan is an abundant and renewable material that may be recovered from the waste products of the seafood processing industry. Farmers can lessen their need for synthetic materials generated from non-renewable sources by employing hydrogels based on chitosan [103]. Chitosan is a safe and environmentally beneficial ingredient for agricultural uses because it is non-toxic to people, animals, or plants. Utilizing chitosan-based hydrogels as seed coating materials does not endanger the environment or public health. Farmers can lessen their reliance on synthetic chemicals and insecticides, which may have negative environmental impacts, by applying hydrogels based on chitosan as seed coatings. The antibacterial qualities of chitosan can also aid in seed protection against infections without artificial fungicides. By encouraging seed germination and seedling growth, hydrogels based on chitosan can enhance soil health and produce healthier plants with more robust root systems. This can support environmentally friendly farming methods that preserve soil fertility and lessen the demand for synthetic fertilizers. Hydrogels based on chitosan provide an environmentally friendly substitute for traditional seed coating materials in farming. Farmers can increase crop output while reducing their environmental impact and encouraging eco-friendly agricultural methods by utilizing the inherent qualities of chitosan.

5.4 Chitosan-Based Hydrogel as Controlled-Release Systems of Fertilizer

Hydrogels based on chitosan can also be used as controlled-release fertilizer solutions in agriculture. A slow-release fertilizer that progressively delivers vital nutrients to plants over an extended period of time can be made by adding nutrients to chitosan hydrogels. The following are some advantages of this controlled-release method for crop growth and soil health:

5.4.1 Enhanced Nutrient Efficiency

Hydrogels based on chitosan can potentially increase plants' nitrogen absorption efficiency. The hydrogel's slow-release characteristics, which guarantee that nutrients are delivered gradually, decrease leaching and runoff. This reduces pollution to the environment and improves plant uptake of nutrients.

5.4.2 Long-Lasting Effects

Plants may access nutrients for a longer period of time because of the controlled-release properties of chitosan-based hydrogels, which give them a steady supply of vital components for development and growth. Throughout the growing season, this can promote healthy plant growth and assist in keeping the soil's nutrient levels at their ideal levels.

5.4.3 Reduced Fertilizer Application Frequency

Farmers can lower the frequency of fertilizer treatments by employing hydrogels based on chitosan as a nutrient transporter. By doing this, you can avoid overfertilization, resulting in nutrient imbalances and environmental harm, while saving time and work.

5.4.4 Improved Soil Structure

Hydrogels based on chitosan may also help strengthen the soil's structure and its ability to hold water. In order to improve the soil's aeration, decrease compaction, and retain moisture, the hydrogel matrix can be used to improve the soil's ability for plant roots to absorb nutrients and water.

5.4.5 Environmentally Friendly

Hydrogel formulations based on chitosan present a sustainable substitute for traditional fertilizers due to their biodegradable nature and environmental friendliness. By limiting nutrient losses and enhancing soil health, adopting chitosan-based controlled-release systems can help lessen the environmental effects of agriculture.

All things considered, chitosan-based hydrogels present a viable way to create controlled-release fertilizer systems that can boost crop yield, encourage sustainable farming methods, and better manage nutrient availability in agriculture. By utilizing the special features of chitosan, a natural polymer, farmers can minimize the environmental concerns associated with typical fertilizer applications while optimizing nutrient delivery to plants.

6 Challenges

Hydrogels derived from chitosan provide several benefits for agriculture, but there are also some drawbacks and difficulties that should be taken into account. For example, when considering traditional synthetic materials, the initial cost of chitosan and its hydrogels can be expensive. This can be a challenge for small-scale farmers or those with tight budgets. The main source of chitosan is the shells of crustaceans, such as shrimp and crabs, which are only sometimes easily found. It can be challenging to find high-quality chitosan, particularly in landlocked or non-seafood countries. Although chitosan-based hydrogels' biodegradability is a benefit, it can also be a drawback in terms of their lifetime and durability. The hydrogels can break down more quickly than synthetic substitutes, necessitating more frequent treatments and possibly decreasing their efficacy over time.

The use of hydrogels based on chitosan may call for specific tools or methods, which could make farming more difficult and expensive. Ensuring that the hydrogels are applied and distributed correctly can take time and effort, reducing their effectiveness. Hydrogels based on chitosan may function and work differently when combined with other agrochemicals, fertilizers, or soil conditions. It is imperative to comprehend the compatibility of chitosan with current farming practices to prevent unfavorable interactions. For usage in agriculture, chitosan-based hydrogels can need regulatory approval, depending on the area.

It might take a lot of time and money to demonstrate the safety and effectiveness of these items while still meeting regulatory requirements. More research is required to optimize the composition, application techniques, and doses of chitosan-based hydrogels for various crops and soil types. The benefits of these hydrogels in agriculture must be maximized by ongoing research and development. Because chitosan-based hydrogels biodegrade quickly, their shelf life may be restricted. Proper handling procedures and storage conditions are required to keep them functioning and stop degradation before use. While chitosan-based hydrogels generally have encouraging advantages for sustainable agriculture, it is imperative to solve the difficulties and constraints related to their application to optimize their potential influence and uptake in agricultural methods. These challenges can be addressed, and the practicality of chitosan-based hydrogels in agriculture has increased with continued research, technology developments, and market backing.

7 Conclusion

Biopolymer-based hydrogels are employed in many different applications worldwide. One of their main uses is utilizing biopolymeric hydrogels for agricultural use. Hydrogels can raise agricultural production by improving plant protection, nutrient availability, and soil moisture levels. Hydrogels based on chitosan are highly desirable materials for use in agriculture because they may be used to deliver active chemicals and regulate their release.

It can be a carrier of nutrients and other agrochemicals and improve the soil's capacity to retain water, making it useful as soil conditioning agents. Crop growth, productivity, and survival are influenced by irrigation methods and water-saving measures. This paper addresses the creation of hydrogels based on chitosan and explores their possible usage in agriculture.

References

1. Abd El-Rehim HA (2006) Characterization and possible agricultural application of polyacrylamide/sodium alginate crosslinked hydrogels prepared by ionizing radiation. J Appl Polym Sci 101(6):3572–3580
2. Abo Elsoud MM, El Kady E (2019) Current trends in fungal biosynthesis of chitin and chitosan. Bull Natl Res Cent 43(1):1–12
3. Adel A, Abou-Youssef H, El-Gendy A, Nada A (2010) Carboxymethylated cellulose hydrogel; sorption behavior and characterization. Nat Sci 8(8):244–256
4. Ahmadi F, Oveisi Z, Samani SM, Amoozgar Z (2015) Chitosan based hydrogels: characteristics and pharmaceutical applications. Res Pharm Sci 10(1):1
5. Ahmed EM (2015) Hydrogel: preparation, characterization, and applications: a review. J Adv Res 6(2):105–121
6. Ahmed S, Ahmad M, Ikram S (2014) Chitosan: a natural antimicrobial agent-a review. J Appl Chem 3(2):493–503
7. Anderson C, De Pablo N, Romo C (1978) Antarctic krill (Euphausia superba) as a source of chitin and chitosan. In: Proceedings of first international conference on chitin and chitosan. MIT Sea Grant Program: Cambridge, MA, USA, pp 54–63
8. Aranaz I et al (2009) Functional characterization of chitin and chitosan. Curr Chem Biol 3(2):203–230
9. Aranaz I et al (2021) Chitosan: an overview of its properties and applications. Polymers 13:3256
10. Ardean C et al (2021) Factors influencing the antibacterial activity of chitosan and chitosan modified by functionalization. Int J Mol Sci 22(14):7449
11. Arikibe JE, Lata R, Kuboyama K, Ougizawa T, Rohindra D (2019) pH-responsive studies of bacterial cellulose/chitosan hydrogels crosslinked with Genipin: swelling and drug release behaviour. ChemistrySelect 4(34):9915–9926

12. Balart R, Garcia-Garcia D, Fombuena V, Quiles-Carrillo L, Arrieta MP (2021) Biopolymers from natural resources. MDPI, p 2532
13. Bhattarai N, Gunn J, Zhang M (2010) Chitosan-based hydrogels for controlled, localized drug delivery. Adv Drug Deliv Rev 62(1):83–99
14. Bósquez-Cáceres MF, Bejar J, Álvarez-Contreras L, Tafur JP (2023) Enhancing electrochemical performance of zinc-air batteries using freeze crosslinked carboxymethylcellulose-chitosan hydrogels as electrolytes. J Electrochem Soc 170(6):060502
15. Che X et al (2024) Application of chitosan-based hydrogel in promoting wound healing: a review. Polymers 16(3):344
16. Chen J et al (2018) Environmentally friendly fertilizers: A review of materials used and their effects on the environment. Sci Total Environ 613:829–839
17. Cheng B, Pei B, Wang Z, Hu Q (2017) Advances in chitosan-based superabsorbent hydrogels. RSC Adv 7(67):42036–42046
18. Croisier F, Jérôme C (2013) Chitosan-based biomaterials for tissue engineering. Eur Polym J 49(4):780–792
19. Dash M, Chiellini F, Ottenbrite RM, Chiellini E (2011) Chitosan—a versatile semi-synthetic polymer in biomedical applications. Prog Polym Sci 36(8):981–1014
20. Dave U, Somanader E, Baharlouei P, Pham L, Rahman MA (2021) Applications of chitin in medical, environmental, and agricultural industries. J Mar Sci Eng 9(11):1173
21. de Alvarenga ES, de Oliveira CP, Bellato CR (2010) An approach to understanding the deacetylation degree of chitosan. Carbohydr Polym 80(4):1155–1160
22. Demitri C, Scalera F, Madaghiele M, Sannino A, Maffezzoli A (2013) Potential of cellulose-based superabsorbent hydrogels as water reservoir in agriculture. Int J Polym Sci
23. Desai R, Pachpore R, Patil A, Jain R, Dandekar P (2021) Review of the structure of chitosan in the context of other sugar-based polymers, chitosan for biomaterials III: structure-property relationships. Springer, pp 23–74
24. Díaz-Montes E, Castro-Muñoz R (2021) Trends in chitosan as a primary biopolymer for functional films and coatings manufacture for food and natural products. Polymers 13(5):767
25. Dima JB, Sequeiros C, Zaritzky N (2017) Chitosan from marine crustaceans: production, characterization and applications. In: Biological activities and application of marine polysaccharides, pp 39–56
26. Dos Santos Z, Caroni A, Pereira M, Da Silva D, Fonseca J (2009) Determination of deacetylation degree of chitosan: a comparison between conductometric titration and CHN elemental analysis. Carbohydr Res 344(18):2591–2595
27. Dutta PK, Dutta J, Tripathi V (2004) Chitin and chitosan: chemistry, properties and applications
28. Eivazzadeh-Keihan R et al (2022) Cross-linked lignin/agarose hydrogels coated with iron oxide magnetic nanoparticles for in vitro hyperthermia cancer therapy. J Mater Res 37(23):4392–4402
29. El Knidri H, Belaabed R, Addaou A, Laajeb A, Lahsini A (2018) Extraction, chemical modification, and characterization of chitin and chitosan. Int J Biol Macromol 120:1181–1189
30. Elieh-Ali-Komi D, Hamblin MR (2016) Chitin and chitosan: production and application of versatile biomedical nanomaterials. Int J Adv Res 4(3):411
31. Fatimi A (2022) Chitosan-based hydrogels: patent analysis. Mater Proc 9(1):1
32. Fu J, Yang F, Guo Z (2018) The chitosan hydrogels: from structure to function. New J Chem 42(21):17162–17180
33. Gao Y, Peng K, Mitragotri S (2021) Covalently crosslinked hydrogels via step-growth reactions: crosslinking chemistries, polymers, and clinical impact. Adv Mater 33(25):2006362
34. Ghorpade VS (2020) Preparation of hydrogels based on natural polymers via chemical reaction and cross-linking. In: Hydrogels based on natural polymers. Elsevier, pp 91–118
35. Gonçalves JO, Esquerdo VM, Sant'Anna Cadaval TR, de Almeida Pinto LA (2019) Chitosan-based hydrogels. In: Sustainable agriculture reviews 36: chitin and chitosan: applications in food, agriculture, pharmacy, medicine and wastewater treatment, pp 147–173

36. Gorgieva S, Kokol V (2012) Preparation, characterization, and in vitro enzymatic degradation of chitosan-gelatine hydrogel scaffolds as potential biomaterials. J Biomed Mater Res Part A 100(7):1655–1667
37. Guan Y-J, Hu J, Wang X-J, Shao C-X (2009) Seed priming with chitosan improves maize germination and seedling growth in relation to physiological changes under low temperature stress. J Zhejiang Univ Sci B 10:427–433
38. Guilherme MR et al (2015) Superabsorbent hydrogels based on polysaccharides for application in agriculture as soil conditioner and nutrient carrier: a review. Eur Polym J 72:365–385
39. Guo L, Zhao ZY, Bai J, Han SY, Zhao X (2012) Preparation and performance evaluation of the analgesic and anti-inflammatory effects of chitosan composite materials. Adv Mater Res 583:171–174
40. Guyot C, Cerruti M, Lerouge S (2021) Injectable, strong and bioadhesive catechol-chitosan hydrogels physically crosslinked using sodium bicarbonate. Mater Sci Eng C 118:111529
41. Hackman R (1954) Studies on chitin I. Enzymic degradation of chitin and chitin esters. Aust J Biol Sci 7(2):168–178
42. Hong F et al (2023) Chitosan-based hydrogels: from preparation to applications, a review. Food Chem X:101095
43. Hoti G et al (2021) Effect of the cross-linking density on the swelling and rheological behavior of ester-bridged β-cyclodextrin nanosponges. Materials 14(3):478
44. Hu C, An R, Han L, Wang X, Shi Y, Ran R (2018) Preparation of high strength double physically cross-linked hydrogels by immersion method—How to avoid uneven soaking. Colloids Surf A: Physicochem Eng Asp 559:74–82
45. Iber BT, Kasan NA, Torsabo D, Omuwa JW (2022) A review of various sources of chitin and chitosan in nature. J Renew Mater 10(4):1097
46. Iftime MM, Ailiesei GL, Ungureanu E, Marin L (2019) Designing chitosan based eco-friendly multifunctional soil conditioner systems with urea controlled release and water retention. Carbohydr Polym 223:115040
47. Ismail H, Irani M, Ahmad Z (2013) Starch-based hydrogels: present status and applications. Int J Polym Mater Polym Biomater 62(7):411–420
48. Jain A, Gupta Y, Jain SK (2007) Perspectives of biodegradable natural polysaccharides for site-specific drug delivery to the colon. J Pharm Pharm Sci 10(1):86–128
49. Joseph SM, Krishnamoorthy S, Paranthaman R, Moses J, Anandharamakrishnan C (2021) A review of source-specific chemistry, functionality, and applications of chitin and chitosan. Carbohydr Polym Technol Appl 2:100036
50. Kaczmarek B, Nadolna K, Owczarek A (2020) The physical and chemical properties of hydrogels based on natural polymers. In: Hydrogels based on natural polymers, pp 151–172
51. Kaminski K, Zazakowny K, Szczubiałka K, Nowakowska M (2008) PH-sensitive genipin-cross-linked chitosan microspheres for heparin removal. Biomacromol 9(11):3127–3132
52. Kashyap PL, Xiang X, Heiden P (2015) Chitosan nanoparticle-based delivery systems for sustainable agriculture. Int J Biol Macromol 77:36–51
53. Kean T, Thanou M (2010) Biodegradation, biodistribution and toxicity of chitosan. Adv Drug Deliv Rev 62(1):3–11
54. Khunmanee S, Jeong Y, Park H (2017) Crosslinking method of hyaluronic-based hydrogel for biomedical applications. J Tissue Eng 8:2041731417726464
55. Kong M, Chen XG, Xing K, Park HJ (2010) Antimicrobial properties of chitosan and mode of action: a state of the art review. Int J Food Microbiol 144(1):51–63
56. Lim JY, Goh L, Otake KI, Goh SS, Loh XJ, Kitagawa S (2023) Biomedically-relevant metal organic frameworkhydrogel composites. Biomater Sci 11(8):2661–2677
57. Liu Y, Mao J, Guo Z, Hu Y, Wang S (2022) Polyvinyl alcohol/carboxymethyl chitosan hydrogel loaded with silver nanoparticles exhibited antibacterial and self-healing properties. Int J Biol Macromol 220:211–222
58. Lu J, Chen Y, Ding M, Fan X, Hu J, Chen Y, Li J, Li Z, Liu W (2022) A 4arm-PEG macromolecule crosslinked chitosan hydrogels as antibacterial wound dressing. Carbohydr Polym 277:118871

59. Łubek-Nguyen A, Ziemichód W, Olech M (2022) Application of enzyme-assisted extraction for the recovery of natural bioactive compounds for nutraceutical and pharmaceutical applications. Appl Sci 12(7):3232
60. Matica A, Menghiu G, Ostafe V (2017) Antibacterial properties of chitin and chitosans. New Front Chem 26(1)
61. Mei J, Zhang H, Li Z, Ou H (2019) A novel tetraethylenepentamine crosslinked chitosan oligosaccharide hydrogel for total adsorption of Cr (VI). Carbohydr Polym 224:115154
62. Michalik R, Wandzik I (2020) A mini-review on chitosan-based hydrogels with potential for sustainable agricultural applications. Polymers 12(10):2425
63. Mignon A, De Belie N, Dubruel P, Van Vlierberghe S (2019) Superabsorbent polymers: a review on the characteristics and applications of synthetic, polysaccharide-based, semi-synthetic and 'smart' derivatives. Eur Polym J 117:165–178
64. Mohamed NA, Al-Harby NF, Almarshed MS (2021) Effective removal of Basic Red 12 dye by novel antimicrobial trimellitic anhydride isothiocyanate-cross-linked chitosan hydrogels. Polym Polym Compos 29:S274–S287
65. Mohan K et al (2022) Green and eco-friendly approaches for the extraction of chitin and chitosan: a review. Carbohydr Polym 287:119349
66. Monteiro OA Jr, Airoldi C (1999) Some studies of crosslinking chitosan–glutaraldehyde interaction in a homogeneous system. Int J Biol Macromol 26(2–3):119–128
67. Mony U, Jayakumar R (2016) Chitin and chitosan as hemostatic agents. In: Encyclopedia of Polymer Science and Technology, pp 1–12
68. Mourya V, Inamdar NN (2008) Chitosan-modifications and applications: opportunities galore. React Funct Polym 68(6):1013–1051
69. Mujtaba M et al (2020) Chitosan-based delivery systems for plants: a brief overview of recent advances and future directions. Int J Biol Macromol 154:683–697
70. Muzzarelli R, Tanfani F, Emanuelli M, Gentile S (1980) The chelation of cupric ions by chitosan membranes
71. No HK, Hur EY (1998) Control of foam formation by antifoam during demineralization of crustacean shell in preparation of chitin. J Agric Food Chem 46(9):3844–3846
72. Novikov VY, Derkach SR, Konovalova IN, Dolgopyatova NV, Kuchina YA (2023) Mechanism of heterogeneous alkaline deacetylation of chitin: a review. Polymers 15(7):1729
73. Okamoto Y et al (2002) Analgesic effects of chitin and chitosan. Carbohydr Polym 49(3):249–252
74. Omer AM, Dey R, Eltaweil AS, Abd El-Monaem EM, Ziora ZM (2022) Insights into recent advances of chitosan-based adsorbents for sustainable removal of heavy metals and anions. Arab J Chem 15(2):103543
75. Pakizeh M, Moradi A, Ghassemi T (2021) Chemical extraction and modification of chitin and chitosan from shrimp shells. Eur Polym J 159:110709
76. Palanivelu SD et al (2022) Hydrogel application in urban farming: Potentials and limitations—a review. Polymers 14(13):2590
77. Paravar A, Piri R, Balouchi H, Ma Y (2023) Microbial seed coating: an attractive tool for sustainable agriculture. Biotechnol Rep 37:e00781
78. Patra SK et al (2022) Prospects of hydrogels in agriculture for enhancing crop and water productivity under water deficit condition. Int J Polym Sci
79. Peers S, Montembault A, Ladavière C (2020) Chitosan hydrogels for sustained drug delivery. J Control Release 326:150–163
80. Pellá MC et al (2018) Chitosan-based hydrogels: from preparation to biomedical applications. Carbohydr Polym 196:233–245
81. Percot A, Viton C, Domard A (2003) Characterization of shrimp shell deproteinization. Biomacromol 4(5):1380–1385
82. Peter S et al (2021) Chitin and chitosan-based composites for energy and environmental applications: a review. Waste Biomass Valoriz 12:4777–4804
83. Qu B, Luo Y (2020) Chitosan-based hydrogel beads: preparations, modifications and applications in food and agriculture sectors–a review. Int J Biol Macromol 152:437–448

84. Racine L, Texier I, Auzély-Velty R (2017) Chitosan-based hydrogels: recent design concepts to tailor properties and functions. Polym Int 66(7):981–998
85. Ranjha NM, Ayub G, Naseem S, Ansari MT (2010) Preparation and characterization of hybrid pH-sensitive hydrogels of chitosan-co-acrylic acid for controlled release of verapamil. J Mater Sci - Mater Med 21:2805–2816
86. Ravishankar K, Dhamodharan R (2020) Advances in chitosan-based hydrogels: evolution from covalently crosslinked systems to ionotropically crosslinked superabsorbents. React Funct Polym 149:104517
87. Reddy N, Reddy R, Jiang Q (2015) Crosslinking biopolymers for biomedical applications. Trends Biotechnol 33(6):362–369
88. Reshad RAI, Jishan TA, Chowdhury NN (2021) Chitosan and its broad applications: a brief review. SSRN 3842055
89. Rinaudo M (2006) Chitin and chitosan: properties and applications. Prog Polym Sci 31(7):603–632
90. Rinaudo M, Pavlov G, Desbrieres J (1999) Influence of acetic acid concentration on the solubilization of chitosan. Polymer 40(25):7029–7032
91. Riseh RS, Vatankhah M, Hassanisaadi M, Varma RS (2024) A review of chitosan nanoparticles: nature's gift for transforming agriculture through smart and effective delivery mechanisms. Int J Biol Macromol 260:129522
92. Riseh RS, Vazvani MG, Kennedy JF (2023) The application of chitosan as a carrier for fertilizer: a review. Int J Biol Macromol 252:126483
93. Romal JRA, Ong SK (2023) Marine polysaccharide-based hydrogels for critical materials selective removal and recovery: a review. Coord Chem Rev 482:215054
94. Román-Doval R, Torres-Arellanes SP, Tenorio-Barajas AY, Gómez-Sánchez A, Valencia-Lazcano AA (2023) Chitosan: properties and its application in agriculture in context of molecular weight. Polymers 15(13):2867
95. Seenuvasan M, Sarojini G, Dineshkumar M (2020) Recovery of chitosan from natural biotic waste. In: Current developments in biotechnology and bioengineering. Elsevier, pp 115–133.
96. Shahidi F, Synowiecki J (1991) Isolation and characterization of nutrients and value-added products from snow crab (Chionoecetes opilio) and shrimp (Pandalus borealis) processing discards. J Agric Food Chem 39(8):1527–1532
97. Shariatinia Z, Jalali AM (2018) Chitosan-based hydrogels: Preparation, properties and applications. Int J Biol Macromol 115:194–220
98. Sharif R et al (2018) The multifunctional role of chitosan in horticultural crops; a review. Molecules 23(4):872
99. Sloeymani Eil Bakhtiari S et al (2021) 3-dimensional printing of hydrogel-based nanocomposites: a comprehensive review on the technology description, properties, and applications. Adv Eng Mater 23(10):2100477
100. Sogias IA, Williams AC, Khutoryanskiy VV (2008) Why is chitosan mucoadhesive? Biomacromol 9(7):1837–1842
101. Tariq Z et al (2023) Significance of biopolymer-based hydrogels and their applications in agriculture: a review in perspective of synthesis and their degree of swelling for water holding. RSC Adv 13(35):24731–24754
102. Tatarusanu SM, Sava A, Profire BS, Pinteala T, Jitareanu A, Iacob AT, Lupascu F, Simionescu N, Rosca I, Profire L (2023) New smart bioactive and biomimetic chitosan-based hydrogels for wounds Care Management. Pharmaceutics 15(3):975
103. Teixeira-Costa BE, Andrade CT (2021) Chitosan as a valuable biomolecule from seafood industry waste in the design of green food packaging. Biomolecules 11(11):1599
104. Tian B, Liu Y, Liu J (2021) Chitosan-based nanoscale and non-nanoscale delivery systems for anticancer drugs: a review. Eur Polym J 154:110533
105. Tolaimate A, Desbrieres J, Rhazi M, Alagui A (2003) Contribution to the preparation of chitins and chitosans with controlled physico-chemical properties. Polymer 44(26):7939–7952
106. Ullah F, Othman MBH, Javed F, Ahmad Z, Akil HM (2015) Classification, processing, and application of hydrogels: a review. Mater Sci Eng C 57:414–433

107. VandeVord PJ et al (2002) Evaluation of the biocompatibility of a chitosan scaffold in mice. J Biomed Mater Res: Off J Soc Biomater Jpn Soc Biomater Aust Soc Biomater Korean Soc Biomater 59(3):585–590
108. Vandepas LE (2018) Chitin the good fight–identification and description of chitin and its genes in Cnidaria
109. Vikhareva IN, Buylova EA, Yarmuhametova GU, Aminova GK, Mazitova AK (2021) An overview of the main trends in the creation of biodegradable polymer materials. J Chem 2021:1–15
110. Wang QZ et al (2006) Protonation constants of chitosan with different molecular weight and degree of deacetylation. Carbohydr Polym 65(2):194–201
111. Wang J, Zhuang, S (2022) Chitosan-based materials: Preparation, modification and application. J Clean Prod 355:131825
112. Whang HS, Kirsch W, Zhu YH, Yang CZ, Hudson SM (2005) Hemostatic agents derived from chitin and chitosan. J Macromol Sci Part C: Polym Rev 45(4):309–323
113. Xue X et al (2022) Fabrication of physical and chemical crosslinked hydrogels for bone tissue engineering. Bioact Mater 12:327–339
114. Yang Q, Dou F, Liang B, Shen Q (2005) Studies of cross-linking reaction on chitosan fiber with glyoxal. Carbohydr Polym 59(2):205–210
115. Yien L, Zin NM, Sarwar A, Katas H (2012) Antifungal activity of chitosan nanoparticles and correlation with their physical properties. Int J Biomater
116. Younes I, Rinaudo M (2015) Chitin and chitosan preparation from marine sources. Structure, properties and applications. Mar Drugs 13(3):1133–1174
117. Zhang S, Liu X, Jin X, Li H, Sun J, Gu X (2018) The novel application of chitosan: effects of cross-linked chitosan on the fire performance of thermoplastic polyurethane. Carbohydr Polym 189:313–321
118. Zhao Y, Park R-D, Muzzarelli RA (2010) Chitin deacetylases: properties and applications. Mar Drugs 8(1):24–46

The manufacturer's authorised representative in the EU is Springer Nature Customer Service Centre GmbH, Europaplatz 3, 69115 Heidelberg, Germany. If you have any concerns regarding our products, please contact ProductSafety@springernature.com

Printed and bound by CPI Group (UK) Ltd, Croydon, CR0 4YY
30/09/2026
02225302-0003